HANDBOOK FOR
HIGHLY CHARGED ION
SPECTROSCOPIC
RESEARCH

HANDBOOK FOR HIGHLY CHARGED ION SPECTROSCOPIC RESEARCH

Edited by
Yaming Zou
Roger Hutton

CRC Press
Taylor & Francis Group
Boca Raton   London   New York

CRC Press is an imprint of the
Taylor & Francis Group, an **informa** business

A TAYLOR & FRANCIS BOOK

CRC Press
Taylor & Francis Group
6000 Broken Sound Parkway NW, Suite 300
Boca Raton, FL 33487-2742

First issued in paperback 2017

ISBN 13: 978-1-138-11656-6 (pbk)
ISBN 13: 978-1-4200-7904-3 (hbk)

Visit the Taylor & Francis Web site at
http://www.taylorandfrancis.com

and the CRC Press Web site at
http://www.crcpress.com

This book is dedicated to the memory of

Professor Indrek Martinson

Contents

Preface

Spectroscopy of highly charged ions is of enormous interest due to its key role in testing quantum electrodynamics (QED), in strong fields and to possible tests on parity nonconservation (PNC), both of which are discussed in this volume. However, highly charged ions also play crucial roles in the physics of hot plasmas, for example, those produced in tokamak fusion devices and in inertial confinement fusion experiments. Much of the diagnostics of matter under such extreme environments relies heavily on spectroscopy and the availability of atomic data. The field of x-ray astronomy hinges almost entirely on the use of spectral lines from highly charged ions to provide information from distant astrophysical plasmas and objects.

Given these fundamental interests and the current rapid developments in fusion and x-ray astronomy, it is clear that the spectroscopy of highly charged ions is a very rich area of research with strong and important connections with many important subfields of physics, for example, nuclear physics. The need for high-quality atomic data is as important now as it has ever been.

Hence we feel that the idea behind this book is very timely. The aim of this book was to bring together a number of the techniques and ideas needed for highly charged ion spectroscopy research.

The book is organized in two parts. Part I brings together techniques of light/ion sources, spectrometers, and detectors and includes also a chapter on coincidence techniques. This part ends with a discussion on how atomic properties change along an isoelectronic sequence. Part II is devoted to investigations of atomic structure and to applications and also to some of the theoretical ideas where precise studies of highly charged ion spectroscopy can be of fundamental significance, for example, QED and PNC.

Yaming Zou
Roger Hutton

Editors

Yaming Zou is a professor of physics and the chair of Modern Physics Institute at Fudan University in Shanghai. She received her PhD in atomic physics from Fudan University. She then did research on collision-based atomic physics, mainly highly charged ion physics, at RIKEN, Lund University, GSI, Freiburg University, and MPI for Nuclear Physics at Heidelberg. She moved back to Fudan University in 2002 and has since been the leader of the Shanghai EBIT project.

Roger Hutton is a professor of physics at the Modern Physics Institute, Fudan University, Shanghai, China. He received his PhD in atomic spectroscopy from the University of Lund, Sweden. Following his PhD, he worked at the Lawrence Berkeley Laboratory studying ion–atom collisions and also came in contact with the very first EBIT at the Lawrence Livermore Laboratory. He has also done research in highly charged ion spectroscopy at the University of Lund and also the RIKEN laboratory in Japan. He moved to Fudan, China, in 2005.

Contributors

Jorge Reyna Almandos
Centro de Investigaciones
    Opticas (CIOp)
Gonnet, La Plata, Argentina

O. Yu. Andreev
Faculty of Physics
Saint Petersburg State University
Petrodvorets, Saint Petersburg,
    Russia

C. Biedermann
Max-Planck-Institut für
    Plasmaphysik
Greifswald und Institut für
    Physik der Humboldt-
    Universität zu
Greifswald, Germany

and

Lehrstuhl Plasmaphysik
Berlin, Germany

Anastasiya Bondarevskaya
Department of Physics
Saint Petersburg State University
Petrodvorets, Saint Petersburg,
    Russia

John T. Costello
School of Physical Sciences and
    National Centre for Plasma
    Science and Technology
    (NCPST)
Dublin City University
Glasnevin, Dublin, Ireland

R. Dux
Max-Planck-Institut für
    Plasmaphysik
Garching, Germany

Alan Hibbert
School of Mathematics and
    Theoretical Physics
Queen's University Belfast
Belfast, Northern Ireland

Roger Hutton
Institute for Modern Physics
Fudan University
Shanghai, China

Ottmar Jagutzki
Institut für Kernphysik der
    Goethe-Universität Frankfurt
Frankfurt, Germany

L. N. Labzowsky
Department of Physics
Saint Petersburg State University
Petrodvorets, Saint Petersburg,
    Russia

and

Saint Petersburg Nuclear Physics
    Institute
Gatchina, Saint Petersburg,
    Russia

Indrek Martinson[*]
Department of Physics
Lund University
Lund, Sweden

Michael Meyer
Université Paris XI
Orsay, Paris, France

Nobuyuki Nakamura
Cold Trapped Ions
    Project
ICORP, JST
Chofu, Tokyo, Japan

Nick Nelms
ESA-ESTEC, Optoelectronics
    Section
Noordwijk, the Netherlands

R. Neu
Max-Planck-Institut für
    Plasmaphysik
Garching, Germany

T. Putterich
Max-Planck-Institut für
    Plasmaphysik
Garching, Germany

R. Radtke
Max-Planck-Institut für
    Plasmaphysik
Greifswald and Institut für
    Physik der Humboldt-
    Universität zu
Greifswald, Germany

and

Lehrstuhl Plasmaphysik
Berlin, Germany

F. B. Rosmej
Sorbonne Universités
Paris, France

and

Ecole Polytechnique
Laboratoire pour Utilisation des
    Lasers Intenses—LULI
Physique Atomique dans les Plasmas
    Denses—PAPD
Palaiseau, France

Zhan Shi
Modern Physics Institute
Fudan University
Shanghai, China

John A. Tanis
Western Michigan University
Kalamazoo, Michigan

Elmar Träbert
Astronomisches Institut
Ruhr-Universität Bochum
Bochum, Germany

and

High Temperature and
    Astrophysics Division
Lawrence Livermore National
    Laboratory
Livermore, California

---

[*] Deceased.

Yang Yang
The Key Laboratory of Applied
 Ion Beam Physics
Educational Ministry
Shanghai, China

and

Shanghai EBIT Laboratory
Fudan University
Shanghai, China

Yaming Zou
The Key Laboratory of Applied
 Ion Beam Physics
Educational Ministry
Shanghai, China

and

Shanghai EBIT Laboratory
Fudan University
Shanghai, China

Part I

Tools and Techniques

1

Light Sources for Atomic Spectroscopy

Jorge Reyna Almandos and Roger Hutton

CONTENTS

1.1 Introduction

The development of appropriate sources for the excitation of atomic spectra is an essential task in atomic spectroscopy. For a long time, classic sources such as electrode-less discharges, arcs, and sparks were used by spectroscopists in the study of the spectra of elements with low ionization degrees. This last source was also used in studies of intermediate and highly ionized atoms, and some examples can be found in the works reported by Sugar et al. [1], Kaufman et al. [2], Churilov et al. [3,4], and Ryabtsev et al. [5]. Such sources basically consist of two electrodes connected to a capacitor which is charged to a high voltage until electric breakdown take place. The differing degrees of ionization produced by this kind of source can be handled by varying the inductance in the circuit. In some cases, by placing an insulator between the electrodes, the spark will be able to slide along the surface of the insulator.

This allows better or more controlled operation, but it does not unfortunately lead to higher stages of ionization [6].

Many studies have been made to develop new spectral sources that permit researchers to obtain new and better data. In this chapter, we will describe some of the different kinds of gas discharge and plasma sources used in atomic spectroscopy, such as hollow cathodes, theta pinch discharges, and capillary and pulsed discharges, which have different applications in the study of plasmas, the development of lasers for the vacuum ultraviolet and X region, the interpretation of spectra from space, the modeling of stellar atmospheres, and to test atomic theoretical models.

1.2 Hollow Cathode

This class of spectroscopic light source is a gaseous discharge, such as a glow discharge, where gases and vapors can be excited at pressures of a few Torr. If the cathode is made hollow (Figure 1.1), when the discharge pressure is reduced, the negative glow fills the hollow cathode (HC).

The material of the cathode, or any metal lining it, is vaporized by bombardment of the gas and excited by electron collisions. In many cases, noble gases, or mixtures of them, are used as carrier gases to obtain changes in the sputtering and the excitation. In this form the cathode material and the filling gas can be studied by changing the current, voltage, and the gas pressure. An important advantage of this discharge is that the electric field in the emission region is very low and consequently the Stark shift and broadening of the spectral lines are small. Another advantage can be obtained by cooling one extreme of the HC in order to reduce the Doppler widths of the lines.

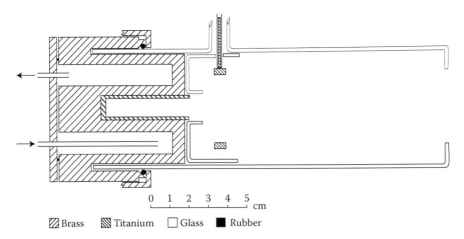

FIGURE 1.1
Cross-section of the HC lamp.

The discharge in the cathode is very stable and may run for hours without any noticeable changes in intensity, an essential feature for Fourier transform spectroscopy, for example. Operating currents usually fall in the region 0.1–2 A, and other discharge conditions are 100–500 V and pressure ranging between 0.05 and 4 Torr, depending on the cathode material and the carrier gas. With these parameters, it is possible to obtain the first degrees of ionization of the cathode material and carrier gas.

Typical dimensions of the HC are in the order of 50 mm length and 10 mm internal diameter.

A symmetric HC discharge tube was also developed where the symmetric design implies that the discharge tube has two anodes (Figure 1.2).

Many works have been made to study different spectra with this kind of discharge making a good use of the above-mentioned advantages, and some examples can be found in the works reported by Valero [7] and Persson [8]. It is also an ideal light source to work in Fourier transform spectroscopy and in recent years was extensively applied to study the spectra of ions of astrophysical interest [9–13].

Pulsed HC discharges have been used to study the spectrum and term system of some elements by discharging a capacitor through the HC. An example of this is the work on Fe II by Johansson et al. [14].

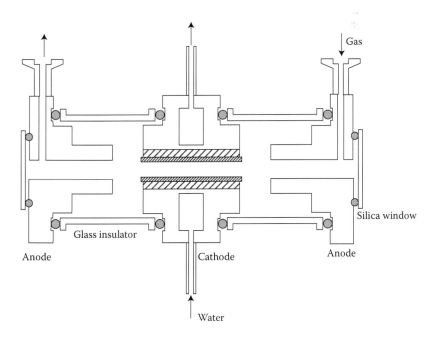

FIGURE 1.2
Cross-section of the double-sided HC lamp.

Studies related with recombination lasers [15] made use of flowing HC discharge, where a fast gas flow can be used to expand the already super-cooled negative glow plasma outside the ionization region, thereby creating a recombining plasma without energetic electrons.

1.3 Pinch Discharges

Due to the efforts made to obtain high density and temperatures of plasmas toward controlled thermonuclear fusion, different kinds of plasma devices were developed which have found subsequent use in the study of the spectra of intermediate and highly ionized atoms.

Light sources using pinch discharges, such as zeta and theta-pinch, were applied to study the spectra of highly charged ions (HCIs) in order to improve the knowledge of the level structure and transitions, including the so-called forbidden lines (for instance, magnetic dipole transitions) that are useful for determining ion densities, ion temperatures, or other plasma characteristics.

In the theta-pinch device, a capacitor bank charged from a high-voltage power supply is discharged over a gas-filled cylindrical tube through an external coil around it. The compression from the generated magnetic field (pinch effect) heats the plasma producing ionization/excitation to several ionization degrees. The ionization stage reached depends on the delivered energy into the tube.

An example of this kind of machine to obtain intermediate and highly ionized spectra is the 2 kJ theta-pinch used by Pettersson [16] to study the spectrum of O III. This device was also used to study the spectra of Ne III [17] and Kr III to Kr VIII [18,19]. In this equipment, which has some differences compared to a previous one built in Uppsala by Bockasten et al. [20] to study N IV, N V, O V, O VI, and Ne VIII spectra, the total capacitance was 7.7 μF and the total inductance was 76 nH. The maximum current at 10 kV discharge voltage was 100 kA, and the period of the damped oscillation was 4.8 μs. The glass tube inside the coil was made of Pyrex and the diameter of the coil was 7 cm with a length of 40 cm. The coil surrounding the discharge tube was connected between the upper and lower plates. A schematic of the discharge circuit is shown in Figure 1.3.

The spark gap used was pressurized with air at about 6 kPa cm^{-2} and the pressure of the filling gas was varied between 1 and 32 mTorr. The gas was preionized using a radio frequency (r.f.) discharge. The repetition rate of the theta-pinch discharge was about 15 per minute at a capacitor bank of 10 kV.

A similar theta pinch was built in Campinas, Brazil, to study noble gas spectra [21–24]. In this case, the internal diameter of the Pyrex tube was 8 cm, the length was 100 cm, the total capacitance was 7.5 μF, and the operation energy was between 1.5 and 3.4 kJ. The calculated electron temperature was 30–200 eV and the density was in the region of 10^{15} cm^{-3}.

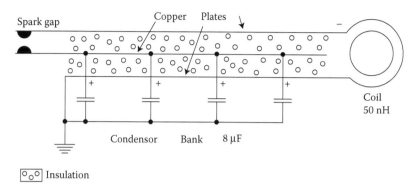

FIGURE 1.3
Schematic of the discharge circuit.

A 40 kJ theta pinch [25], where the electron temperature was 200–300 eV and the density was 10^{16}–10^{17} cm^{-3}, was used to study high ionization degrees in Ar, Co, and Ni.

A 30 kJ theta pinch was used to excite argon ion spectra from Ar V to Ar X [26]. Xe VII and Xe VIII spectra were studied by Roberts et al. [27] and Knystautas et al. [28] using a 50 kJ theta pinch with electron temperature at or around 100 eV and electron density of 5×10^{15} cm^{-3}.

Other pinch devices were developed which have found use in the spectroscopy of highly ionized atoms. Some examples are the toroidal plasma device ZETA which was used to study transitions in noble gases [29,30], the gas-liner pinch with electron density 1018 cm^{-3}, and electron temperature of 7 eV, used to study Stark broadening of Ar IV lines [31] and also employed for Stark broadening measurements of multiply ionized species [32–34].

1.4 Capillary Discharges

Capillary discharges as sources to study highly ionized spectra of noble gases have been used by Schonheit [35] and Léveque et al. [36] to study the Ar IV to Ar VIII emission spectra. Also the generation of soft x-ray and XUV radiation in capillary plasmas was analyzed several years ago by Bogen et al. [37]. Electron–ion recombination processes have been indicated as playing an important role in achieving population inversion in these systems, and the possibility to obtaining soft x-ray amplification in a capillary discharge has motivated several experimental studies [38,39].

As an example of these devices [40], pure Ar gas, or a mixture of Ar and H$_2$, was injected into a capillary channel 4 mm in diameter and 12 cm tube length. The capacitor was charged by a seven-stage 700 kV Marx generator, and the capillary tube was excited by discharging the capacitor. The capillary is in the axis of a 3 nF circular parallel plate capacitor containing ethylene glycol as dielectric. The current pulses used had an amplitude of ~40 kA and a half period of 60 ns. The gas pressure was on the order of 700 mTorr. With these parameters, line amplification at 46.9 nm was observed with a population inversion suitable for lasing.

A modified device using a fast capillary discharge with inductive storage was also used for lasing effects investigations by Antsiferov et al. [41] and Ryabtsev et al. [42]. In this experimental device, high current rates could be reached with much lower voltages.

Studies on xenon emission spectra in the EUV region originating from Xe IX, Xe X, Xe XI, and Xe XII ions using capillary discharge plasmas were made by Klosner and Silfvast [43] in a 6-mm long 1-mm diameter capillary channel, where the pressure of the gas was varied from 0.05 to 1 Torr and the gas was excited by two 0.25-μF capacitors charged at voltages to ~5 kV. The estimated peak electron temperature was ~50 eV.

The influence of the prepulse current on the 46.9-nm Ne-like Ar laser emission using a capillary discharge was recently reported by Tan et al. [44].

1.5 Other Kind of High Current Discharges

The pulsed laser-tube-like source (without end mirrors) was extensively used by the Group at La Plata in the study of Ne, Ar, Kr, and Xe spectra in intermediate and high degrees of ionization [45–55], related with the interest of spectroscopic data from rare gases due to application in astrophysics, laser physics, fusion diagnostic, and so on.

The light source consisted of a Pyrex tube ending in quartz windows with an internal diameter from 3 to 6 mm and distance between electrodes varies from 20 to 120 cm approximately. The electrodes are made of tungsten covered with indium. At one side of the tube, there is an inlet connected via a pressure reduction system to the bottles of noble gases. The pressure range is varied between 1 and 300 mTorr approximately. Gas excitation is produced by discharging through the tube a bank of low-inductance capacitors between 1 and 280 nF, charged up to 20 kV.

The typical current observed by using a Rogowsky coil shows a damped sinusoidal period of 1.2 μs, having peak values between 1 and 4 kA. A schematic of the electronic circuit is shown in Figure 1.4. To obtain better stability in the discharge, the triggered spark gap shown in Figure 1.4 was replaced by a coaxial hydrogen thyratron [52].

When this light source is adapted to study spectra in the VUV region, one end of the tube is connected to a vacuum spectrograph through a nylon flange adaptor. The other end has a window for observing the discharge and alignment of the tube [56].

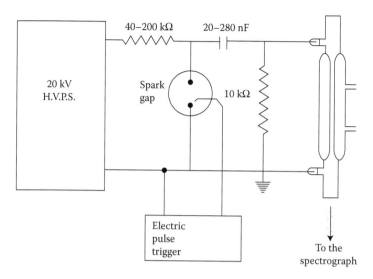

FIGURE 1.4
Schematic of the electronic circuit.

As an example, when the tube has 3 mm internal diameter and 80 cm length, a xenon gas pressure of 45 mTorr, a capacitor of 30 nF giving a current with maximum up to 2 kA, and a pulse swinging period of 1.2 µs, the electron density is estimated to be 10^{16} cm^{-3} and the electron temperature is in the order of 1 eV [57,58]. These parameters correspond to low xenon ionization degrees, but the electron temperature can reach about one order of magnitude higher when Xe IX ions are produced in this kind of source.

1.6 Advances in Light Source Techniques

1.6.1 Beam Foil Spectroscopy

A major development in light source for atomic spectroscopy occurred in the early 1960s when both Bashkin [59] and Kay [60] reported light from ion beams that had passed through thin carbon foils. This was the birth of what was to be found close to a revolution in the study of HCIs, the so-called technique of beam foil spectroscopy (BFS). A schematic diagram of an experimental setup for BFS is shown in Figure 1.5.

Basically an ion beam, from an accelerator, is sent through a thin foil, usually of carbon, where the ions are ionized/excited. The degree of ionization depends on the velocity of the ions and the thickness of the foil. A major factor in determining the importance of this technique in the studies of atomic structure is that the beam velocity is almost always in the region of a few mm/ps for

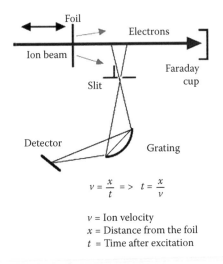

$$v = \frac{x}{t} \Rightarrow t = \frac{x}{v}$$

v = Ion velocity
x = Distance from the foil
t = Time after excitation

FIGURE 1.5
Schematic setup for a typical BFS experiment. (Courtesy of Elmar Träbert.)

very HCIs to mm/ns for lower charge states. As a matter of fortuosity, atomic lifetimes are very often in the picosecond range for HCIs and in the nanosecond range for less HCIs. This means that if the beam were viewed through a perpendicular slit, in particular, the entrance slit of a monochromator, then the decay of excited levels would occur over easy-to-measure distances. Measurement of atomic lifetimes using BFS will be covered in Chapter 10, and so will not be discussed here (see also a recent review [61]). BFS obviously played a crucial role in lifetime measurements, as basically there was no technique previously available, and all lifetime data, for anything other than neutral or possible single charged ions, came from theoretical calculations. However, BFS also played an important role in spectroscopy of HCIs. BFS had one property that was both its greatest advantage (for spectroscopy) and greatest disadvantage (for lifetime measurements). That is, that all atomic energy levels appear to be populated by the beam-foil excitation method. For spectroscopy, this property meant that full spectral analyses could be carried out. However, for lifetime work, this leads to severe problems due to level repopulation through cascades. The population of all levels was put to good use in spectroscopic terms in, for example, a study of ground-state quantum electrodynamics (QED) for He-like O via a measurement of the ionization potential. The problems caused by cascading were solved by a technique known as ANDC (arbitrary normalized decay curves) which was an idea by Curtis et al. [62] and implemented in a computer code (CANDY) by Engström [63]. One of the most interesting results from lifetime work taking into account cascade repopulation was for the resonance transitions in Na-like ions [64]. Here, a systematic discrepancy was found between cascade-corrected measurements and all calculated lifetime values at that time from any of the available computer codes. This was soon found to be due to the fact that it was not good enough to treat the $2p^6$ shell is totally closed. Once the shell was "opened," that is, core polarization was included in the calculation, an agreement with the experimental results was obtained [65].

1.6.1.1 Electron Cyclotron Resonance Ion Source-Based Spectroscopy

Electron cyclotron resonance (ECR) ion sources were pioneered in the mid-1970s and early 1980s by Geller, Bliman, and coworkers in France. In fact, the construction of the first ECR ion source dedicated to atomic physics research was initiated already back in 1976. The basic idea behind ECR ion sources is that in a minimum B-type magnetic field, electrons can be made to absorb energy from a microwave field. This absorption of energy goes to kinetic energy, and the hot electrons can then ionize atoms inside the magnetic bottle. It is, however, interesting to try to imagine where the first free electron comes from. Such ion sources are not capable of producing extreme HCIs as in the case of EBITs but they can produce a much larger number of ions.

The ECR plasma can be used as a light source and is indeed used this way by a group in Paris [66]; such a work was first done by Jean Jacques Bonnet and Samuel Bliman already in the late 1970s. However, most of the spectroscopic uses of ECRs have been with extracted ion beams. As examples of typical beams from a modern (commercial) ECR ion source beams of Ar^{8+}, extracted at an energy of around 10 keV/q, can reach many hundreds of mAs.

A very interesting method for using ECR beams for spectroscopy comes from the process of electron transfer in slow ion–atom collisions. Initial work on such spectroscopy was reported by Bliman in the mid-1980s [67]. In the case of Ar^{8+}, for example, collisions at 10 keV/q (i.e., 80 keV) are considered as slow on the atomic scale. In this respect, slow means that the velocity of the ion is slow compared to the velocity of the electrons in the atomic target. The main process in such collisions is that a number of electrons can be transferred from the target atom to the ion. If the target atoms are hydrogen or helium, then only one or two electrons can be transferred if the gas pressure is kept low enough to ensure single collision conditions. The term "single collision conditions" needs some explanation, it is not a fixed target pressure but depends on the cross-section of the process involved. In other words, the pressure should be chosen such that the process with the largest cross-section has the condition of occurring under single collision conditions (that the ion makes one collision while getting across the gaseous target). Given single collision conditions and electron capture as the dominant process, rather simple spectra can be obtained. In the example here, Ar^{8+} on He, then only Ar^{7+} and Ar^{6+} can result. Ar^{7+} is Na-like and a well-known spectrum, making this an ideal way to study the spectrum of Mg-like Ar (Ar^{6+}). This method of spectroscopy was well demonstrated for a slightly more complex spectrum, that of the $3d^9\,4s^2\,^2D$ levels of Cu-like Kr [68]. Lines from these levels were considered of importance in x-ray laser development. In the work of Hutton et al. [68], a discussion of line widths, and so on, was taken up and it would appear that such spectroscopy should be of wide interest. However, the reality is that this method of generating spectra never really fulfilled its promise. Having said that, a number of interesting results came from ECR-based spectroscopy, although these results were not the intention of the experiments at the time. In the mid-1980s, there were many discussions with respect to the nature of double-electron capture from helium targets [69]. Two processes were discussed. One was a sequential capture, that is, first one electron was transferred and then the other. This was expected to lead to fairly symmetric states, such as $1s^24l4l'$ for O^{6+} on He. However, it was clear that some of the lines in the Auger spectrum were the result of populating highly asymmetric states, such as $1s^23lnl'$ when Ne could be quite large, and this process was known as correlated capture. It was later understood that before making the distinction between correlated and uncorrelated capture, a good understanding of configuration interacting between the $4l4l'$ and $3lnl'$ states was needed.

For some value of n, there would be almost energy degeneracy between $4l4l'$ and $3lnl'$ states and large configuration mixing could occur. Different approaches to couple the Auger spectra with the X-VUV spectra needed further theoretical atomic structure calculations to support the configuration interaction and mixing. The fluorescence yield for each level made it possible to relate the main Auger decay to the weak radiative decays of these doubly excited states. Such effects were also seen in optical spectra produced by HCIs traversing a microcapillary target [70]. Hence one was forced to consider strong electron correlation effects for HCIs which were often neglected. A spectacular result of such correlation was demonstrated in experiments by Schuch et al. [71], where a beam of very highly charged ions, albeit from an EBIT, was impinged on a metal target. There was a surge in that experimental direction after the release of the first results in *Physical Review Letter* [72] introducing the concept of "hollow atoms."

1.6.1.2 Laser-Produced Plasma Spectroscopy

As discussed in Chapter 9, laser-produced plasmas have played a key role in the development of the spectroscopy of HCIs. By firing a high-power laser onto a solid target, very hot and dense plasmas could be obtained. The high plasma density leads to a very bright spectra and these sources seemed an idea for the spectroscopy of HCIs. Two drawbacks in this method were later identified and these along with the high costs of large laser facilities have led to a lull in such spectroscopic efforts. One drawback is that the plasma consisted of many charge states leading to problems with satellite lines. Another problem was found in the calibration of such spectra. As the ions in the plasma are moving quite fast and at different velocities for different charge states, it was found to be difficult to get good calibrations for high accuracy wavelength measurements. With the advent of electron beam ion traps, which produce HCIs in a more controlled manner and also at much lower temperatures, there seems little interest in laser-produced plasmas for pure spectroscopy. Laser-produced plasmas have found other uses, for example, in inertial confinement fusion, and such things are discussed in Chapter 12. Here spectroscopy mainly plays the role of a diagnostic tool although some very interesting effects on atomic structure in very dense plasmas have been observed.

1.6.1.3 Tokamak Spectroscopy

Tokamaks are of course not developed as spectroscopic light sources but as devices for producing nuclear fusion and related studies; however, they have provided some interesting spectroscopic data over the years. Chapter 11 will describe the use of atomic data in the diagnostics of fusion plasmas, so here we will just mention a few examples of tokamaks being used as spectroscopic light sources. A recent review of the use of tokamaks to provide new spectroscopic data can be found in [73]. Clearly, one of the milestones of tokamak-based spectroscopy was the observation of forbidden lines from HCIs by Suckewer and Hinnov [74]. Forbidden lines had been known long before these observations, mainly from astrophysical spectroscopy, but they were difficult to observe under laboratory conditions. The unique environment of high temperature and low density did not exist in many laboratory light sources until quite recently. This topic is better covered in Chapter 9.

1.6.1.4 Electron Beam Ion Traps

The first spectra from an electron beam ion trap (EBIT) were reported more than 20 years ago. Since then, EBITs have come to dominate the study of spectroscopy of HCIs. This subject will be dealt with in detail in Chapter 2, so only some highlights will be mentioned here. A useful reference to EBIT-based spectroscopy is the proceedings from the conference "20 Years of EBIT Spectroscopy," which was held in Berkeley in 2006 [75]. Among the highlights of first observations and very high accuracy measurements using EBITs that should be mentioned are: (a) the first $M3$ decay [76], (b) the first observation of magnetic sensitive lines [77], (c) resolution of Zeeman line splitting for highly ionized argon [78], (d) very precise lifetime measurements for HCIs [79], and (e) spectroscopy of very highly charged ions. These five categories will be discussed in a little detail in the following.

(1) Before the introduction of EBITs as a light source for studies of HCI spectroscopy, the idea of observing such an exotic object as an $M3$ transition would probably have been laughed at. Indeed, the first observation of such a transition, in Ni-like uranium, without doubt, took some thinking about before it was possible to identify such a line. Relatively new additions to the range of spectrometers/detectors available at EBITs are the microcalorimeters at the EBITs in Livermore (Livermore micro) and NIST (NIST micro). Using an EBIT/calorimeter combination, not only is it possible to observe the $M3$

decay in Ni-like Xe, but it is also possible to measure the $3d^9 4s\,^3D_3$ level lifetime [80]. Ni-like ions are very interesting as all levels belonging to the first excited configuration are forbidden to decay via $E1$ transitions (see Figure 1.6).

A measurement of the $^3D_3$ lifetime in Ni-like Xe performed in [80] gave a lifetime result which disagreed with the then currently available theoretical results. There followed new calculations increasing the amount of electron correlation, but there was no convergence to the experimental lifetimes [80]. This discrepancy was later found to be due to the hyperfine interaction mixing the $^3D_3$ level with the $^3D_2$ (and also the $^1D_2$) level [81]. This result showed that the lifetime of the $^3D_3$ level did not have a unique value for ions with a nuclear spin. The ideas presented in [81] were very quickly confirmed by using pure Xe isotopes (^{132}Xe and ^{129}Xe) in separate experiments [82].

(2) Figure 1.7 shows a spectrum recorded at the Livermore Electron Beam Ion Trap, EBIT, of the spectral region containing lines from the decay of the $2p^5 3s\,^3P_j$ lines of Ne-like argon [77].

The line from the $2p^6\,^1S_0 - 2p^5 3s\,^3P_0$, marked as B in Figure 1.7, is strictly forbidden to decay via a one photon transition in the absence of a nuclear spin. However, the external magnetic field present in the EBIT light source opens up an extra decay channel, the so-called Zeeman-induced transition. The $(2p^5_{1/2}3s)_{j=0}$ level will be mixed with the $(2p^5_{1/2}3s)_{j=1}$ level by the magnetic field and hence the $^3P_0$ level will have an open $E1$ decay channel. It turns out that the decay rate associated with this Zeeman-induced transition scales as the square of the magnetic field. The ratio of this line intensity to that of a close by, in terms of wavelength, allowed line then shows a monotonic increase as a function of the magnetic field strength. The work so far on magnetic sensitive lines has been done using an EBIT light source which generally operates with a

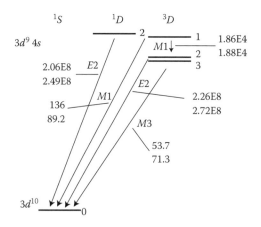

FIGURE 1.6
Schematic diagram of the energy and decay properties levels of the first excited configuration of Ni-like Xe.

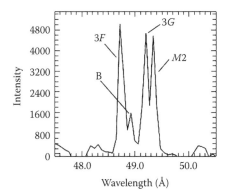

FIGURE 1.7
This shows the magnetically induced line, B, for the strictly forbidden $2p^{6\,1}S_0 - 2p^5 3s^3 P_0$ transition in Ne-like Ar.

magnetic field around 3 T, which is very relevant to the situation at tokamaks. Unfortunately, this interesting effect has so far only been seen for Ne-like Ar [77]; however, the effect should exist for all Ne-like ions and the relative strength of the magnetic-induced line depends on energy level separations and the competition from the $M1$ decay of the $^3P_0$ level to the $^3P_1$ level of the same $2p^5 3s$ term. However, other cases where an appreciable effect may be expected and perhaps elements of other sequences such as Ar-like ions deserve study. In Ar-like ions, the $3p^5 3d\ ^3P_0$ is the lowest excited level and hence there are no competing decay channels.

(3) Perhaps the highest resolution spectroscopy done on HCIs has been done in the visible spectral region on Ar^{13+}, B-like Ar, where Zeeman splitting for the $1s^2 2s^2 2p\ ^2P_{1/2} - {}^2P_{3/2}$ $M1$ ground-state transitions was observed [78]. We must remember that all EBITs operate with a magnetic field, usually on the order of a few Tesla, and this has its consequences, another of such will be discussed later. It is very rare that an ionic system is confined to a region with such a well-defined magnetic field as in an EBIT. Magnetic field accuracies in the trap region of an EBIT are generally precise on the 10^{-4} T level.

Figure 1.8 shows the Zeeman components of the $^2P_{1/2} - {}^2P_{3/2}$ ground-state $M1$ transition in B-like argon [78].

(4) Way back in the 1940s when Edlén first understood the origin of a number of unidentified spectral lines in spectra of the solar corona as being forbidden transitions in HCIs, it was not possible to observe such lines in laboratory light sources. It is relatively easy to observe forbidden transitions in HCIs using an EBIT and even to measure the lifetime of the upper energy level. As discussed in Chapter 2, in an EBIT a cloud of HCIs is generated through successive electron impact collisions with atoms in the interaction region of the drift tube assembly. The charge state distribution is determined mainly by the energy of the electron beam. The ions are trapped along the beam direction by applying

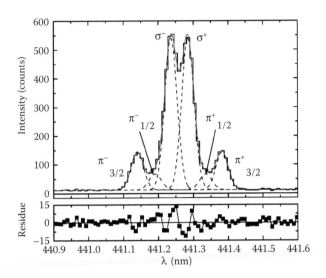

FIGURE 1.8

This shows the Zeeman components of the $^2P_{1/2} - {}^2P_{3/2}$ ground state $M1$ transition in B-like Argon. (From Draganic et al., *Phys. Rev. Lett.* 91, 183001, 2003. With permission.)

suitable voltages to the drift tubes. Radial trapping is done in the first instance by the space charge of the electron beam. To measure atomic lifetimes using an EBIT, the charge state of interest is first created by using a well-chosen electron beam energy. The electron beam is then switched off, and photons from the transition of interest are recorded as a function of time after switching off the beam. One thing of importance here is that the lifetime of the upper level should be compared to the time required to switch off the electron beam and also short compared to the time taken for the ions to move away from the trapping volume. There is a secondary trapping mechanism, the so-called magnetic trap mode where the EBIT behaves in a way similar to a Penning trap (PB mag trap mode [83]). If the lifetime under investigation fits into this category, then it appears that very accurate measurements can be made. There are of course a number of systematic effects that can be investigated but final lifetimes with error bars of the order 0.1% can be obtained. As an example, the $M1$ decay rate of the $3s^23p\ {}^2P_{3/2}$ upper level of the ^2P ground term in Al-like Fe has been measured to be 16.726(−0.010/+0.020) ms [79]. The decay of this particular level gives rise to the famous corona green line.

(5) In the past, a routine method for studying spectroscopy of HCIs was through laser-produced plasmas. However, in a work on Na-like spectroscopy using laser-produced plasmas at a number of laser centers, an interesting trend was found when comparing observed and calculated energy levels was found [84]. This problem is illustrated in Figure 1.9 [85].

There is a clear divergence in the difference (Energy$_{obs}$ − Energy$_{cacl}$) for the $3s_{1/2} - 3p_{3/2}$ transition energy in Na-like ions when comparing results from

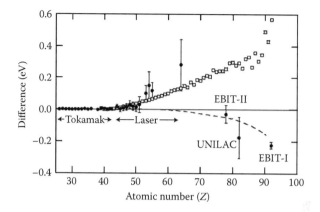

FIGURE 1.9

The difference (Energy$_{obs}$ − Energy$_{cacl}$) for the $3s_{1/2} - 3p_{3/2}$ transition energy in Na-like ions. (The calculation is from Seely et al., *At. Data Nucl. Data Tables* 47, 1, 1991.)

laser-produced plasmas and other experimental techniques, in particular, with EBIT results. This divergence could lead to very difficult problems for theoretical development. The measurement marked EBIT-II was done in Livermore in 1991 [86] and cast doubt on the reliability of the results from laser-produced plasmas. Since then, the 1991 EBIT results have been corroborated by an accelerator-based experiment (BFS) for Na-like Pb at the GSI UNILAC [87] and a higher Z EBIT result for Na-like U, also from Livermore [88]. A similar situation, that is, a divergence between EBIT results and those from laser-produced plasma, has also been observed for Cu-like ions. One point that comes to mind, and shows another of the advantages EBITs offer to atomic spectroscopy, is that it is very much easier to introduce calibration lines into EBIT spectra. Spectral calibration lines can be obtained by a number of methods, including: (i) by making use of the fact that the ions in an EBIT move relatively slowly and hence by using an external (fixed) spectral lamp, such as an argon lamp or an HC, or (ii) introducing a calibration element into the EBIT. Both these methods have been used with great success in past EBIT experiments.

1.6.1.5 Astrophysical Plasmas

The first wavelengths associated with forbidden transitions were measured using solar corona spectra way back in 1869. Of course, it was not understood at that time that these wavelengths corresponded to forbidden transitions. The origin of these wavelengths remained a mystery until the mid-1940s when these were identified as forbidden transitions in HCIs of calcium, iron, nickel, and so on [89]. We usually think about atomic spectroscopy as a means of providing data for astrophysical plasma diagnostics, however, just as in the case of tokamaks mentioned a few sections previously, there are exceptions.

A more recent example of where an astrophysical plasma was used to provide atomic data is the work of Brage et al. [90]. In that work, a planetary nebula (NGC3918) is used as a light source to provide a measure of the hyperfine-induced decay rate of the $2s2p\,^3P_0$ decay in Be-like N. This turns out to be one of the longest lifetimes ever measured. More details on the use of astrophysical plasmas as light sources for atomic spectroscopy, especially in comparison of spectra with laboratory light sources, are given in Chapter 9.

Acknowledgment

We would like to thank Prof. Samuel Bliman for sharing his memories of the early days of ECRs and ECR-based spectroscopy.

References

1. Sugar et al., *Phys. Scripta* 26, 419–421, 1982.
2. Kaufman et al., *JOSA B* 4, 1924–1926, 1987.
3. Churilov et al., *Phys. Scripta* 57, 500–502, 1998.
4. Churilov et al., *Phys. Scripta* 61, 420–430, 2000.
5. Ryabtsev et al., *Phys. Scripta* 72, 377–384, 2005.
6. Martinson, *Rep. Prog. Phys.* 52, 157–225, 1989.
7. Valero, *JOSA* 58, 484–489, 1968.
8. Persson, *Phys. Scripta* 3, 133–155, 1971.
9. Forsberg, Ph.D. Thesis, Lund University, 1987.
10. Quinet et al., *Phys. Rev. A* 49, 2446–2452, 1994.
11. Whaling et al., *J. Quant. Spectr. Radiat. Transf.* 53, 1–22, 1995.
12. Kramida et al., *Eur. Phys. J.* D37, 1–21, 2006.
13. Aldenius et al., *Mon. Not. R. Astron. Soc.* 370, 444–452, 2006.
14. Johansson et al., *Phys. Scripta* 18, 217–265, 1978.
15. Wernsman et al., *IEEE J. Quantum Electron* 26, 1624–1632, 1990.
16. Pettersson, *Phys. Scripta* 26, 296–318, 1982.
17. Persson et al., *Phys. Rev. A* 43, 4791–4823, 1991.
18. Bredice et al., *JOSA B* 5, 222–235, 1988.
19. Trigueiros et al., *Phys. Scripta* 34, 164–166, 1986.
20. Bockasten et al., *Ark. Fys.* 32, 437–446, 1966.
21. Trigueiros et al., *Nucl. Inst. Meth. Phys. Res. A* 280, 589–592, 1989.
22. Trigueiros et al., *JOSA B* 14, 2463–2468, 1997.
23. Pagan et al., *JOSA B* 12, 203–211, 1995.
24. Cavalcanti et al., *J. Phys. B* 29, 6049–6053, 1996.
25. Gabriel and Paget, *J. Phys. B* 5, 673–685, 1972.
26. Fawcett et al., *Phys. Scripta* 18, 315–322, 1978.
27. Roberts et al., *JOSA* 69, 1620–1622, 1979.

28. Knystautas et al., *JOSA* 69, 1726–1727, 1979.
29. Fawcett et al., *Proc. Phys. Soc. London Ser.* 78, 1223–1226, 1961.
30. Fawcett et al., *J. Phys. B* 13, 2711–2716, 1980.
31. Hey et al., *J. Phys. B* 22, 241–249, 1989.
32. Ackermann et al., *Phys. Rev. A* 31, 2597–2609, 1985.
33. Bottcher et al., *Phys. Rev. A* 36, 2265–2271, 1987.
34. Bottcher et al., *Phys. Rev. A* 38, 2690–2693, 1988.
35. Schonheit, *Optik* 23, 409–434, 1966.
36. Léveque et al., *J. Physique* 45, 665–670, 1984.
37. Bogen et al., *JOSA* 58, 203–206, 1968.
38. Roca et al., *Phys. Rev. E* 47, 1299–1304, 1993.
39. Lemoff et al., *JOSA B* 13, 180–184, 1996.
40. Roca et al., *Phys. Rev. Lett.* 73, 2192–2195, 1994.
41. Antsiferov et al., *Phys. Scripta* 62, 127–131, 2000.
42. Ryabtsev et al., *J. Phys. IV* 11, 317–319, 2001.
43. Klosner and Silfvast, *JOSA B* 17, 1279–1290, 2000.
44. Tan et al., *Phys. Rev. A* 75, 0438081-5, 2007.
45. Gallardo et al., *J. Quant. Spectrosc. Radiat. Transfer* 95, 365–372, 2005.
46. Gallardo et al., *Spect. Lett.* 40, 879–891, 2007.
47. Bredice et al., *Phys. Scripta* 51, 446–453, 1995.
48. Raineri et al., *Phys. Scripta* 45, 584–589, 1992.
49. Raineri et al., *J. Quant. Spectrosc. Radiat. Transfer* 102, 391–395, 2006.
50. Raineri et al., *Atomic Data and Nuclear Data Tables* 94, 140–159, 2008.
51. Raineri et al., *J. Phys. B* 42, 205004 (6pp), 2009a.
52. Raineri et al., *Phys. Scripta* 79, 025302 (8pp), 2009b.
53. Reyna Almandos et al., *Phys. Scripta T* 134, 014018 (6pp), 2009.
54. Saloman, *J. Phys. Chem. Ref. Data* 33, 765–921, 2004, and references therein.
55. Saloman, *J. Phys. Chem. Ref. Data* 36, 215–386, 2007, and references therein.
56. Gallardo et al., *Appl. Opt.* 28, 4513–4515, 1989.
57. Di Rocco et al., *J. Quant. Spectrosc. Radiat. Transfer* 35, 443–446, 1986.
58. Di Rocco et al., *J. Quant. Spectrosc. Radiat. Transfer* 40, 513–517, 1988.
59. Bashkin, *Nucl. Instrum. Meth.* 28, 88, 1964.
60. Kay, *Phys. Lett.* 5, 36, 1963.
61. Träbert, *J. Phys B* 43, 074034, 2010.
62. Curtis et al., *Phys. Lett.* 344, 159, 1871.
63. Engström, *Nucl. Instrum. Meth.* 202, 369, 1982.
64. Hutton et al., *Phys. Rev. Lett.* 60, 2469, 1988.
65. Theodosiou and Curtis, *Phys. Rev. A* 38, 4435, 1988.
66. Martins et al., *Phys. Rev. A* 80, 032501, 2009.
67. Bliman et al., *J. Phys. B* 20, 5127, 1987.
68. Hutton et al., *Phys. Scripta* 48, 569, 1993.
69. Stolterfoht et al., *Phys. Rev. Lett.* 57, 74 1986.
70. Morishita et al., *Phys. Rev. A* 70, 012902, 2004.
71. Schuch et al., *Phys. Rev. Lett.* 70, 1073, 1993.
72. Briand et al., *Phys. Rev. Lett.* 65, 159, 1990.
73. Martinson and Jupén, *J. Chin. Chem. Soc.* 48, 469, 2001.
74. Suckewer and Hinnov, *Phys. Rev. Lett.* 41, 756, 1978.
75. Beiersdorfer, 20 years of EBIT spectroscopy, *Can. J. Phys.* 86, 2008.
76. Beiersdorfer, *Phys. Rev. Lett.* 67, 2272, 1991.

77. Beiersdorfer et al., *Phys. Rev. Lett.* 90, 235003, 2002.
78. Draganic et al., *Phys. Rev. Lett.* 91, 183001, 2003.
79. Brenner et al., *Phys. Rev. A* 75, 032504, 2007.
80. Träbert et al., *Phys. Rev. A* 73, 022508, 2006.
81. Yao et al., *Phys. Rev. Lett.* 97, 183001, 2006.
82. Träbert et al., *Phys. Rev. Lett.* 98, 263001, 2007.
83. Beiersdorfer et al., *Rev. Scientific Instrum.* 67, 3818, 1996.
84. Seely et al., *At. Data Nucl. Data Tables* 47, 1, 1991.
85. Beiersdorfer, *Phys. Scripta*, T120, 40, 2005.
86. Cowan et al., *Phys. Rev. Lett.* 66, 1150, 1991.
87. Simionovici et al., *Phys Rev A* 48, 3056, 1993.
88. Beiersdorfer et al., *Phys Rev A* 67, 052103, 2003.
89. Edlén, *Zeitschrift fur Astrophysik* 22, 30, 1943 and Edlen, *Arkiv Matem. Astron. Fysik* 28B(1), (1942).
90. Brage et al., *Phys. Rev. Lett.* 89, 281101, 2002.

2

Electron Beam Ion Traps: Principles and Applications to Highly Charged Ion Spectroscopy

---

Yang Yang and Yaming Zou

CONTENTS

2.1 Introduction

The history of spectroscopic studies of highly charged ions (HCIs) goes back to the 1930s. At that time, HCIs were produced using vacuum spark techniques. With such techniques, the charge state distribution was rather broad and not easy to control, but the precise spectroscopic studies of vacuum spark led to the first identification of HCIs in astrophysical plasma, for example, highly charged Fe in the solar corona [1]. These are very important studies as they dispelled the ideas concerning the elements coronium and nebulium and revealed the corona temperature to be several millions of degrees instead of several thousands of degrees believed at the time. Since then, many techniques for producing HCIs have been developed, among which the most successful one was the beam-foil (BF) method. The BF technique was developed independently by Bashkin [2] and Kay [3] and pioneered by several groups. In this technique, an ion beam from an accelerator passes through a thin foil (usually a carbon foil), where the ions are stripped and excited. Higher incident ion beam energies will lead to higher average charge state distributions of the ions passing through the foil. Other techniques, such as various spark forms (sliding, vacuum, etc.) and laser-produced plasmas, have also made important contributions to the knowledge of HCI spectra and structure. In recent years, the interest for HCIs has been growing due to the efforts to produce controllable fusion power. HCIs from wall materials would degrade the fusion plasma operation, as the photon radiation of the HCIs exhausts large amounts of the energy of the plasma, and the radiation power increases rapidly with increasing atomic number of the HCIs. On the other hand, the photons emitted from the HCIs provide valuable diagnostics of the plasma temperature, plasma density, and plasma movement. Being both a blessing and a menace for the success of controlled fusion experiments, detailed knowledge of the HCIs involved in fusion plasmas is imperative. Laser plasma spectra often suffer from many charge states, satellite line contamination, and Doppler shift and broadening. BF spectroscopy of HCIs requires access to large accelerator facilities and hence expensive.

A more convenient and cheaper method to produce HCIs in a controlled manner was suggested by Plumlee [4] in the 1950s. He suggested using an electron beam to create HCIs, at the same time to trap them. The creation process being that of electron impact ionization, whereas the trapping would be provided by the space charge field of the electron beam. Although this seems to be a rather straightforward idea, the practical application of this to produce the first electron beam ion source (EBIS) took more than 10 years [5]. Then in 1986, the first electron beam ion trap (EBIT) was developed at the Lawrence Livermore National Laboratory (LLNL) based on the currently available EBISs at that time, along with some further improvements by Marrs et al. [6]. The original design of the LLNL EBIT allowed operation up to electron beam energy of around 40 keV. However, it was consequently redesigned to operate at much higher electron beam energies (up to 200 keV) in 1993. The power of this technique was demonstrated in 1994 when bare uranium was detected in the LLNL EBIT via radiative recombination (RR). This experiment was done with the electron beam energy set at 198 keV [7]. After the groundbreaking developments at LLNL, EBITs have been set up in Oxford [8], NIST [9], Tokyo [10], Berlin [11], Freiburg [12] (since relocated to Heidelberg), Dresden [13], Shanghai [14], Vancouver [15], Stockholm [16], Belfast [17], and Darmstadt [18].

Before the development of EBITs, HCIs of charge state above 30 only were available from large accelerators and hence not convenient for easy access and study. As was shown in LLNL, EBITs can produce almost any charge state of any element and capable of acting both as HCI light sources and ion sources. As HCI ion sources, EBITs can provide low-velocity ions with much higher charge states compared with electron cyclotron resonance ion sources and with much lower cost compared with high-energy accelerators. As HCI light sources, they can basically provide light from emission states of any charge state of any element in the periodic table. Because the HCIs produced in an EBIT are moving at much lower velocities compared to those produced in heavy ion accelerators, the spectra suffer much less from Doppler shifts and spectral line broadening. These are very good characteristics for precise spectroscopic research. And because of the flexibility in producing various ions with EBITs, such devices are very powerful tools for studies along isoelectronic and isonuclear-charge sequences aimed to reveal the underlying physics behind many physical properties. In an EBIT, tenuous plasmas can be formed from basically any element, as long as the element can be injected into the trap region. The way an EBIT works gifts it with the property of a tunable and almost mono-energy electron beam. The range the tunable electron beam energy can cover spans from a few hundreds of eV to few hundreds of keV. This property makes it possible to use EBITs, especially when combined with spectroscopic techniques, for detailed studies of processes in hot plasmas, that is, disentanglement studies of the various processes in hot plasmas and to assist plasma diagnostics for temperature, density, electromagnetic field, and ion motion.

2.2 Electron Beam Ion Trap

As is well known for two-particle collisions, only when the mass of one partner is negligible compared to the other, can the kinetic energy of the lighter partner be transferred efficiently to its heavier partners' internal energy. On the other hand, in collisions between two particles of comparable mass, a large part of the kinetic energy must remain as "center of mass motion" so as to make both energy and momentum conservation rules happy. In the later case, the efficiency of energy transfer is much lower, hence, colliders in which zero center of mass motion is obtained were invented to overcome this problem. Fortunately, in the case of EBIT, the collisions belong to the former case, that is, light electrons and heavy atoms/ions, and hence the energy transfer efficiency is very high.

The core of an EBIT is an area where ions are trapped and undergo successive collisions with electrons, with ions, and with neutral atoms or molecules of the residual gas. This was transplanted from EBIS devices [5], which were the first successful application of Plumlee's idea [4]. The main structure of an EBIS is shown in Figure 2.1. The principal features of an EBIS are: an electron gun to produce the electron beam; a (superconducting) magnetic solenoid to produce a (strong) magnetic field for compressing the electron beam to a higher current density; a cryogenic system to maintain the superconduction (if a superconducting magnetic solenoid is employed) and to produce the needed ultrahigh vacuum in the trap region; drift tubes to produce a potential well for axial trapping of the ions and to supply the high voltage (relative to the cathode of the electron gun) required to accelerate the electron beam to

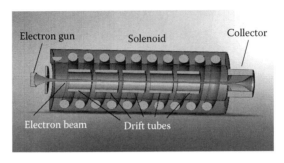

FIGURE 2.1
The main parts of an EBIS.

the needed high energy; the electron beam to provide radial trapping of ions and at the same time to collide with atoms and ions to successively strip their electrons and to excite them; an electron collector to collect the electrons after they have finished their job; and an ion extraction assembly to remove the HCIs from the trap region, to inject them into an accelerator, or to introduce them to a collision chamber for other collision experiments. An EBIS is mainly used as an ion source, and the ion intensity is the major concern. Hence a long drift tube assembly (around 1 m) is most commonly chosen in EBIS design, as the number of trapped ions and consequently extracted ions increases with increasing drift tube length. Unfortunately, at the same time, the long drift tube introduces plasma instability inside the trap region. These instabilities in some sense prohibit getting very high charge states of the trapped ions. Marrs et al., in developing their EBIT, shrank the trap region to only a few centimeters, which made evaporation cooling more efficient and so effectively solved the problem of plasma instabilities, hence removed many of the obstacles hindering the production of extremely high charge states. On top of this, they used a pair of superconducting Helmholtz coils instead of a solenoid to produce a both high and uniform magnetic field inside the trap region. This construction immediately made it possible to directly view the trap region of the EBIT through observation ports on the central drift tube using various diagnostic devises. This made EBIT not only an excellent source for HCIs, but also an excellent light source. A sketch of the main parts of an EBIT is shown in Figure 2.2.

2.2.1 Electron Gun

One of the most important parts in an EBIT is an electron gun. In many EBITs, spherical concave cathode Pierce-type guns are used. Normally, it contains a spherically concave cathode, a focus electrode, an anode, an exit electrode (also called second anode), and bucking coils. The electric field between the cathode and the anode extracts the electrons by providing them a longitudinal velocity. The focus electrode is to correct the fringing effect of this field. The

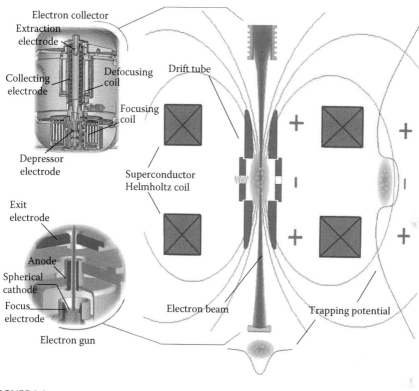

FIGURE 2.2
A sketch of the main parts of an EBIT.

exit electrode is to further adjust the electron beam for better quality. The main part of the electron gun is usually shielded by soft iron from the main magnetic field of the EBIT, and the bucking coils are to further offset the magnetic field at the cathode, so as to get a more easily controllable electron beam. An example is shown in the left-hand side lower inset in Figure 2.2, which is a sketch of the electron gun employed at the Shanghai EBIT. In the Shanghai EBIT, the cathode is a standard series barium tungsten dispenser cathode with Sc coating, from HeatWave Labs, Inc. (http://www.cathode.com). It works at 950–1200°C to overcome the working function of the cathode surface for emitting electrons. The perveance of the electron gun in the Shanghai EBIT is around $0.45\,\mathrm{A/V^{3/2}}$.

2.2.2 Drift Tube Assembly

The electron beam emitted by the gun is then accelerated by the high voltage between the drift tub assembly and the cathode. When the electrons arrive into the central drift tube region, they gain the kinetic energy required for specific tasks. The drift tube assembly in an EBIT normally contains three sections, as shown in Figure 2.2. In some cases, when EBIT and EBIS are combined, it could contain many sections. Bias voltages are applied on the two end sections, making them more positive relative to the central one, so a potential well is formed along the axis to trap positively charged ions. Radial trapping of the ions is done by the electron beam space charge combined with the axial strong magnetic field which will be described in Section 2.2.3. So when an EBIT is working, positively charged ions are trapped in the central drift tube, along the electron beam.

Normally the size of the ion cloud, also called the EBIT plasma, is lower than $100\,\mu m$ in diameter and few centimeters long.

2.2.3 Helmholtz Coils

The electron beam current density from an electron gun in an EBIT is usually around a few amperes per cm^2. However, it turns out that this is not enough as all the reaction rates of ion electron collisions are proportional to the current density instead of current only. So the electron beam is compressed by a strong magnetic field provided by a pair of Helmholtz coils. A strong compression, leading to an increase in the current density of three orders of magnitude (in the central drift tube region), can be achieved by using superconducting Helmholtz coils. The coils have to be designed carefully in order to get both a strong and uniform magnetic field in the center of the central drift tube. The magnetic field strength in that region is usually a few Tesla. As an example, in the Shanghai EBIT, the Helmholtz coils are made from Cu:SC = 1:1 superconducting wire from Oxford Instruments. The wires are fixed with a special technique to protect against displacement through quenching of the magnetic field. The highest field strength at the Shanghai EBIT could reach 5.5 T. The uniformity of the field along the electron beam is around 10^{-4} within the center trap region. The superconducting coils are immersed in a liquid helium vessel which is shielded by two cryogenic screens, at 20 and 80 K, respectively, refrigerated by a CVI CGR-511 2-stage refrigerator, to reduce the consumption of liquid helium.

2.2.4 Electron Collector and Ion or Atom/Molecule Injection

The electron collector in an EBIT collects the electrons after they pass the drift tubes. Some EBITs also equipped with ion extraction assemblies to extract ions for collision studies, for retrapping, and so on. The left-hand side upper inset in Figure 2.2 shows the structure of the electron collector in the Shanghai EBIT. It contains a beam position monitor system, focusing and defocusing coils, depressor electrodes, collecting electrodes, and extraction electrode (for ion injection and extraction). The whole collector is magnetically shielded. The beam monitor is to monitor the electron beam position, to ensure it is traveling along the central axis of the EBIT system. The focusing coil is to make sure the electron beam can properly enter the collector assembly, the defocusing coil is to help spread the electron beam onto the collecting electrode, and the depressor electrode is to prevent second electrons (eject from the collector caused by the electron beam hitting it) coming back to the drift tube region. Gas injection, Knudson cell, metal vapor vacuum arc (MEVVA) ion source, and sometimes laser ion source are used to inject atoms, molecules, or low-charge ions into an EBIT. Low-charge ions are usually produced by MEVVA or laser ion sources and introduced through the electron collector to the central drift tube region, while neutral species are introduced through observation ports on the central drift tube. After atoms or molecules are introduced into the EBIT, which become positively charged ions as soon as they are hit by the electron beam, and subsequently get trapped in the central drift tube. In the Shanghai EBIT, there are eight observation ports on the central drift tube to facilitate beam diagnostics, gas injection, and spectroscopy from visible through VUV up to the hard x-ray region.

A more realistic drawing of the Shanghai EBIT is shown in Figure 2.3. For high-energy EBITs, usually the electron collector and the electron gun are sitting on a same high (negative)-voltage platform to reduce the electric power consumption and hence to reduce the heat production. The drift tube assembly is sitting on a separate high (positive)-voltage platform. The capability of the high voltage difference between these two platforms roughly decides the upper limit of the electron beam energy of an EBIT.

Let us now consider the requirements on the electron beam energy of an EBIT. In principle, as long as the energy gained by the electrons, through acceleration on their way from the electron gun cathode to the central drift tube, exceeds the binding energy of the electron to be striped from an ion (or atom), the ion (or atom) can be further ionized. In an atom or ion, inner electrons are bound more tightly by the nuclear charge, and so need more energy to get ionized. The innermost shell is called K-shell (principal number $n = 1$) in which only up to two electrons can coexist. The neighboring outer shell is the L-shell ($n = 2$), in which up to eight electrons can exist and are less tightly bound. Then from inside to outside, there are $M, N, O, \ldots$ shells. Along an isoelectronic sequence, electrons in the same shell are bound tighter for heavier ions (with higher atomic number Z). So, with a higher energy electron beam, the shell accessible for ionization gets deeper (in other words, closer to the nucleus) or heavier ions along a given isoelectronic sequence can be ionized. For example, when the electron beam energy exceeds the binding energy of the only electron in a hydrogen atom, 13.6 eV, the H atom can be ionized to become a bare nucleus. However, for a hydrogen-like uranium ion, the ground-state binding energy of the electron is 131.82 keV. Only when the electron beam energy exceeds 131.82 keV, does it become possible for bare U to appear. However, it does not mean that with an electron beam energy of 132 keV, the trap will be full of bare U ions. Even with much higher energy, say around 200 keV, only a few bare U were observed in the Livermore Super EBIT. The reasons will be discussed in Section 2.3.

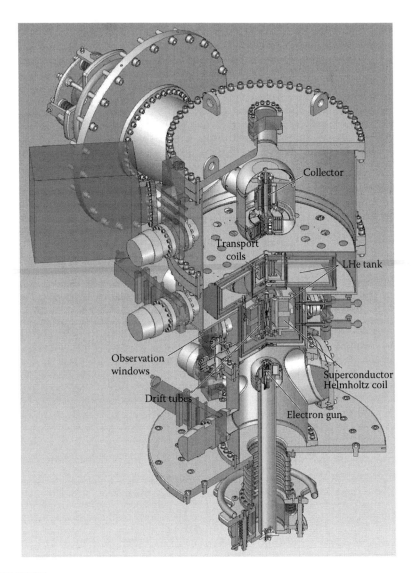

FIGURE 2.3
A drawing of the Shanghai EBIT.

2.3 The Ions Trapped in an EBIT

When ions are trapped inside the central drift tube region of an EBIT, they undergo various collisions with electrons, with ions, and with residual gas atoms (or molecules). During collisions with electrons, they can be ionized by direct and indirect ionizations, excited by direct and indirect excitations, and can capture electrons through resonant and nonresonant recombinations. Ionization and excitation are similar processes; in the former case, electrons are excited to continuous states, whereas in the later case to bound states. Ionization contributes to increase the ion charge, while excitation does not. However, recombination always lowers the ion charge. Although (cryogenic) EBITs are run under very high vacuum, residual gas never totally disappears. Collisions of the trapped ions with the residual gas atoms (or molecules) will lead to charge exchange (CX), that is, the ions capture some loosely bound electrons from the atoms (molecules). The CX cross-sections of HCIs are rather large and hence very efficient in lowering the charge states. Even for modest charge state, like Ar^{16+}, the CX cross-section is about five orders of magnitude higher than the electron impact ionization cross-section at around threshold. The charge state distribution in the trap is determined by the competition of the above-mentioned processes. To get very high charge states, high electron beam energy is necessary, and at the same time it is important to suppress CX processes by keeping the EBIT under a vacuum as high as possible. On the other hand, to increase the electron beam density, so as to increase the ionization rate, will also help.

During collisions with energetic electrons, ions are continuously heated. When the ions obtain sufficient kinetic energy to overcome the trapping potential, they will escape from the trap. The characteristic time for this process depends on the depth of the trap, the electron beam energy and density, ion species, and so on. The kinetic energy transferred from the electron beam to the ions will be redistributed among the ions through ion–ion collisions. The characteristic time for the redistribution is on the order of millisecond. Based on this, evaporative cooling is realized to keep the trapped ions longer before they get hot enough to escape. The trapping potential, which equals to the trapping voltage multiplied by the ion charge, is different for different charged ions. Higher charged ions are trapped more tightly, even with the same bias on the drift tubes. Although higher charge states have higher heating rates from the collisions with the energetic electron beam, they also have higher collision rates with other ions to redistribute the obtained kinetic energy. In a short while, lower charged ions will get enough kinetic energy and escape from the trap carrying away the "heat" they acquired. If lower charged ions are continuously supplied, they can continuously obtain "heat" from higher charged ions and take the "heat" away. In this way, the trap can be cooled down, and the higher charged ions can then accumulate "heat" much slower and so stay inside the trap longer. This is the principle of evaporative cooling. To realize this, light element atoms (molecules) are introduced into the trap continuously, as soon as they collide with the electron beam and become ionized, and they can cool the higher charged ions inside the trap in the way described above. By virtue of this technique, it is possible to keep HCIs trapped in an EBIT for a few hours.

A detailed understanding of the factors determining charge state distributions, ion temperatures, and so on in an EBIT device is very important in planning and understanding experiments. Adjusting operating parameters such as electron beam energy, electron beam current, magnetic field strength, axial trapping potential, injection density, and so on can be facilitated by such an understanding. This requires considerable knowledge of the physical processes taking place in the trap region of an EBIT. Taking into account the processes of electron impact ionization, dielectronic recombination (DR), RR, ion–atom collisions involving multiple CX (up to four-electron transfer), and ion escape caused by too high kinetic energy, the rate equation controlling the evolution of the ion density of a specific charge state in an EBIT trap can be expressed as [19]

$$
\frac{dn_i}{dt} = R^{ion}_{i-1 \to i} n_{i-1} - R^{ion}_{i \to i+1} n_i + R^{DR}_{i+1 \to i} n_{i+1} - R^{DR}_{i \to i-1} n_i + R^{RR}_{i+1 \to i} n_{i+1}
$$

$$
- R^{RR}_{i \to i-1} n_i + \sum_{j=1}^{4} R^{CX}_{i+j \to i} n_{i+j} - \sum_{j=1}^{4} R^{CX}_{i \to i-j} n_i - R^{axesc}_i n_i
$$

$$
- R^{resc}_i n_i + R^{source}_i, \tag{2.1}
$$

where $R^{ion}_{i \to j}$ is the effective ionization rate from charge state q_i to q_j, $R^{DR}_{i \to j}$ the effective DR rate, $R^{RR}_{i \to j}$ the effective RR rate, $R^{CX}_{i \to j}$ the effective CX rate, R^{axesc}_i and R^{resc}_i are, respectively, the axial and radial escape rates, and R^{source}_i the source rate caused by CX between ions with continually injected neutral atoms (of the same element species) given by

$$
R^{source}_i = \begin{cases} \sum_{j=i}^{z} n_j R^{CX}_{j \to j-i}, & i \leq 4, \\ 0, & i > 4. \end{cases} \tag{2.2}
$$

The effective rate for electron impact ionization is determined by the electron impact ionization cross-section $\sigma^{ion}_{i \to i+1}$, the geometrical electron ion overlap factor $f(e, i)$, and the electron beam density j_e/e,

$$
R^{ion}_{i \to i+1} = \frac{j_e}{e} \sigma^{ion}_{i \to i+1} f(e, i). \tag{2.3}
$$

A semiempirical formula [20] can be employed to evaluate the ionization cross-section from charge state q_i to q_{i+1}:

$$
\sigma^{ion}_{i \to i+1} = \sum_j \frac{a_{ij} \zeta_j}{E_e I_j} \ln \frac{E_e}{I_j} \left\{ 1 - b_{ij} \exp\left(-c_{ij}\left[\frac{E_e}{I_j} - 1\right]\right) \right\}. \tag{2.4}
$$

In which E_e is the electron beam energy, ζ_j the number of electrons in the jth subshell, I_j the binding energy, and a_{ij}, b_{ij}, c_{ij} are constants. The parameters $a_{ij} = 4.5 \times 10^{-14}$ cm^2 eV^2, $b_{ij} = 0$, and $c_{ij} = 0$ will lead to reasonably good results for ion charge states above 3, but two or three times underestimation is expected for lower charge states [21]. The electron beam energy in EBIT is usually high enough so that relativistic correction is needed $\sigma^{rion}_{i \to i+1} = \xi \sigma^{ion}_{i \to i+1}$ [22]:

$$\xi = \left(\frac{\tau + 2}{\varepsilon + 2} \right) \left(\frac{\varepsilon + 1}{\tau + 1} \right)^2 \left(\frac{(\tau + \varepsilon)(\varepsilon + 2)(\tau + 1)^2}{\varepsilon(\varepsilon + 2)(\tau + 1)^2 + \tau(\tau + 2)} \right)^{3/2}. \tag{2.5}$$

Here ε and τ are the electron beam energy and target binding energy, respectively, in units of the electron rest mass.

DR is a resonant process. In this process, a free electron is captured by an ion, and at the same time a bound electron in the ion is promoted, forming a multiply excited intermediate state situated above the autoionization threshold. The process is finally completed by stabilization through emitting one or more photons, so as to reduce the ion energy to below its ionization limit. DR processes are generally labeled using an inverse Auger process notation. For instance, a process in which an electron is captured, whereas a bound electron is excited from the K-shell, results in an autoionization state in which one of the two active electrons is in the L-shell, and the other in an n shell (n can be L, M, N, O, etc.). This process is labeled as KLn DR. DR process only happens when the excess energy in capturing the free electron matches the excitation energy for promoting the bound electron in the ion. In an EBIT, the electron beam energy is almost monochromatic with a small spread, and hence resonant processes, such as DR, can be neglected basically. Such processes must be considered specifically at the resonance energies.

In contrast, RR is a nonresonant process that can occur at any electron energy. In this process, a free electron is captured by an ion, and the excess energy is released by photon emission. The excess energy equals the sum of the kinetic energy of the free electron and the ionization energy of the final state of RR. The effective rate of this process in an EBIT plasma which can be written as

$$R^{RR}_{i \to i-1} = \frac{j_e}{e} \sigma^{RR}_{i \to i-1} f(e, i) \tag{2.6}$$

and the RR cross-section σ^{RR} can be described by a semiempirical formula given by Kim and Pratt [23]:

$$\sigma^{RR}_{i \to i-1} = \frac{8\pi}{3\sqrt{3}} \alpha \lambda_e^2 \chi \ln \left[1 + \frac{\chi}{2(n_v)_{eff}^2} \right], \tag{2.7}$$

in which α is the fine structure constant, λ_e electron Compton wavelength,

$$\chi = \frac{2Z_{eff}^2 I_H}{E_e}, \tag{2.8}$$

$$Z_{eff} = \frac{1}{2}(Z + q_i), \tag{2.9}$$

and

$$(n_v)_{eff} = n_v + (1 - W_{n_v}) - 0.3, \tag{2.10}$$

where Z is the atomic number, n_v the principal quantum number of the valence shell of the recombined electron, and W_{n_v} the number ratio of the unoccupied states over the total states of the valence shell.

In all of the above-mentioned processes involving electron ion collisions, the electron beam and ion cloud over lap factor $f(e, i)$ is a matter that needs to be noticed. Very often it is assumed to be 1, which means that the ion cloud and electron beam are 100% overlapped. But it is not true. Assume a radial Boltzmann distribution for the ion cloud of charge state q_i:

$$n_i(r) = n_i(0) \exp\left(\frac{-eq_i \varphi(r)}{kT_i}\right) \tag{2.11}$$

and a Gaussian shape for the electron beam radial distribution:

$$n_e(r) = n_e(0) \exp\left(-\frac{r^2}{r_e^2} \ln 2\right), \tag{2.12}$$

$n_i(0)$ and $n_e(0)$ denote the densities of the ion cloud and of the electron beam along the central axis, respectively, and the potential function

$$\varphi(r) = \varphi_e(r) + \varphi_i(r) \tag{2.13}$$

accounts for radial trapping by both the electron beam space charge and radial screening from ions closer to the beam center. Ignoring small variation along the Z-axis, the overlap factor between the electron beam and ion cloud can then be expressed as

$$f(e, i) = \frac{\int_0^{r_{dt}} 2\pi L\, dr \int_0^{r_{dt}} 2\pi r^2 L\, dr\, n_e(0) \exp(-r^2 \ln 2/r_e^2) n_i(0) \exp(-eq_i \varphi(r)_i/kT_i)}{\int_0^{r_{dt}} 2\pi L r_1\, dr_1 n_e(0) \exp(-r_1^2 \ln 2/r_e^2) \int_0^{r_{dt}} 2\pi L r_2\, dr_2 n_i(0)} \\ \times \exp(-eq_i \varphi(r_2)_i/kT_i) \tag{2.14}$$

where L and r_{dt} are the length and radius of the drift tube, respectively.

The electron beam is not involved in the CX process in an EBIT. The effective rate of the CX process is

$$R^{CX}_{i \to i-r} = n_0 \sigma^{CX}_{i \to i-r} \bar{v}_i, \tag{2.15}$$

in which n_0 is the density of the neutral atoms and $\bar{v}_i$ the mean velocity of the ions with charge q_i. In the work done by Selberg et al. [24], semiempirical scaling rules for transfer of a well-defined number of target electrons from various target atoms were obtained, and a good agreement with experiments was seen. According to the scaling rule, the CX cross-section with r electron transfer, for an ion of charge q_i is

$$\sigma_{i \to i-r}^{CX} = \frac{(2.7 \times 10^{-13})q_i r}{I_1^2 I_r^2 \sum_{j=1}^{N} (j/I_j^2)},$$

(2.16)

where I_j is the ionization potential (in eV) of the jth electron, N the number of outer shell electrons, and the cross-section is given in unit of cm^2.

As was mentioned above, the ions trapped in an EBIT are heated all the time by the collisions with the electrons of the electron beam, and then distribute the heat through collisions with other ions. Actually ion–ion collisions tend to reduce the thermal difference among the ions. When the ions are thermally energetic enough, they can overcome the trap potential and escape from the trap. The escape rate can be described as [25]

$$R_i^{esc} = -n_i v_i \left[\frac{e^{-\omega_i}}{\omega_i} - \sqrt{\omega_i}(\mathrm{erf}(\omega_i) - 1) \right],$$

(2.17)

$$\omega_i = \frac{q_i e V_\omega}{kT_i},$$

(2.18)

where V_ω is the trap depth and T_i the temperature of the ions of charge state q_i. The total Coulomb collision rate is

$$v_i = \sum_j v_{ij}^{eff}.$$

(2.19)

The effective collision rate

$$v_{ij}^{eff} = \frac{4}{3}\sqrt{2\pi}n_j \left(\frac{q_i q_j e^2}{4\pi\varepsilon_0 M_i} \right)^2 \left(\frac{M_i}{kT_i} \right)^{3/2} \ln \Lambda_{ij} f(r_i, r_j),$$

(2.20)

in which M_i is the mass of the ion q_i and $\ln \Lambda_{ij}$ the ion–ion Coulomb logarithm. Now the overlap between the q_i and q_j ion clouds has to be considered. The overlap factor is then

$$f(r_i, r_j) = \frac{\int_0^{r_{dt}} 2\pi L \, dr \int_0^{r_{dt}} 2\pi L r^2 \, dr \, n_i(0) \exp(-eq_i\varphi(r)_i/kT_i)n_j(0)}{\int_0^{r_{dt}} 2\pi L r_1 \, dr_1 \, n_i(0) \exp(-eq_i\varphi(r_1)_i/kT_i) \int_0^{r_{dt}} 2\pi L r_2 \, dr_2 \, n_j(0)} \times \frac{\times \exp(-eq_j\varphi(r)_j/kT_j)}{\times \exp(-eq_j\varphi(r_2)_j/kT_j)}.$$

(2.21)

Here $n_i(0)$ and $n_j(0)$ are the ion densities along the central axis, and $\varphi(r)_i$ and $\varphi(r)_j$ are the effective radial potentials experienced by the two ion species.

The ion temperature in an EBIT is determined by electron collision heating, energy exchange among ions, and ion escape. The rate equation for ion temperature evolution is then

$$\frac{d}{dt}(n_i k T_i) = \left[\frac{d}{dt}(n_i k T_i)\right]^{\text{heat}} + \left[\frac{d}{dt}(n_i k T_i)\right]^{\text{ex}} - \left[\frac{d}{dt}(n_i k T_i)\right]^{\text{axesc}}$$

$$- \left[\frac{d}{dt}(n_i k T_i)\right]^{\text{resc}}. \tag{2.22}$$

The contribution from electron heating [26] is

$$\left[\frac{dE_i}{dt}\right]^{\text{heat}} = \left[\frac{d}{dt}(n_i k T_i)\right]^{\text{heat}} = \frac{4m_e}{3M_i} v_{ei} n_i E_e, \tag{2.23}$$

in which the electron–ion Coulomb collision frequency is

$$v_{ei} = 4\pi \frac{n_e}{v_e^3} \left(\frac{q_i e^2}{4\pi \varepsilon_0 m_e}\right)^2 \ln \Lambda_i, \tag{2.24}$$

where v_e is the electron velocity, ε_0 the permittivity of free space, q_i is the ion charge state, and $\ln \Lambda_i$ the electron–ion Coulomb logarithm. The contribution to the temperature evolution from collisions between the ions of charge states q_i and q_j is [25]

$$\left[\frac{dE_i}{dt}\right]^{\text{ex}} = \left[\frac{d}{dt}(n_i k T_i)\right]^{\text{ex}} = \sum_j 2v_{ij}^{\text{eff}} n_i \frac{M_i}{M_j} \frac{k(T_j - T_i)}{(1 + M_i T_j / M_j T_i)^{3/2}}. \tag{2.25}$$

Escaping ions take away energy, which can be expressed [27] for both axial and radial escape by

$$\left[\frac{dE_i}{dt}\right]^{\text{esc}} = \left[\frac{d}{dt}(n_i k T_i)\right]^{\text{esc}} = \left[\frac{2}{3} n_i v_i e^{-\omega_i} - \frac{dn_i}{dt}\right] k T_i \tag{2.26}$$

with different potential expressions in ω_i for the two escapes. Here v_i is the total Coulomb collision rate described above in Equation 2.19.

Using the above formulas, the time evolution of charge state distribution and ion temperature of the EBIT plasma can be calculated. The following is an example of a calculation done for Dy ions in an EBIT trap when the electron beam energy is 150 keV, the beam current 150 mA, the drift tubes are biased to produce an axial trap of 300 eV in depth, and under a magnetic field of 3 T. At time zero (when the trap was just closed), a certain amount of neutral Dy atoms were sent into the EBIT trap in order to have an initial density of neutral Dy of $N_{Dy} = 10^9 \, \text{cm}^{-3}$. For the purpose of seeing how the coolant gas works, Ne gas was injected continuously to keep a constant density of neutral Ne of $N_0 = 10^4 \, \text{cm}^{-3}$ in the EBIT. Figures 2.4 and 2.5 show the time evolution of the charge state distributions of Dy and Ne ions, respectively. From Figure 2.4,

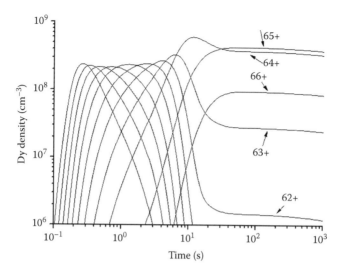

FIGURE 2.4

The time evolution of the charge state distributions of Dy ions.

we can see that in the early stages, lower charge states are dominate, most of them grow up with time at the beginning, and consequently decline when higher charge states start growing up. Equilibrium establishes after the highest charge states have appeared. This is quite obvious, as electrons in an ion are stripped sequentially by the electron beam, and it takes longer to strip

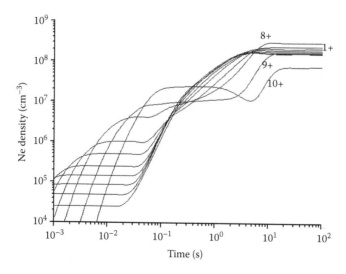

FIGURE 2.5

The time evolution of the charge state distributions of Ne ions.

away a higher number of electrons and hence to reach the higher charge states. With the condition described above, equilibrium is established about 30 s after the trap is closed and at the equilibrium, charge state of 65+ is most favorable, 64+ the second, 66+ is the highest charge state reached and also the third favorable. We notice that when higher charge states increase, the lower charge states decrease. This is because there is no more supply of Dy atoms after time zero. This is not true for the Ne case shown in Figure 2.5. Although lower charge states also appear earlier than higher charge states in Figure 2.5, they do not decline with the appearance of the higher charge states. Instead, they experience a relatively constant period and then increase again until equilibrium is established. Higher charge states behave in a similar way but with a time delay. For the higher charge state, more delay is seen. When equilibrium is reached, all the charge states keep more or less constant with the time. It is worth to notice that the Ne atoms are injected constantly. The constant density of Ne ions means that the escape rate of the Ne ions should balance the injection. For the equilibrium charge state distribution of Ne ions, the most favorable charge state is 8+, the second and third favorable are 1+ and 2+, respectively, instead of 9+ or 7+. Bare Ne ions exist only as a small (but not negligible) fraction. Favoritism of the lower charged Ne ions roots from the constant injection of neutral Ne atoms, which leads constant production of fresh low-charged Ne ions. Otherwise, with such a high electron beam energy, one would expect that bare Ne ions would be the most probable charge state.

In the time period between 1 and 10 s, there is a dip in the Ne^{10+} ion density. During this same time period, the drop off of the middle high-charge state Dy ions is steeper than the drop off of the less highly charged Dy ions in the time period earlier. The cooling effects of middle charge state ions to the higher charge states should take the responsibility for the special behavior in this time period. After this period, the cooling duty transfers to lower charge states, mostly the singly charged Ne ions. This can be proven (see Figure 2.6) by the equilibrium temperature of the Ne^{1+} of 300 eV, same as the trapping potential for Ne^{1+}. Figure 2.6 shows the temperature evolution of both Dy ions and Ne ions. We can see from Figure 2.6 that the equilibrium temperatures of Ne^{q+} ions or Dy^{p+} ions are all lower than q times 300 eV or p times 300 eV.

We can expect that the constant injection of neutral Ne atoms finally contributes as a constant supply of Ne^{1+} ions, and the constant escape of the Ne^{1+} ions takes away thermal energy at a constant rate from the EBIT trap, which would balance the constant heat load from electron beam impact. So we can expect a constant temperature of the ions inside the EBIT trap when the equilibrium is reached. This is confirmed also by the results of the simulation which are shown in Figure 2.6. Although the highly charged Dy ions have higher temperatures, as shown in Figure 2.6, their trapping potential is also higher. Thanks to the coolant activities of Ne^{1+}, highly charged Dy ions can stay inside the EBIT trap for a long time.

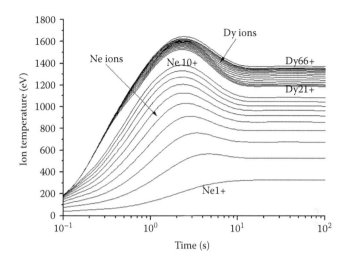

FIGURE 2.6
The temperature evolution of both Dy ions and Ne ions.

2.4 HCI Spectroscopy at EBITs

When ions are trapped in an EBIT, they undergo numerous collision processes with electrons, with ions, and with atoms or molecules. Many of these processes result in photon emission. By properly adjusting the electron beam energy, electron beam current, magnetic field strength, and coolant density, the charge states balance in the EBIT plasma can be controlled. By properly tuning the electron beam energy, the excitation states of the ions in the EBIT plasma can also be controlled to some degree. EBIT plasma usually has a cylindrical shape, 2 cm long, and 70–100 μm in diameter being quite typical. Hence the EBIT plasma is basically a line source, very suitable for grating and crystal spectrometers for high-resolution spectroscopic studies. A sketch of the setup for spectroscopic experiments at an EBIT is shown in Figure 2.7. Spectrometers can either view the EBIT plasma directly or be coupled through lenses or other focusing elements to the EBIT.

2.4.1 HCI Spectroscopy for Fundamental Physics

As EBIT can basically produce and trap ions of any charge states of any elements and is an ideal light source for spectroscopic studies of ions along an isoelectronic sequence. Due to this property, one can access many interesting physical phenomena. An isoelectronic sequence is a sequence of ions with a similar electronic configuration, for example, along the H sequence, we have H, He^{1+}, Li^{2+}, Be^{3+}, B^{4+},..., U^{91+},..., and for the He sequence He,

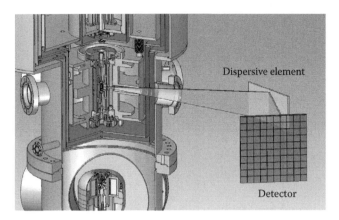

FIGURE 2.7
A sketch of setup for spectroscopic experiments at an EBIT.

Li^{1+}, Be^{2+}, B^{3+}, ..., U^{90+}, ... The physical processes giving contributions to the final energy or lifetime of an atomic energy level evolve differently along an isoelectronic sequence. Hence, the behavior of a physical quantity along the sequence can often reveal the underlying physics. For example, along the H sequence, the energy intervals between the main shells are proportional to the square of the atomic numbers Z, whereas the energy intervals between subshells inside a main shell are linearly proportional to Z. The spin-orbit interaction, which arises from relativistic effects, is proportional to the fourth power of Z and also the Lamb shift, which is a manifestation of quantum electrodynamic (QED) effects, scales roughly with the fourth power of Z. Hyperfine structure scales with the third power of Z, and so on. Table 2.1 lists the Z dependence for the probabilities of allowed and forbidden transitions [28], hyperfine interaction, QED effects, relativistic effects, and parity violation effects along the H isoelectronic sequence. All the transition rates in Table 2.1 increase as a function of the atomic number, and most of the forbidden transition rates increase faster than those of the allowed transitions (this of course is embedded in the definition of allowed and forbidden transitions). Hence, for very-highly charged ions, many forbidden transitions are comparable in rates to allowed transitions or in some cases even exceed the rates of allowed transitions. Different forbidden transitions are induced by different physical effects. Spin forbidden transitions come from mixing of wave functions of different spin states by the spin-orbit interaction. Parity violation transitions come from the mixing of different parity wave functions by the weak interaction arising from the small overlap of electron and nuclear wave functions. Through spectroscopic studies of various forbidden transitions, one can test theories of not only atomic physics, but also parity violation, nuclear structure, and so on. Examples of spectroscopic studies of forbidden transitions at EBIT are shown in [29,30]. In the work of [29], magnetic

TABLE 2.1

The Z-dependence of the Probabilities of Allowed and Forbidden Transitions, Hyperfine Interaction, QED Effects, Relativistic Effects and Parity Violation Effects along the H Isoelectronic Sequence

E1 ($\Delta n = 0$)	Z
E1 ($\Delta n \neq 0$)	Z^4
M1 ($\Delta n = 0$)	Z^3
M1 ($\Delta n \neq 0$)	Z^6
M1 (within fine structure)	Z^{12}
E2 ($\Delta n = 0$)	Z
E2 ($\Delta n \neq 0$)	Z^6
E2 (within fine structure)	Z^{16}
2E1	Z^6
E1M1	Z^6
Hyperfine splitting	Z^3
QED effect	Z^4
ESO	Z^4
Parity violation	Z^5

dipole transitions of titanium-like xenon and barium ions were studied. The transition was $3d^4\ ^5D_2 - {}^5D_3$ within the ground configuration. In their experiments, light emitted from the NIST EBIT was imaged by a pair of lenses to the entrance slit of an $f/3.5$ Ebert scanning monochromator, and detected by a blue sensitive photomultiplier. Their results show poor agreements with *ab initio* calculations. In [30], magnetic dipole transitions of highly charged argon and barium ions were studied using the Oxford EBIT. Light from the Oxford EBIT was also imaged by a pair of lenses, but onto the entrance slit of a 1 m focal length, $f/10$ Monospek plane-grating spectrometer, a charge-coupled device (CCD) was used as a detector instead of a photomultiplier. The results for the same transitions of the same ions in [29,30] agree with each other, with higher accuracies for the later measurements. Discrepancies with theoretical predictions persist. Hence these measurements highlight the need for improving calculations for many electron atoms.

QED effects have been stringently tested by ultrahigh precision spectroscopy for H atoms. However, higher-order effects start to become influential for heavier ions, giving significant contribution to the self-energy term for $Z > 40$ [31]. With increasing Z, the overlap of electronic and nuclear wave functions becomes substantial, and the electrostatic distribution of the nucleus not only affects the eigenvalues of the Dirac equation, but also has implications for the Lamb shift. This contribution will exceed that from vacuum polarization for $Z > 70$ [32] and reach the same order of magnitude as the self-energy term. These effects can only be studied through experiments using very heavy

HCIs. There are many good works done on related aspects since the first EBIT was set up in LLNL, the examples of which are shown in [33–35].

The hyperfine interaction has been mostly studied through its effect on spectral line splitting. For $Z > 50$ H-like ions however, the ground-state splitting is already large enough so that transitions among these levels fall into the range suitable for very high-resolution spectroscopic studies, that is, the visible region. Precise studies of hyperfine splitting can lead to important knowledge concerning nuclear electric and magnetic distributions. These properties are very important to not only nuclear physics, but also to a precise understanding of QED effects. EBIT is an excellent machine for performing such studies, as can be seen in [36–38]. In [36], a direct measurement of the spontaneous emission due to hyperfine splitting of the ground state of hydrogen-like $^{165}Ho^{66+}$ was made, at the Super EBIT in LLNL. They measured the transition between the $F = 4$ and $F = 3$ levels of the $1s\ ^2S_{1/2}$ of $^{165}Ho^{66+}$ ions. The transition wavelength is around 5726 Å, in the visible range, so they used a prism spectrograph instead of a monochromator for higher efficiency, directly viewing the EBIT plasma at an electron beam energy of 132 keV and a current of 285 mA. Holmium was introduced into the EBIT via low-charge states Ho ions produced by an MEVVA source. They employed a CCD to record the photons after wavelength dispersion by the prism spectrograph. Their work leads to a renewed nuclear dipole magnetic moment for the ^{165}Ho nuclide. Using the same apparatus, the same group obtained the nuclear magnetization distribution radii of ^{185}Re and ^{187}Re nuclides, by studying the $F = 3$ to $F = 2$ hyperfine transitions in the $1s$ ground states of the two isotopes [37]. They found that these radii are substantially larger than their nuclear charge distribution radii.

Also the finite overlap of the wave functions of electrons and the atomic nucleus couples the states of the inner-shell electrons and nuclear particles, which can influence the rates of nuclear decays or even induce unexpected decay channels. A successful example of such a study can be seen in the work by Jung et al. at GSI [39]. Here the totally forbidden β decay of the neutral ^{163}Dy atom to ^{163}Ho was observed with bare ^{163}Dy ions.

2.4.2 EBIT-Based Spectroscopy in Assisting Plasma Diagnostics

Many laboratory and astrophysical plasmas involve a Maxwell–Boltzmann energy distribution of electrons colliding with an ensemble of ions. To understand the behavior of plasmas from their emission spectra needs accurate electron–ion collision rate coefficients, which include cross-sections for various processes, such as excitations, ionizations, recombinations, and so on. The cross-sections are needed in the whole energy range at a certain plasma temperatures. Theoretical calculations need to make approximations in almost all the cases and may not always lead to accurate results. Experimental studies using laboratory plasma sources, such as laser plasmas, tokamaks, and θ pinch devices, often suffer from limited accuracies caused by complications such as density effects, radiative transfer, ion abundance, field gradients and temperature effects, and so on, and the combinations of all these effects simultaneously.

At LLNL, a special setup was devised to simulate a Maxwell–Boltzmann electron distribution for clean and optically thin EBIT plasmas, in order to study collisions between ions of specific charge states with electrons of a Maxwell–Boltzmann energy distribution [40]. In their work, they developed a system to sweep the electron beam, in energy and current as a function of time, to simulate conditions in hot plasma. The current is swept to keep the electron density constant during the energy sweep, so as to keep the same effect of the space charge distribution. While the energy sweeping is done by converting the number distribution to time distribution, for example, for a higher number fraction at a given energy the time duration in sweeping was set to a longer value according to a Maxwell–Boltzmann distribution. The energy sweep covered the majority of the energy range of the electrons at a given temperature, with a minimum and a maximum energy cut off according to their EBIT operation conditions. Because EBIT operating parameters can be easily changed, a wide range of temperatures can be simulated. Ion density effects are generally unimportant, as the EBIT plasma is usually optically thin. By measuring the emitted photons at the same time as sweeping electron beam energy and current, they could automatically integrate the collision cross-sections with the desired electron energy distribution. The observed properties of ions in such experiments are directly dependent on the relevant rate coefficients [40].

The spectroscopy of HCIs, in particular, He-like and Li-like ions, is very useful for plasma diagnostics. Their spectra show, in addition to the strong resonance lines, numerous satellites which are produced by DRs and inner-shell excitations. Since the intensity ratios between the resonance lines and their dielectronic satellites depend only on the electron energy distribution, they are sensitive measures of electron temperature. The intensity ratios between the resonance lines and their inner-shell excited satellites are proportional to the relative density of the neighboring charge states. These satellites then provide information of charge balance and impurity ion transport. From the experimental point of view, the satellite spectra of He-like ions of medium Z elements are very attractive, because the separation between the resonance line and satellites are sufficiently large to allow for clean identifications. At the same time, the wavelength range of the spectrum is small enough, so that the sensitivity of the instrument is constant over the entire range. The measurements of the satellites-to-resonance line ratios are therefore reliable.

In a work done at TEXTOR-94 [41], satellite spectra of He-like Ar from a discharge with auxiliary heating by neutral-beam injection was measured, as shown in Figure 2 in [41]. In that work, the intensity ratio of the line q, from the Li-like $1s^2 2s\ ^2S_{1/2} - 1s2s2p\ ^2P_{1/2}$ transition, over the line w, from the He-like $1s^2\ ^1S_0 - 1s2p\ ^1P_1$ transition, was used to deduce the charge state ratio of Ar^{15+}/Ar^{16+}, believing that the satellite line q was produced by impact excitation of the ground state of Li-like Ar $1s^2 2s \, {}^2S_{1/2}$. However, the result was about four times higher than theoretical predictions. A few years later, the $n = 2 - 1$ spectral emission pattern of He-like Ar, together with the associated satellite emissions, was studied using high-resolution x-ray spectroscopy at the Berlin EBIT [42]. In the work [42], a mono-energetic electron beam was swept in energy between 1.9 and 3.9 keV (from below the resonance energy for dielectronic satellites to above the direct excitation threshold for $n = 1 - 2$), and spectra were recorded at each electron beam energy. In this way, the satellites caused by dielectronic resonance are separated from those caused by direct excitations, as they happen at different electron energies. Their results showed that the line q could be caused by both direct excitation of the ground-state Li-like Ar $1s^2 2s \, {}^2S_{1/2}$ and by DR of the ground-state He-like Ar $1s^2 \, {}^1S_0$. This questions the applicability of using the q/w line ratio to deduce the corresponding charge state ratio, that is, Li-like/He-like, directly.

X-ray emissions from He- and Li-like Ar ions are also used in diagnostics for inertial confinement fusion plasmas. Ar is usually added as a dopant to the fuel. The electron temperature is deduced from the intensity ratio of the transitions between $n = 3$ and 1 levels of H-like and He-like Ar, $I_{\text{H}\beta}/I_{\text{He}\beta}$, while the density is derived from Stark broadening of the K_β line. But the satellites from Li-like $1s3lnl' - 1s^2nl'$ transitions could distort the profile of the K_β line. They can cause asymmetry of the line profile and broadening of the line width. This would mislead the temperature and density interpretation, if the satellite contributions were not properly taken into account. These contributions were usually estimated by theoretical models, with only the $n = 2$ and 3 satellites being taken into consideration [43,44]. In 1995, the contributions of higher-n satellites to the K_β line of He-like Ar were measured [45], using a high-resolution crystal spectrometer to analyze the x-ray emission of the Ar ions trapped and excited in the LLNL EBIT. The electron beam energy was swept to hit the *KLM*, *KMM*, *KMN*, and *KMO* dielectronic resonances, as well as through the excitation threshold of $n = 1 - 3$. Their results showed that the resonance strength of $n = 4$ satellites was larger than that of $n = 3$, while the resonance strength of $n = 5$ was nearly equal to that of $n = 3$. This means that the contribution of the satellites from $n \geq 4$ was considerably larger than expected from standard scaling procedure. This result has lead to reassessment of line profile calculations of the He-like K_β line used in density diagnostics in laser fusion.

DR is an important process in high-temperature plasmas. It affects the charge state balance and energy-level populations as well as the plasma temperature. As mentioned above, resolvable satellite lines caused by DR are often used for electron temperature determinations, whereas unresolvable satellite lines disturb the determination of line shape, line intensity, and line position, consequently corrupting the determination of temperature, density, and ion movement in plasmas. Therefore, whether the influence of DR being good or bad to plasmas, it plays an important role. Its occurrence strength is vital for accurate modeling of high-temperature plasmas. DR studies are also important for testing atomic structure and atomic collision theories since such resonances carry information on QED, relativistic effects, electron correlations, and so on. There are many experiments done at EBITs for DR studies, including resonance strengths and resonance energies of various ion species. Examples are shown in [46–55]. In the work [53–55], resonance strengths and energies were studied for KLL DR of He-like up to O-like Xe ions at Shanghai EBIT. A schematic drawing of the experimental setup is shown in Figure 2.8.

In these experiments, the electron beam was compressed to a diameter (fullwidth at half-maximum) of 70 μm, under a magnetic field of 2–3 T. The energy of the electron beam was scanned through the KLL resonances, in the range 19.5–23.8 keV. During the experiments, Xe gas was continuously injected and the x-rays from the ion cloud in the EBIT were detected by a high-purity Ge detector in a direction perpendicular to the electron beam. By recording the

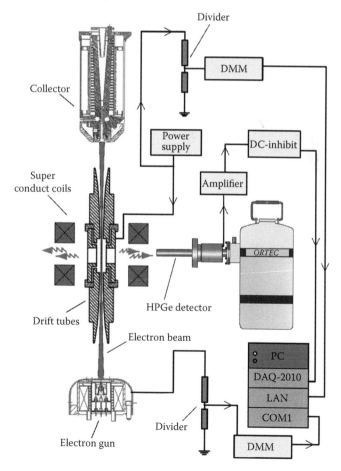

FIGURE 2.8
Schematic drawing of the experimental setup used in studying DR.

x-ray spectrum at each step of electron beam energy, two-dimensional spectra were obtained. An example is shown in Figure 2.9. This kind of spectrum is also called scatter plot, with the x-axis for electron beam energy, y-axis for photon energy (vice versa), and color shade depicts the intensity. The bright sports in Figure 2.9 show the events of KLL DR processes of He-like, Li-like, ... to O-like Xe ions. The rather uniform events in the inclined lines are from RR processes to the Xe ions, from the up most inclined line to lower, they are RR to $n = 2, 3, 4, \ldots$. It is not difficult to understand that the photon energy gets lower and the energy spread gets narrower for RR to higher n orbitals (n is the principal quantum number), and the energy interval between the RR to the neighboring orbitals also gets smaller for higher n, because the photon energy from an RR equals the kinetic energy of the free electron recombined, plus the ionization energy of the final state of the corresponding RR process. Projecting the events in a scatter plot onto its photon energy axis will lead to a photon emission spectrum, whereas projecting to the electron energy axis will lead to a spectrum of excitation function. By properly analyzing these spectra, the relevant DR strengths and energies can be obtained.

In the experiments in [53,54], the electron beam energy was scanned very slowly, in the so-called steady-state mode. This way, the EBIT plasma can reach equilibrium at each step of electron beam energy. This is good for resonance energy measurements, as EBIT is a complicated system, and an RC delay is not avoidable. The steady-state measurements [54], with the help of the high-precision high-stability high-voltage deviders developed in [56], lead to very high precisions of the DR resonance energy determinations, at an average

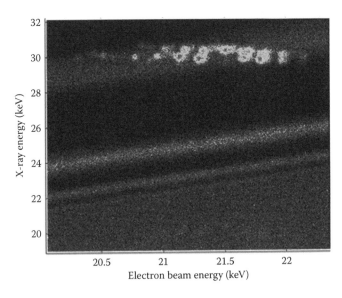

FIGURE 2.9
An x-ray scatter plot of Xe ions in the electron beam energy range 20–22.4 keV.

of 0.03%. But in the steady-state mode, one has less control over the charge states of the ions in the EBIT plasma. In fact, the charge state balance changes when scanning through a DR resonance in the steady-state mode, as the ions have enough time to recombine in a significant number. This fact makes the data analysis extremely complicated and introduces more uncertainties in deducing the DR strengths. A typical uncertainty in DR strengths determination of the middle high Z ions, using a steady-state mode as shown in [53] is around 20%. To improve the precision in DR strength determination, work [55] used a fast scan technique. The timing scheme employed in [55] started with 4 s charge breeding time (cooking time) at the electron beam energy of 23.0 keV, to obtain the desired charge state distribution (see Figure 2.10). The beam energy was then ramped linearly down through the KLL DR resonances to 19.8 keV and back up in 25 ms. The beam energy was then kept at 23.0 keV for another 75 ms before the next ramping in order to maintain the charge state distribution. After 80 ramping-maintaining circles, the trap was dumped to prevent accumulation of other heavy ions, such as tungsten and barium, which could be sputtered from the cathode of the electron gun and make their way into the trap. The trap was then refilled with fresh ions and another cooking-ramping-maintaining circle was started. The x-ray energy, electron beam energy, and the time were all recorded simultaneously, event by event. In this way, the charge state distribution basically remained constant during the measurements, only 1.5% change for the highest resonance (the worst case) according to the estimation in [55]. This work lead to an average precision of 9% in the DR strength determination, improved obviously from the results in [53].

In the past few decades, EBITs have been producing definite values of electron impact excitation, ionization, and resonance excitation cross-sections, as well as dielectronic resonance strengths. Some results have demonstrated shortcomings of theoretical data. For example, systematic measurements of $n = 3$ to 1 line emission from He-like ions at the LLNL EBIT [57] showed significant disagreement between the measured and the calculated ratios of the $1s3p$ $^3P_1 - 1s^2 \, ^1S_0$ intercombination and the $1s3p \, ^1P_1 - 1s^2 \, ^1S_0$ resonance lines for low and medium Z ions. In [56], systematic disagreement was also reported for the measured and the calculated ratios of the $2p^5 3d \, ^3D_1 - 2p^6 \, ^1S_0$ intercombination

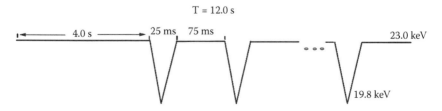

FIGURE 2.10
The timing scheme for fast scan in the DR strength measurements in [55].

and $2p^53d\ {}^1P_1 - 2p^6\ {}^1S_0$ resonance lines in the Ne isoelectronic sequence. On top of these, EBIT spectroscopy has been used for disentangling the emissions from tokamak plasmas, as shown in [58].

EBIT spectroscopy is a useful tool for producing accurate data to test theories of fundamental physics, for disentangling plasma processes and assisting plasma diagnostics.

References

1. B. Edlén, *Arkiv Math. Astron. Fysik* 28B(1), 1942.
2. S. Bashkin, *Nucl. Instrum. Meth.* 28, 88, 1964.
3. L. Kay, *Phys. Lett.* 5, 36, 1963.
4. R. H. Plumlee, *Rev. Sci. Instrum.* 28, 830, 1957.
5. E. D. Donets, In I. G. Brown (Ed.), *The Physics and Technology of Ion Sources*. New York: Wiley, 1989, p. 245.
6. R. E. Marrs et al., *Phys. Rev. Lett.* 60, 1715, 1988.
7. R. E. Marrs, S. R. Elliott, and D. A. Knapp, *Phys. Rev. Lett.* 72, 4082, 1994.
8. J. D. Silver, A. J. Varney, H. S. Margolis, P. G. E. Barid, I. P. Grant, P. D. Groves, W. A. Hallet, A. T. Handfort, P. J. Hirst, A. R. Holmes et al., *Rev. Sci. Instrum.* 65, 1072, 1994.
9. C. A. Morgan, F. G. Serpa, E. Takacs, E. S. Meyer, J. D. Gillaspy, J. Sugar, J. R. Roberts, C. M. Brown, and U. Feldman, *Phys. Rev. Lett.* 74, 1072, 1994.
10. F. J. Currell, J. Asada, K. Ishii, A. Minoh, K. Motohashi, N. Nakamura, K. Nishizawa, S. Ohtani, K. Okazaki, M. Sakurai et al., *J. Phys. Soc. Jpn.* 65, 3186, 1996.
11. C. Biedermann, A. Forster, G. Fussmann, and R. Radtke, *Phys. Scripta T* 73, 360, 1997.
12. J. R. Crespo, A. Dorn, R. Moshammer, and J. Ullrich, *Phys. Scripta T* 80, 502, 1999.
13. V. P. Ovsyannikov and G. Z. Lornack, *Rev. Sci. Instrum.* 70, 2646, 1999.
14. X. F. Zhu, Y. Liu, X. Wang et al., *Nucl. Instrum. Methods B* 235, 509, 2005.
15. M. Froesel, C. Champagne, J. R. Crespo López-Urrutia, S. Epp, G. Gwinner, A. Lapierre, J. Pfister, G. Sikler, J. Ullrich, and J. Dilling, *Hyperfine Interact.* 173, 1572, 2006.
16. S. Bohm, A. Enulescu, T. Fritioff, I. Orban, S. Tashenov, and R. Schuch, *J. Phys: Conf. Ser.* 58, 303, 2007.
17. H. Watanabe and F. Currell, *J. Phys.: Conf. Ser.* 2, 182, 2004.
18. S. Geyer, O. Kester, J. Pfister, A. Sokolov et al., *GSI Rep.: At. Phys.* 19, 290, 2007.
19. Y. F. Liu, K. Yao, R. Hutton, and Y. Zou, *J. Phys. B: At. Mol. Opt. Phys.* 38, 3207, 2005.
20. W. Lotz, *Z. Phys.* 216, 241, 1968.
21. B. M. Penetrante et al., *Phys. Rev. A* 43, 4861, 1991.
22. M. Gryzinski, *Phys. Rev. A* 138, 322, 1965.
23. Y. S. Kim and R. H. Pratt, *Phys. Rev. A* 27, 2913, 1983.
24. N. Selberg et al., *Phys. Rev. A* 54, 4127, 1996.
25. V. P. Pastukhov, *Nucl. Fusion* 14, 3, 1974.
26. B. M. Penetrante et al., *Phys. Rev. A* 43, 4861, 1991.

27. E. D. Donets and V. P. Ovsyannikov, *Sov. Phys. JETP* 53, 466, 1981.
28. C. F. Fischer, T. Brag, and P. Joensson, *Computational Atomic Structure*. Bristol and Philadelphia: Institute of Physics Publishing, 1997.
29. C. A. Morgan, F. G. Serpa, E. Takács et al., *Phys. Rev. Lett.*, 74, 1716, 1995.
30. D. J. Bieber, H. S. Margolis, P. K. Oxley, and J. D. Silver, *Phys. Scripta T* 73, 64, 1997.
31. W. R. Johnson and G. Soff, *At. Data Nucl. Data Tables* 33, 405, 1985.
32. P. Indelicato, *X-ray and Inner Shell Processes*, X-90 AIP Conference Proceedings, Vol. 215, New York, 1990, p. 591.
33. P. Beiersdorfer, M. H. Chen, R. E. Marrs, and M. A. Levine, *Phys. Rev. A* 41, 3453, 1990.
34. T. E. Cowan, C. L. Bennett, and D. D. Dietrich et al., *Phys. Rev. Lett.*, 66, 1150, 1991.
35. P. Beiersdorfer, H. Chen, D. B. Thorn, and E. Träbert, *Phys. Rev. Lett.* 95, 233003, 2005.
36. J. R. Crespo López-Urrutia, P. Beiersdorfer, D. W. Savin, and K. Widmann, *Phys. Rev. Lett.* 77, 826, 1996.
37. J. R. Crespo López-Urrutia, P. Beiersdorfer, B. B. Birkett et al., *Phys. Rev. A* 57, 879, 1998.
38. P. Beiersdorfer, A. Osterheld, J. H. Scofield, J. R. Crespo López-Urrutia, and K. Widmann, *Phys. Rev. Lett.* 80, 3022, 1998.
39. M. Jung, F. Sosch et al., *Phys. Rev. Lett.*, 69, 2164, 1992.
40. D. W. Savin, P. Beiersdorfer, S. M. Kahn, B. R. Beck, G. V. Brown, M. F. Gu, D. A. Liedahl, and J. H. Scofield, *Rev. Sci. Instrum.* 71, 3362, 2000.
41. G. Bertschinger, A. Biel, the TEXTOR-94 team, O. Herzog, J. Weinheimer, H. J. Kunze, and M. Bitter, *Phys. Scripta T* 83, 132, 1999.
42. C. Biedermann and R. Radtke, *Phys. Rev. E* 66, 066404, 2002.
43. R. C. Mancini, C. F. Hooper et al., *Rev. Sci. Instrum.* 63, 5119, 1992.
44. B. A. Hammel, C. J. Keane, M. D. Cable et al., *Phys. Rev. Lett.* 70, 1263, 1993.
45. A. J. Smith, P. Beiersdorfer, V. Decaux, K. Widmann, K. J. Reed, and M. H. Chen, *Phys. Rev. A* 54, 462, 1996.
46. R. Ali, C. P. Bhalla, C. L. Cocke, and M. Stockli, *Phys. Rev. Lett.* 64, 633, 1990.
47. Y. Zou, J. R. Crespo López-Urrutia, and J. Ullrich, *Phys. Rev. A* 67, 042703, 2003.
48. P. Beiersdorfer, T. W. Phillips, K. L. Wong, R. E. Marrs, and D. A. Vogel, *Phys. Rev. A* 46, 3812, 1992.
49. B. E. O'Rourke, H. Kuramoto, Y. M. Li, S. Ohtani, X. M. Tong, H. Watanabe, and F. J. Currell, *J. Phys. B* 37, 2343, 2004.
50. D. A. Knapp, R. E. Marrs, M. A. Levine, C. L. Bennett, M. H. Chen, J. R. Henderson, M. B. Schneider, and J. H. Scofield, *Phys. Rev. Lett.* 62, 2104, 1989.
51. D. A. Knapp, R. E. Marrs, M. B. Schneider, M. H. Chen, M. A. Levine, and P. Lee, *Phys. Rev. A* 47, 2039, 1993.
52. N. Nakamura, A. P. Kavanagh, H. Watanabe, H. A. Sakaue, Y. Li, D. Kato, F. J. Currell, and S. Ohtani, *Phys. Rev. Lett.* 100, 073203, 2008.
53. W. D. Chen, J. Xiao, Y. Shen et al., *Phys. Plasmas* 14, 103302, 2007.
54. W. D. Chen, J. Xiao, Y. Shen et al., *Phys. Plasmas* 15, 083301, 2008.
55. K. Yao, Z. Geng, J. Xiao, Y. Yang, C. Chen, Y. Fu, D. Lu, R. Hutton, and Y. Zou, *Phys. Rev. A* 81, 022714, 2010.
56. W. D. Chen, J. Xiao, Y. Shen, Y. Q. Fu, F. C. Meng, C. Y. Chen, Y. Zou, and R. Hutton, *Rev. Sci. Instrum.* 79, 123304, 2008.

57. P. Beiersdorfer, G. V. Brown, M. F. Gu, C. L. Harris, S. M. Kahn, S. H. Kim, P. A. Neill, D. W. Savin, A. J. Smith, S. B. Utter, and K. L. Wong, *Proceedings of the International Seminar of Atomic Processes in Plasmas*, Nagoya, Japan, Vol. 25, 2000.

58. T. Pütterich, R. Neu, C. Biedermann, R. Radtke, and ASDEX Upgrade Team, *J. Phys. B* 38, 3071, 2005.

3

Spectroscopic Instruments

Roger Hutton, Zhan Shi, and Indrek Martinson

CONTENTS

3.1 Introduction

The term "spectrometer" refers to any energy-resolving instrument; however, in this chapter we will concentrate on photon energy-resolving instruments or, to be more specific, wavelength-dispersive instruments. The basic theme of this chapter concerns techniques for the study of highly charged ions (HCIs), and one aspect of this should be their spectra. Spectra from HCIs can cover very large ranges in photon energy, or wavelength, that is, from over 100 keV photons for H-like resonance lines in very-highly charged ions to less than 1 eV photons for hyperfine transitions. For example, the 1s hyperfine splitting gives a transition at $3858.2260 \pm 0.30\,\text{Å}$ (3.2 eV) in $^{203}\text{Tl}^{80+}$ [1], whereas the 1s to 2p resonance transitions have energies more than 90 keV (simple Z scaling). Another example is provided by transitions in highly ionized iron. The resonance transitions in He-like iron, Fe XXV (from $n = 2$ to $n = 1$), occur in the x-ray region, at 1.85 Å, or 6.7 keV. These lines have been observed in solar flares and in tokamaks [2]. In Al-like Fe XIV, there is a forbidden M1 transition within the $3s^2 3p^2 P$ ground term, a strong line in the solar corona, at 5303 Å. Already in 1945, Edlén [3] had identified 23 such corona lines, in highly charged Ar, Ca, Fe, and Ni, with wavelengths ranging from 3328 to 10,797 Å. The unit Å is not an official SI unit but is named after the Swedish physicist Ångström and is equal to 10^{-10} m, that is, 0.1 nm.

Excited energy levels in an HCI, A^{q+}, decay by electronic transitions, resulting in the emission of electromagnetic radiation. However, nonradiative transitions to the ground state, or to some excited state in the next ion $A^{(q+1)+}$ by electron emission, are sometimes also energetically possible. To determine excitation energies and energy differences in ions, both photon and electron spectroscopies are therefore applicable. In this chapter, the emphasis is placed, as said above, on photon spectroscopy of multiply ionized atoms. This is often called classical spectroscopy, which may sound a bit old-fashioned. However, it can often be the only possible way for experimental studies of spectra and structures of HCIs, for instance, when investigating solar and stellar spectra or hot thermonuclear fusion plasmas.

As the energy, and so the wavelength, region covers a very large range, a variety of spectrographs and spectrometers will be needed. For instance, in He-like boron, B IV, 141 spectral lines have been observed between 43 and 4813 Å [4]. In a famous and pioneering paper, Edlén and Swings [5] studied the Fe III spectrum between 500 and 6500 Å and observed 1500 spectral lines in this region. They classified all these lines which resulted in 320 new energy levels for Fe III, which are of great importance in astrophysics.

It will be seen in the following that a number of optical techniques will be needed to cover such large ranges in photon energies. Basically, three wavelength regions can be defined where different techniques must be used. They are (i) from infrared down to about 1850 Å, (ii) 1850 Å down to around 10 Å, and (iii) below 10 Å. Each of these regions and their peculiarities will be discussed below.

Before discussing the three regions defined above, a few general remarks about photon spectrometers and other instruments used in photon spectroscopy are called for.

3.2 General about Spectroscopic Instruments

The group of the so-called classical spectroscopic instruments can be divided between those with prisms and those with gratings. Of these, the latter are dominant, especially when HCIs are being studied because of their limited useful wavelength region for prism-based instruments. There are also several types of interferometers, often used in the so-called Fourier transform spectroscopy. Such instruments are very important for accurate atomic and molecular studies, but so far have seen limited applications in the case of multiply ionized atoms.

For the instruments to be discussed here, we can distinguish some special chief outlines. In a typical spectrometer, the light to be studied enters the instrument through an entrance slit in the focal plane of a spherical mirror or grating. In the case of the mirror, the divergent bundle of light will be made parallel by the collimator lens or spherical mirror and sent to a grating as dispersing element and then to a focusing mirror which forms an image of the entrance slit in the focal plane. The rays leave the grating with angles which depend on the wavelengths, and hence the position of the image of the entrance slit is a function $x(\lambda)$ of the wavelength.

Photographic emulsion was the dominating detector for spectroscopic instruments for many years. It has several good properties thanks to which it can still be used. Among the good properties are the integrating capability, the two-dimensional registration with high resolution, storage of the signal directly in the detector with high information density, and practically unlimited storage time. The data storage capacity of a photographic plate should not be mocked. Twenty years ago, or so, data were recorded on 9 in. floppy discs but who can read such things now! Among the disadvantages, the most serious one is the highly nonlinear response to light intensity. Moreover, the sensitivity is lower than that for photoelectrical detection and the emulsions are not generally sensitive to single-photon events. This method of photon detection misses any time variation of the light intensity, and photographic emulsion can thus be described as a detector of illumination. Photographic plates have been almost exclusively replaced by either micro-channel plates (MCPs), detectors (see Chapter 6), or charge-coupled device (CCD)-based detectors (see Chapter 5).

There are several designations and names for spectroscopic instruments which are sometimes used without any accurate distinction. A spectrograph usually means an instrument with a photographic plate in the spectral image plane, sometimes recording a large range of wavelengths simultaneously. Nowadays, photographic plates are often replaced by electronic array detectors as mentioned in the previous paragraph. The term "monochromator" means that a single-exit slit is used to select a spectral line or a narrow spectral band. We can also have a scanning monochromator, by rotating the disperser (for instance, a grating) which can thus also be called as a scanning spectrometer.

Spectroscopic instruments for different wavelength ranges are constructed according to different principles. Thus, spectrographs for photographic registration can only be used for wavelengths below 11,000 Å. Furthermore, because of reduced reflectivity, at low angles of incidence, for all materials and for wavelengths shorter than 1000–1500 Å (see Figure 3.1), reflections should usually be minimal, that is, only one. Below 300 Å, one really must apply grazing incidence methods for this single reflection as the reflectivity for all materials increases with the incidence angle. Readers may demonstrate this to themselves using the XOP software [6]. XOP is a freely downloadable set of software which performs the calculation of the reflectivity of many materials as a function of incidence angle, photon energy, surface roughness, and so on. Because of transmission problems, it is also necessary to eliminate air (and thus measure in vacuum) when radiation below 2000 Å is observed, that is, the vacuum UV region. Figure 3.2 shows an example calculated using XOP. It is for the reflection of 1000 Å photons from a gold surface as a function of angle (given in milliradians) and measured from the surface (i.e., not from the normal). The surface is assumed to be perfectly smooth, but roughness can be dealt with by the software.

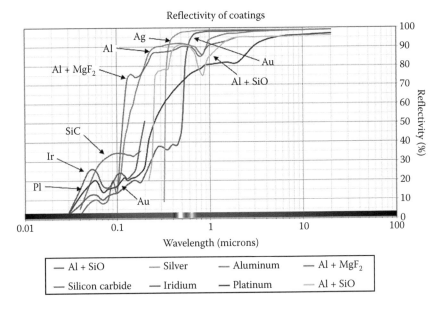

FIGURE 3.1
Reflectivity for a number of common mirror coatings as a function of wavelength (μm) at normal incidence, that is, light incident along (or close to the normal of the optical component). (Courtesy of McPherson Inc., Massachusetts, USA [7].)

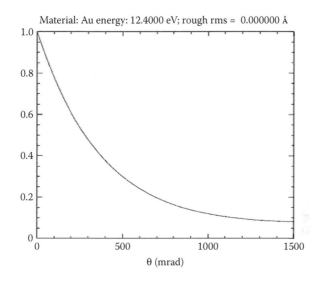

FIGURE 3.2
Reflection of 1000 Å photons from a gold surface as a function of angle (given in milliradians) and measured from the surface (i.e., not from the normal). The surface is assumed to be perfectly smooth in this case.

Not only does the wavelength region influence the construction of a spectroscopic instrument, but also the width of the region in which one would like to work with a certain instrument. Those who want to study the hyperfine structure of a certain spectral line will often select a laser spectroscopic method. For term analysis, where many spectral lines are needed, a variety of instruments may be required. Out of interest, the region below 1850 Å and down to around 100 Å is often called the vacuum ultraviolet region. There are three underlying reasons for this change at around 1850 Å: (i) the absorption of quartz, (ii) the absorption of oxygen, and (iii) more of historic interest, the absorption of the gelatin in standard photographic plates.

3.3 Diffraction Gratings

With the exception of interferometers, almost all spectroscopic instruments use reflective gratings as the dispersive element. However, the first gratings to be produced were of the transmission type and were constructed by the American astronomer D. Rittenhouse in 1785. Rittenhouse (1732–1796) was viewing a distant light source through a fine silk handkerchief. He then made up a square of parallel hairs and found that red light was bent more than blue light. In the early nineteenth century, the German physicist Joseph von Fraunhofer (1787–1826) discovered the dark lines in the Sun's spectrum, now known as Fraunhofer lines. He was also the first scientist to extensively use diffraction gratings, which he himself made on glass or metal surfaces. It is considered that his work set the stage for the development of spectroscopy. Somewhat more modern diffraction gratings were first developed by the American physicist Henry A. Rowland (1848–1901). He also invented the concave grating, which could replace prisms and plane gratings and revolutionized spectrum analysis. Around 1882, he managed to construct a machine which could engrave as many as 20,000 lines/in. (785 lines/mm) for diffraction gratings. He then ruled on spherical concave surfaces, thus eliminating the need for additional lenses and mirrors in spectrometers and used them to develop exact spectrometry. Later developments were made at the Massachusetts Institute of Technology (MIT) and also by the company Bausch and Lomb and others [8].

From 1965, a new technique of grating production has been developed. This is the holographic method, which produces gratings by recording interference fringes on photosensitive materials. This procedure results in the grove spacing being absolutely constant, and the gratings are therefore free of ghosts. The efficiency is, however, lower than that of ruled gratings.

From the mid-1980s, a new form of diffraction grating became available, the so-called aberration-corrected flat-field grating. These were originally produced by Hamamatsu using a mechanical ruling device but more recently have been manufactured using holographic techniques, for example, those produced by Jobin Yvon [9]. Note that Yvon, along with a number of other companies, provided excellent online guides and details of grating design, use, and technology.

3.4 Spectrometer Slits

Before discussing the effect of slits on the performance of a spectroscopic instrument, it is interesting to think about why spectral features are most often lines.

A line is called a line just because it is the line-shaped image of the line-shaped slit in the focal plane of the spectrometer. And that is how it looked on a photographic plate. It is interesting to notice how things change with time. For some time, and in almost all "modern" instruments, a spectral line was just the photon or light intensity distribution as a function of wavelength. But currently, most instruments use CCDs or other imaging detectors, so now the lines look like lines again, in the original sense. One can also ask whether other shapes for the entrance aperture could be used. Newton used a circular opening in his first prism experiment, but as soon as one thinks of trying to separate spectral features, like, for example, Fraunhofer did, then you need the smallest possible extension of the input aperture in the dispersion direction. But it does not matter if the aperture is wide in the perpendicular direction. A very small dot would be difficult to see directly with the eye in the early instruments and also later when photographic plates began to be used. However, circular apertures are fine for use in combination with CCD detectors, except now lines become spots. The use of circular apertures has great implications, for example, in the design of echelle-based spectrometers (see Section 3.7.2).

3.5 Entrance Slit

Whether a spectrometer has two slits or one is depending on whether it is designed to be a monochromator or a spectrograph. In either case, the width of the slit (or slits) plays an important role in defining the characteristics of the instrument. The most obvious function is that the entrance slit will limit the amount of light, not only intensity but more importantly angular direction, entering the spectrometer.

The width of the entrance slit may also be the defining factor for the resolution of the spectrometer. Basically, the optics in a spectrometer is designed to image the entrance slit, usually in a one-to-one manner on the spectral image plane. If the width of the slit is smaller than the diffraction limited image of the optics, then the slit does not define the resolution and the resolution will be diffraction-limited. If the width of the entrance slit is greater than the diffraction-limited image, then the resolution is defined by the width of the entrance slit.

The instrument profile of a spectrograph is given by the convolution of the slit function and the diffraction profile defined by the aperture of the dispersive element (grating). In the case of a monochromator, the instrument profile must include the exit slit function as part of the convolution. Although the role of exit slits has been greatly reduced in recent years, due to the increasing use of multichannel detection techniques, it is still useful to look at the role of an exit slit. One of the roles of an exit slit is to limit the amount of light falling onto the detector, in the case of single-channel detection. A little argument will show that the optimal width of the exit slit is exactly the same width as the entrance slit.

First consider the case where diffraction can be ignored and look at the influence of the entrance and exit slits. If we assume that the entrance slit has a width a_1 and luminance (see below for a definition) B, then we can describe the slit as a rectangle. We can also assume that the exit slit can be defined by a width a_2 and a transmission $= 1$ over the slit width and 0 outside. If we ignore diffraction and aberration, then the convolution can be seen as a scanning of the exit slit over the entrance slit. Hence the flow of light onto a detector will be proportional to the total area presented by the two slits and this area changes during the scanning (convolution). There are two situations to consider: (i) the entrance slit is narrower than the exit slit, and (ii) the entrance slit is the broadest. However, the optimal appears when $a_1 = a_2$ and where the slits lead to a triangular instrument profile.

Luminance describes the amount of light that passes through or is emitted from a particular area, and falls within a given solid angle. It is an invariant in geometrical optics, that is, if a light bundle is hard focused, then it will appear as though the spot is brighter, but the solid angle will also be larger, so the luminance in constant (except for absorption at lenses, etc.). This is an important concept in optics and is similar to the idea of emittance in ion beam optics.

3.6 Aberrations

No discussion of spectrometers would be complete without a mention of aberrations, as in the end it is often these that limit resolution or light collection. The easiest way to approach aberration studies for a particular instrument is through ray tracing. Here free downloadable software such as Zmax or Shadow can be used. The most common aberrations when discussing spectrometers are spherical aberration and astigmatism. Below we will discuss these aberrations from Seidel theory. The basis of this is to take the next higher-order approximation to the theory of geometrical optics. Geometrical optics, or Gaussian theory of image formation, works in the region when the so-called paraxial rays are considered, that is, $\sin \alpha = \tan \alpha = \alpha$. The next order of approximation is given by letting $\sin \alpha = \alpha - \alpha^3/3!$ and $\tan \alpha = \alpha + \alpha^3/3!$ Doing this allows the construction of a third-order theory of image formation which takes into account nonparaxial rays. The deviations from perfect imaging will be the Seidel aberrations, of which spherical aberration and astigmatism are just two of the terms. There are five Seidel terms, each representing a different form of aberration and each term is a complex function of the position in the object and image planes, the radius of curvature of the optical element, and also the refractive index (if needed). The five Seidel terms were first defined and developed in the original article by Seidel in 1856 [10]. Spherical aberration refers to the fact that even light from an infinitely distant object will not be focused to a point by a spherical mirror. This is caused by the fact that light rays are reflected with a larger angle by the outer parts of a mirror compared to those from the inner mirror region.

Like all aberrations, this is best studied using ray-tracing techniques. However, an interesting measure of the importance of spherical aberration is the circle of least confusion. Here, a comparison can be made between the size of the circle of least confusion and the size of the image due to diffraction. This can be done by comparing the two terms below:

$$\Delta_{SA} = h^3/8f^2, \text{ which gives the size of the circle of least confusion and}$$
$$\Delta_R = \lambda f/2h, \text{ which gives the diffraction-limited image size.}$$

Here, h is the height of the light rays from the optical axis (i.e., the mirror radius), f the focal length, and λ the wavelength.

As an example, a mirror with $f = 3\,\text{m}$ and $h = 6.5\,\text{cm}$ gives $\Delta_{\text{SA}} = 4\,\mu\text{m}$, whereas a mirror with the same radius but $f = 1.5\,\text{m}$ has $\Delta_{\text{SA}} = 16\,\mu\text{m}$. For $5000\,\text{Å}$ wavelength photons, the values of Δ_{R} are 12.5 and $6\,\mu\text{m}$ for the 3 and 1.5 m mirrors, respectively. Hence the spatial resolution of the 3 m mirror is limited by diffraction, whereas for the 1.5 m mirror it is by spherical aberration.

Another aberration that should be mentioned as of importance to spectrometer is astigmatism. Astigmatism leads to a point object not being imaged as a point. There will be two positions where the image will be a straight line and the lines are perpendicular to each other. The distance between the two focus positions is given by the Seidel theory as hy^2, where y is the distance of the object from the optical axis. This aberration limits the useful height of spectrometer entrance slits, for example. Seidel theory was developed for normal, or close to normal incidence. The size of most aberrations increase as a function of the incidence angle, which is particularly true for astigmatism. Grazing incidence spectrometers are very limited in both entrance slit height and grating width by aberrations. For a full discussion of this, see the work of Mack et al. [11]. In particular, astigmatism will limit the useful grating size and hence the light collection angle for very high incidence angles. This has to be weighted against the increase in reflectivity as a function of incidence angle for wavelengths below 1000–1500 Å when choosing spectrometer geometry (see Section 3.8). Before leaving this short discussion on aberrations, we should mention coma. This aberration comes about when imaging a point that does not lie on the optical axis. The importance of coma can be judged by the Seidel expression h^2y, where h and y are as above. Coma is particularly irritating in spectroscopic instruments because it will lead to asymmetric line shapes and all the problems incorporated with this. A full account of aberrations and their influence on optical systems can be found in the book by Welford [12]. As mentioned a number of times in this chapter, the way to estimate the importance of aberrations on the performance of an optical instrument is through ray-tracing.

3.7 Plane Grating Mounts

In spectrometers that employ plane gratings, the grating only performs the task of dispersing the light depending on wavelength. Focusing, and so on is done by auxiliary optical components. Hence wavelength dispersion and focusing can be optimized independently and this leads to plane grating spectrometers usually having higher light gathering properties. Plane grating spectrometers are usually more flexible than instruments using curved gratings as the grating can be changed without changing the optical properties.

Hence, it is often recommended to use plane grating instruments whenever it is possible. A number of mountings for plane gratings exist and some will be discussed below.

3.7.1 Czerny-Turner Mounting

One of the biggest advantages of this mounting is that coma can be totally eliminated, at least in the meridian plane.

The basic optics of this instrument is quite straightforward and it is relatively easy to understand how coma can be eliminated. The coma introduced by the collimating mirror, C, is equal and opposite in sign to that introduced by the focusing mirror, E (see Figure 3.3).

There are other mounts for plane gratings, for example, the Ebert and Eagle mounting, which can be described in the book by Samson and Ederer [13].

3.7.2 Echelle Mounting

In recent years, echelle gratings have become more popular in the construction of spectroscopic instruments. The main reasons for this are the developments in two-dimensional photon detectors such as MCPs and CCD-based detectors. An echelle grating acts more like an interferometer than a grating and works in a very high order of diffraction (see Figure 3.4).

Although echelle gratings offer very high wavelength dispersion, they give rise to very low divergence. Hence the most common way to use echelle gratings is with a cross-disperser, which can be either a grating or prism. For work below, the transmission cut-off of quartz at about 1850 Å gratings must be used. An interesting solution for an echelle cross-disperser geometry was given in 1958 by Harrison [8]. In Section 3.8, we will develop the idea of the Rowland circle. However, there is a second solution to the imaging properties of concave gratings, the so-called Wadsworth solution. This

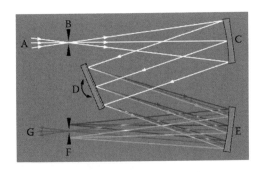

FIGURE 3.3
This shows one of the schemes for using a plane grating, the so-called CzernyTurner mounting.

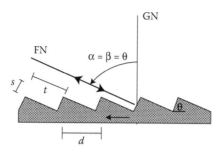

FIGURE 3.4
This shows the operation of an echelle grating. GN is the grating normal, FN the face normal, and d is the groove spacing.

geometry is very suited to work with an echelle grating. As will be shown in the section below, this solution requires the object to be at an infinite distance from the grating and the spectrum will be imaged along the direction of β (diffraction angle) $= 0$ and on a curved surface given as $r' = R/(1 + \cos \alpha)$, where R is the radius of curvature of the grating and α the angle of incidence. The mounting of an echelle grating requires parallel light impinging on the grating from an angle. The reflected and dispersed light will also be a parallel beam, hence the match with the Wadsworth mounting of the cross-disperser. The final spectrum will be two-dimensional with the cross-disperser throwing the different orders from the echelle ideally by just the height of the entrance slit. An example of such an instrument is described in the following section.

As we will see in the following section, the development of the focusing properties of a concave grating led to a number of conditions that must be fulfilled. One being the grating equation which tells in which direction a given wavelength will be focused, but not where. The following equation tells where the light at a given wavelength will be focused:

$$\cos \alpha \left(\frac{\cos \alpha}{r} - \frac{1}{R} \right) + \cos \beta \left(\frac{\cos \beta}{r'} - \frac{1}{R} \right) = 0,$$

where r is the object distance, r' the image distance, and R the radius of curvature.

One solution is to let the light source be at an infinite distance, that is, $r = \infty$, which is the same as using parallel light to fall on the echelle and also β (the diffraction angle) $= 0$. The spectrum will then fall on a surface given by $r' = R/(1 + \cos \alpha)$. This is known as the Wadsworth mounting of a concave grating.

As an example, we will consider the design of an echelle-based spectrometer where the wavelength resolution should be around 0.1 Å and the working range from 1500 to 6000 Å. Also we will consider that the instrument should be compact for convenience. A further important parameter in the design of echelle spectrometers is the height of the object, as this defines how far each echelle order needs to be "thrown" to get good order sorting. In the example here, the object was an ion beam of 3 mm in diameter. Hence the entrance slit was chosen to be 3 mm in height. Using CCD detectors, there is no longer any reason why a slit is needed and circular apertures can be used. Such an instrument will be briefly mentioned below but we can mention here that it is very important to match the spectrometer to the properties of both the light source and detector. Most likely the optical components are going to be the cheapest part of any spectrometer.

For example, here 1 m radius of curvature optics are chosen. The line density of the echelle grating is fixed once the design criteria of the instrument are specified. It is surprising how little the choice is to the designer in fact. The free spectral range of an echelle grating is given by $\lambda^2/(2d \sin \theta)$, where d is the groove separation and θ the blaze angle. The plate factor, that is, the number of Å per mm, along an echelle order is $\lambda/R \tan \theta$, where R is the radius of curvature of the concave grating (and most likely the collimating concave mirror too). The plate factor is directly related to the required resolution and the properties of the detector. If we choose, for example, 2 Å/mm at 4000 Å, then the blaze angle is determined as we specified $R = 1$ m. The blaze angle should be 63°. If we choose the maximum length of the echelle order contains 4000 Å, this determines the echelle line spacing to be around 31 lines/mm. These parameters are very close to those of some commercially produced echelle gratings, for example, blaze angle of 63.26° and 31.6 lines/mm. The order sorting concave grating is now basically fixed by the detector dimensions. The instrument was required to work up to 6000 Å, so based on CCD chip of 25 mm length, we can require the echelle order containing 6000 Å to be, say 21 mm, in length and a free spectral range of 63 Å. We now require the plate factor of the order sorter to be λfrs/slit height. This leads to a groove density of around 1100 lines/mm at 6000 Å and the spectrometer is completely specified.

3.8 Concave Grating Mounts

The geometry of plane grating mountings requires at least three reflections, collimation, dispersion, and finally focus. Such a spectrometer operating at short wavelength and at normal incidence, that is, the incident angle being less than around 10°, would have very bad efficiency due to the low reflectivity of all materials for wavelengths shorter than 1000–1500 Å (see Figure 3.1). For example, 20% of 20% of 20% would be a reflectivity of 0.8%. Hence mountings of gratings where the number of reflections can be minimized and preferably only one are highly useful, which shows the importance of concave grating. For wavelengths below 300 Å, the one "allowed" reflection must be at grazing incidence as reflectivity increases at higher angles of incidence. Of course, there is a price to pay for the increased efficiency and that is the inability to compensate for aberrations. In particular, astigmatism is high for concave grating-based spectrometers.

Most mountings of concave gratings are based on the so-called Rowland circle geometry which will be introduced below. A full derivation of the results that will be presented here can be found in the work of Beutler [14].

3.8.1 Rowland Circle

The imaging properties of the concave grating can be understood via the following arguments. First, we define a co-ordinate system as in Figure 3.5: $A(x, y, 0)$ is the position of an object in the $z = 0$ plane and $B(x', y', 0)$ is the position of the image after diffraction by the grating. $P(u, w, l)$ is any point on the grating surface and O is the centre of the grating. We then set up equations to describe the optical path from A to B via P. This path length is then minimized according to Fermats principle and this will lead to a set of relations defining the imaging properties of the grating:

$$(AP)^2 = (x - \zeta y)^2 + (y - w)^2 + (z - l)^2,$$
$$(BP)^2 = (x' - \zeta)^2 + (y' - w)^2 + (z' - l)^2.$$

We then change co-ordinates and use the relation defining the surface of the grating:

$$x = r \cos \alpha \quad \text{and} \quad x' = r \cos \beta,$$
$$y = r \sin \alpha \quad \text{and} \quad y' = r' \sin \beta.$$

The equation for the grating surface is $(\zeta - R)^2 + w^2 + l^2 = R^2$.

The optical path $F = AP + PB$ should then be subject to Fermats principle and for an image both $\delta F / \delta w$ and $\delta F / \delta l$ should be zero for all values of w and l.

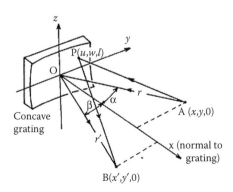

FIGURE 3.5
Schematic diagram of optical system.

However, it is not important in this case that all light paths from A to B are equal, but that all waves reaching B, from A, are in phase. For a grating, there is the possibility for different path lengths as there are reflections from very discrete areas, namely the grating rulings. This can be taken into account by allowing the light path to change by an integer number of wavelengths as w (the distance along the grating perpendicular to the grooves) changes by one grating ruling, that is,

$$F = AP + PB + \frac{wm\lambda}{a},$$

where m is the diffraction order, a the grating spacing (1/line density), and λ the wavelength.

Using the above relations, F can be written as a function of the variables $r, r', \alpha, \beta, \alpha', \beta', w$, and l.

The last part of the development requires F to be written as an expansion in various powers of w and l. The terms in the expansion can then be grouped together and useful relations can be obtained. There are a number of ways to do such an expansion but the one developed by Beutler in his 1945 paper leads to the most useful expressions. For a full derivation, the reader is directed to Beutlers paper.

Here $AP = \Sigma F_i$ and $PB = \Sigma F_{i'}$. The first two terms, that is, $i = 1$ and 2 will be

$$F_1 = r - w \sin \alpha,$$

$$F_2 = \frac{1}{2} w_2 \left(\frac{\cos^2 \alpha}{r} - \frac{\cos \alpha}{r} \right) + \text{higher-order terms},$$

$$F'_1 = r' - w \sin \beta,$$

$$F'_2 = \frac{1}{2} w_2 \left(\frac{\cos^2 \beta}{r'} - \frac{\cos \beta}{r} \right) + \text{higher-order terms}.$$

We can now examine the condition for a focus, that is, look at $\delta F / \delta w$ and $\delta F / \delta l = 0$ for all positions on the grating. Any deviation from zero will represent aberrations, which, in principle, can be calculated.

The first term gives

$$F_1 = r - w \sin \alpha + r' - w \sin \beta + \frac{wm\lambda}{a}.$$

If

$$\frac{\delta F}{\delta w} = \frac{\delta F}{\delta l} = 0,$$

then

$$- \sin \alpha - \sin \beta + \frac{m\lambda}{a} = 0.$$

This is nothing else, but the grating equation.

In the special case of plane gratings, $R = \infty$ and also $r = r' = \infty$, so the above equations are always zero and the only relation to think about is the grating equation and the plane grating cannot introduce any aberrations.

The grating equation tells in which direction an image at a given wavelength can be expected, but not where the focus will be. The position of the focus comes from the next order of the expansion.

$$F_2 = 0.5\, w^2 \left(\frac{\cos^2 \alpha}{r} - \frac{\cos \alpha}{R} + \frac{\cos^2 \beta}{r'} - \frac{\cos \beta}{R} \right)$$

$$+ 0.5\, w^2 \left(\frac{\sin \alpha}{r} \left(\frac{\cos^2 \alpha}{r} - \frac{\cos \alpha}{R} \right) + \frac{\sin \beta}{r'} \left(\frac{\cos^2 \beta}{r'} - \frac{\cos \beta}{R} \right) \right).$$

Now consider the first term and put its derivative w.r.t. w and l equal to 0. Again $\delta F / \delta l = 0$ by default. Then $\delta F / \delta w = 0$ implies either $w = 0$ or

$$\frac{\cos^2 \alpha}{r} - \frac{\cos \alpha}{R} + \frac{\cos^2 \beta}{r'} - \frac{\cos \beta}{R} = 0,$$

which leads to

$$\cos \alpha \left(\frac{\cos \alpha}{r} - \frac{1}{R} \right) + \cos \beta \left(\frac{\cos \beta}{r'} - \frac{1}{R} \right) = 0.$$

One solution to the above is given by $r = R \cos \alpha$ and $r' = R \cos \beta$.

If this solution is chosen, then even the higher-order terms in x.8 are zero. This solution represents a circle with radius $R/2$ and points on the circle are given by (r, α) and (r', β). This is the well-known Rowland circle. There are, of course, terms that depend on l and cross-terms that depend on both l and w that do not vanish and lead to aberrations. However, the overall imaging properties are not.

There is another useful solution to the above imaging equations. A second solution leads to the Wadsworth mounting as discussed in Section 3.7.2. The Wadsworth solution requires the light source to be at an infinite distance from the grating, that is, $r = \infty$, which can be represented by parallel light impinging on the grating. The image is then on the grating axis, that is, $\beta = 0$.

There are basically two types of spectrometers based on the Rowland circle. These operate in different wavelength regions, as discussed above, and are based on either normal or grazing incidence geometry. A discussion of these two types of instruments follows.

3.9 Normal Incidence Spectrometers

From the formulation above, it is clear that if an entrance slit is positioned on the Rowland circle, then the spectrum will also be imaged on the Rowland circle. The image where the incidence angle equals the diffraction angle is the zeroth-order image, or mirror reflection ($\sin \alpha + \sin \beta = n\lambda/d$ and $n = 0$). This image contains light of all wavelengths and is often the strongest image produced by the spectrometer and hence a very useful calibration mark. We will mention this in more detail under the discussion of grating blaze angles. In principle, any position along the Rowland circle can be chosen for an entrance slit.

However, as with most optics, aberrations increase for off-axis imaging. Hence, if the incidence angle is kept small, the spectral image will not be distorted significantly by aberrations. In a classic design by Paul McPherson, the incidence angle was chosen to be $7.5°$. Most commercial normal incidence spectrometers use this angle (or something very close) and unless the gratings are aberration-corrected, the opening angle is usually limited to less than $f/10$. Normal incidence instruments are relatively easy to construct and in the old days would have an entrance slit and a photographic plate bent to match the Rowland circle. In principle, a very wide wavelength range can be covered by a large enough plate, but in practice this would lead to problems with emulsions having different sensitivities for different wavelength regions. Also the geometrical size of such instruments can become very large, leading to problems of stabilities, and so on. The largest normal incidence instrument was built at the Argonne National Laboratory and had a diameter of 9 m. This geometry is not very convenient for using photoelectric detectors, such as photomultipliers, and so on. Hence the scanning design introduced by McPherson (see Figure 3.6) in 1963 made a huge impact on the usefulness of the normal incidence mounting.

Several other mountings for concave gratings are possible, as discussed in the book by Thorn et al. [15].

(a)   (b)

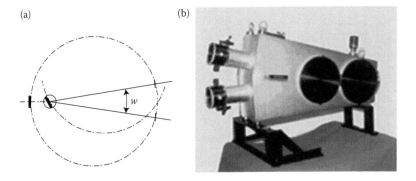

FIGURE 3.6
(a) shows schematically how by moving and rotating the grating the Rowland circle can be moved to keep the focus at the exit slit, and (b) shows how this leads to a much more compact spectrometer.

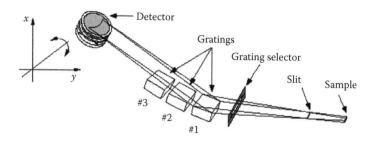

FIGURE 3.7
Layout of the optics.

3.10 Grazing Incidence Spectrometers

Grazing incidence spectrometers work exactly as one would imagine from the name, with the incident light falling at a very high (grazing) angle with respect to the grating. Typically this angle is between 85° and 89°, and the higher the angle, the better the reflectivity. One can then ask why not always use grazing incidence? The answer can be found above in the discussion on aberrations. As an example, we will consider the efficiency of a 1 m normal incidence spectrometer compared to a 1 m grazing incidence instrument.

Another classic problem with the grazing incidence geometry is clear from Figure 3.7, in which the angle of the detector to the incoming photons will also be grazing and lead to reflection off the detector surface. This can be a problem if, for example, a CCD detector is used in the image plane. A solution to this problem will be discussed in Section 3.12.2. A second problem associated more with grazing incidence spectrometers, although in principle it is a more general problem, is that of curved line shapes in the image plane (see Figure 3.7). This can, of course, be taken care of using software for any given spectrometer as the line curvature is known based on the exact geometry of the instrument.

3.11 What Influences the Efficiency of a Spectrometer?

The overall efficiency of a spectrometer is influenced by a number of effects/parameters, an example being the reflective properties of the optics. For wavelengths under 2000 Å, it is advised to minimize the number of reflections and optimally use only one. As said earlier, this is because the reflective properties of all materials drop rapidly for shorter wavelengths. The rapid drop in reflectivity can be combated by using the optic at a high incidence angle. An illustration of this is shown in Figure 3.2 (GI-gold) where the reflectivity of gold is plotted as a function of incidence angle for a wavelength of 400 Å. One could then ask "why not always work using grazing incidence optics?" The answer, as indicated earlier, is that grazing optics suffer from large aberration properties and hence limit the light collection angles that can be used (see [11]). Hence it may well be that a normal incidence spectrometer can be more efficient than a grazing incidence spectrometer ever for quite short-wavelength light, indeed normal incidence spectrometers have been used down to around 250 Å.

A further influence on the efficiency of a spectrometer is the so-called grating blaze. Blaze will be explained below, however, if a grating is not blazed most of the diffracted light ends up in the zeroth-order, that is, the directly reflected light and hence contains no spectral information. Blaze is therefore defined as the concentration of a limited region of the spectrum into any order other than the zero order. Blazed gratings are manufactured to produce maximum efficiency at designated wavelengths. A grating may, therefore, be described as "blazed at 250 nm" or "blazed at 1 μm," etc., by appropriate selection of groove geometry.

The blaze condition is achieved by controlling the groove shape to be right-angled triangles, as shown in Figure 3.8. The groove profile is most often calculated for under the Littrow condition for reflection, that is, the input and output rays propagate along the same axis:

$$\sin \alpha + \sin \beta = \frac{n\lambda}{d}.$$

However when blazed, $\alpha = \beta = \omega$, where ω is the blaze angle.

Hence, $2 \sin \omega = n\lambda/d$ gives the blaze angle.

As a general approximation, for blazed gratings the strength of a signal is reduced by 50% at two-thirds the blaze wavelength, and 1.8 times the blaze wavelength.

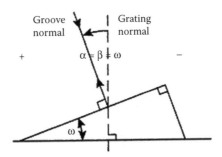

FIGURE 3.8
The principle of the grating blaze angle.

3.11.1 Aside

A major break through in grating design/manufacture occurred in the early 1980s. This was the invention of variable-spacing gratings. Such gratings could lead to flat-field spectral imaging and combined with the soon-to-be developed CCD and multichannel plate detectors has led to very powerful spectrometers of particular interest at synchrotron facilities. This breakthrough will be described below.

3.12 The Aberration-Corrected Flat-Field Spectrograph

Based on their development of numerically controlled ruling engines, in the early 1980s, Harada and Kita reported the availability of mechanically ruled aberration-corrected concave gratings, which are still widely used in vacuum ultraviolet spectroscopy nowadays.

The discussion below follows the work of Harada et al. from the early 1980s.

3.12.1 Principles of Flat-Field Spectrograph

For any spectrograph, the diffracted light varies with wavelength following the so-called grating equation:

$$\sin \alpha + \sin \beta = \frac{m\lambda}{d}.$$

If the groove spacing d is constant, for a given wavelength λ and diffraction order m, the angle of diffraction β is determined only by the angle of incidence α. If the slit of the spectrograph and the grating is placed on the Rowland circle, whose diameter is equal to the radius of curvature of the grating surface, then the diffracted light has to fall on the Rowland circle, as shown in Figure 3.9a. However, if the groove spacing varies, which is true in mechanically ruled aberration-corrected concave gratings, the angle of diffraction β now depends both on the angle of incidence α and the groove spacing d. So the angle of diffraction β can be regulated through proper ruling of the grooves. Through proper design of the groove spacing, aberration-corrected concave gratings can be made, as shown in Figure 3.9b.

3.12.2 Design of Aberration-Corrected Concave Gratings through Fermat's Principle

Figure 3.10 shows a typical optical set up of a grazing incidence flat field spectrometer. As shown in Figure 3.5, the light emitting from $A(x, y, 0)$ is diffracted at $P(u, w, l)$ and focused at $B(x', y', 0)$. The light path function F can be expessed by

$$F = \langle AP \rangle + \langle PB \rangle + nm\lambda.$$

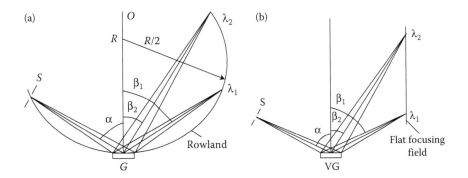

FIGURE 3.9
Concave gratings with (a) invariable and (b) variable line space.

The above equation can be derived following the throught that the light diffracted at two adjacent grooves is enhanced only when the light path difference equals $m\lambda$. And n is the groove number counted from the center of the grating.

According to Fermat's principle, the real light path has to satisfy

$$\frac{\partial F}{\partial w} = 0 \quad \text{and} \quad \frac{\partial F}{\partial l} = 0.$$

In the papers [17,18], the authors used a simpler way to rule the grating, which in contrast to their 1980 ruling process, in which aside from the groove spacing they also included the groove tilting angle θ [16]. In the following discussion, only the groove spacing varies, and the tilting angle is always zero.

The variation of the groove spacing is defined as

$$\sigma = \frac{\sigma_0}{1 + (2b_2/R)w + (3b_3/R^2)w^2 + (4b_4/R^3)w^3 + \cdots},$$

where σ_0 is the nominal spacing of the grooves, or the groove spacing at the center of the grating. B_j ($j = 2, 3, 4, \ldots$) are the ruling parameters.

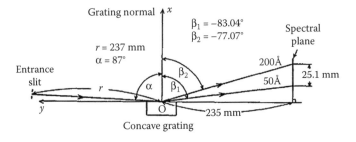

FIGURE 3.10
Schematic and design specifications for the flat-field spectrograph using a mechanically ruled aberration-corrected concave grating: 1200 groves/mm; $R = 5649$ mm; 50×30 mm^2 ruled area.

Combining the above several equations, the light path function as a power series of w and l can be expressed as

$$F = r + r' + wF_{10} + w^2F_{20} + l^2F_{02} + w^3F_{30} + wl^2F_{12} + w^4F_{40}$$
$$+ w^2l^2F_{22} + l^4F_{04} + T(w^5).$$

In the above equation, every F_{ij} term can be expressed by the term $F_{ij} = C_{ij} + (m\lambda/\sigma_0)M_{ij}$.

For the detailed expression of each term, C_{ij} and M_{ij}, please refer to the literature [16]. In general, C_{ij} is the term corresponding to the conventional grating, M_{ij} is the term arising from the varied spacing grooves. F_{10} is related to the dispersion of the grating, F_{20} to the horizontal focal condition, F_{02} to astigmatism, F_{30} to coma-type aberration, and other F_{ij} term to higher-order aberrations.

To achieve completely stigmatic image focusing, the condition $F_{ij} = 0$ have to be satisfied. However, it is impossible to satisfy the stigmatic condition for the total wavelength range in which the grating is used, so in practical design one only needs to make the key term of F_{ij} zero or minimum.

We can obtain, for example, for $F_{10} = 0$ and $F_{20} = 0$,

$$r' = \frac{rR\cos^2\beta}{r[\cos\alpha + \cos\beta - 2(\sin\alpha + \sin\beta)b_2] - R\cos^2\alpha}.$$

When some of the grating and mounting parameters such as σ_0, α, r, and the position of the flat detector surface are predetermined, the above equation which gives the combination of R and b_2 will minimize the deviation between the horizontal focal curve $F_{20} = 0$ and the detector plane within the diffraction domain of β. Coma and spherical aberrations can be reduced by choice of proper values of b_3 and b_4 to minimize the values of F_{30} and F_{40}, respectively, within the diffraction domain of β.

Acknowledgment

Many of the ideas presented in this chapter are based on a compendium called "Spectroscopiska Instrument" written in Swedish by Professor Ulf Litzen, Lunds Universitet, Reprocentralen, in 1985. We are very grateful for his co-operation in this current project.

References

1. P. Beierersdorfer et al., *Phys Rev. A* 64, 032506, 2001.
2. P. Doschek, in *Autoionization*, edited by A. Temkin. New York: Plenum Press, 1985, pp. 171–256.

3. B. Edlén, *Mon. Not. R. Astron. Soc.* 105, 323, 1945.
4. A. E. Kramida, A. N. Ryabtsev, J. O. Ekberg, I. Kink, S. Mannervik, and I. Martinson, *Phys. Scripta* 78, 025302, 2008.
5. B. Edlén and P. Swings, *Astrophys. J.* 95, 532, 1942.
6. *XOP X-ray Software*. European Synchrotron Radiation Facility. http://www.esrf.eu/computing/scientific/xop2.1/
7. McPherson, Inc. http://www.mcphersoninc.com/
8. G. R. Harrison, *Proc. Am. Phil. Soc.* 102(5), 483–491, 1958.
9. Jobin   Yvon.   http://www.horiba.com/scientific/products/optics-tutorial/diffraction-gratings/
10. L. Seidel, *Astr. Nach.* 1856, 289, 1840.
11. J. E. Mack, J. R. Stehn, and B. Edlén, *J. Opt. Soc. Am.* 22, 245, 1932.
12. W. Welford, *Aberrations of Optical Systems*. Taylor & Francis, 1986.
13. J. A. Samson and D. L. Ederer, *Vacuum Ultraviolet Spectroscopy II, Vol. 32 in Experimental Methods in the Physical Sciences*. San Diego, CA: Academic Press, 1998. ISBN 0-12-475979-30.
14. H. G. Beutler, *J. Opt. Soc. Am.* 35, 311 1945.
15. A. Thorn, U. Litzen, and S. Johansson, Spectrophysics, Principles and Applications. Berlin, Heidelberg, New York: Springer, 1999. ISBN 3-540-65117-9.
16. T. Harada and T. Kita, *Appl. Opt.* 19(23), 3987, 1980.
17. T. Kita et al., *Appl. Opt.* 22(4), 512, 1983.
18. T. Harada et al., *Appl. Opt.* 38(13), 2743, 1999.

4

Crystal Spectrometers

Nobuyuki Nakamura

CONTENTS

4.1 Introduction

Since the transition energy as well as the energy level of an atomic system can be scaled as Z^2 and n^{-2} (where Z denotes the atomic number and n the principal quantum number) [1], most transitions fall in the x-ray range for highly charged heavy ions. Figure 4.1 shows examples of the Z dependence of transition wavelengths. As seen in Figure 4.1, not only $\Delta n \neq 0$ transitions, but also $\Delta n = 0$ transitions can fall in the x-ray range for heavy ions. Thus, x-ray spectroscopy is one of the most important methods in studying highly charged ions. There are two ways to analyze the energy (wavelength) of an x-ray photon. One is called the energy-dispersive method by which quantities

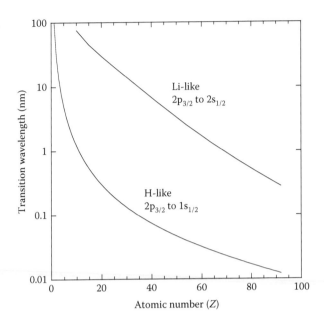

FIGURE 4.1
Atomic number dependence of transition wavelengths. Theoretical values by Johnson and Soff [2] and Chen and Reed [3] are used. (From M. H. Chen and K. J. Reed, *Phys. Rev. A* 47, 1874, 1993. With permission.)

proportional to the photon energy are measured. For example, for semiconductor detectors, the number of electron–hole pairs produced in the semiconductor through the interaction with an x-ray photon is measured. The resolution $E/\Delta E$ of a semiconductor detector is usually limited by the statistical variability of the number of electron–hole pairs to the order of 10^2. Another way to analyze the x-ray energy (wavelength) is called the wavelength-dispersive method in which diffraction by a grating or a crystal is used. Generally, a grating is used for soft x-rays ($>50\,\text{Å}$), and a crystal for hard x-rays ($<50\,\text{Å}$). The resolution of the wavelength-dispersive method with a crystal is typically on the order of 10^3–10^4. Recently new types of energy-dispersive instruments show remarkable development. For example, the resolution of an x-ray micro-calorimeter [4] reaches the value comparable to that of the wavelength-dispersive spectrometer. However, the technique for such a detector is still state of the art so that it is rather difficult to acquire such detectors without collaborating with groups involved in research and development of these devices. In addition, the effective size of such a detector is generally very small (typically less than $1\,\text{mm}^2$). Thus, crystal spectrometers are still the most important tools for the high-resolution spectroscopy of hard x-rays. In this chapter, various types of crystal spectrometers are described.

4.2 Basic Consideration

4.2.1 Principle

The principle of a crystal spectrometer is based on the Bragg diffraction, which is shown in Figure 4.2, where an x-ray with a wavelength λ is incident on a crystal surface with a lattice constant d. The x-ray is reflected when the following condition is met:

$$n\lambda = 2d \sin \theta, \tag{4.1}$$

where θ is the incident angle (measured from the crystal surface) and is also equal to the reflection angle, and n the order of reflection. On the other hand, when this equation is not met, the x-ray is not reflected but absorbed in (or penetrates) the crystal. This is Bragg's law, which is the most and almost only important equation for crystal spectrometers. Since d is a constant, the wavelength can be related directly to the angle θ. If the incident x-rays are in the form of parallel light, then all of x-rays are incident on the lattice plane with the same angle so that all of them are reflected if Bragg's condition is met (Figure 4.2b). Consequently, none of the x-rays will be reflected if Bragg's condition is not met. On the other hand, if the incident x-rays are in the form of divergent light, such as light diverging from a point, then the incidence angle is dependent on the position of the crystal where the photon hits so that only a portion of the x-rays are reflected (Figure 4.2c).

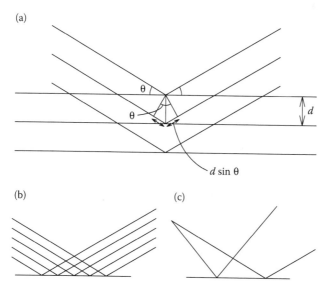

FIGURE 4.2
Bragg diffraction. (a) Bragg diffraction (principle), (b) diffraction of a parallel light, and (c) diffraction of a dispersive light.

4.2.2 Basic Structure of a Spectrometer

A common visible spectrometer consists of (1) an entrance slit, (2) a lens (or a mirror) for converting the divergent light to a parallel light, (3) a dispersion element (such as a grating, a prism, etc.), (4) a lens (or a mirror) for focussing the reflected parallel light onto a exit slit (or the surface of the imaging detector), and (5) a detector. However, since neither lenses nor mirrors are useful for x-rays, except in special cases, an x-ray spectrometer consists of a slit, a crystal, and a detector only. Some spectrometers do not need a slit depending on the type of crystals or the shape of the source.

Since the transmission of x-rays ranging from 1 to 10 keV in air is low, the x-ray path in the spectrometer must be vacuum or He atmosphere. For example, the transmission of a 5 keV x-ray in air is only less than 1% when the path length is 1 m. Although the transmission increases with the photon energy, it is still only about 50% even for a 10 keV x-ray. On the other hand, the transmission in a He atmosphere (with a path length of 1 m) is larger than 90% for x-rays with a photon energy larger than 2 keV.

Both vacuum and He atmosphere spectrometers have advantages and disadvantages. For He-based spectrometers, no vacuum vessel is needed as He can flow through the system during operation. Thus, a lightweight vessel can be made at a low price because it can be made of, for example, acrylic resins sealed with adhesive tapes. However, because the x-ray light source is inside a vacuum chamber, for most cases in spectroscopy of highly charged ions, the source and the spectrometer must be divided by a window which can stand the 1 atmospheric pressure difference. Beryllium is usually used as a window because of its high transmittance for x-rays, but the absorption by the beryllium window is considerable for photon energies below 3 keV. For example, the transmission of a 3 keV photon is only about 30% for a 0.3-mm thick beryllium window. On the other hand, for a vacuum spectrometer, it is not needed to use such a window in principle. A window is often used also for a vacuum spectrometer because it is rather difficult to obtain ultra-high vacuum inside the spectrometer; however, in this case, the window can be very thin as it does not need to withstand 1 atmosphere pressure difference and hence the absorption by the window is negligible. One of the disadvantages of a vacuum spectrometer is that it must be constructed from stainless steel (or titanium), which is heavy and expensive. Another drawback with vacuum-based spectrometers is that in some designs it is difficult to change the Bragg angle without breaking the vacuum.

4.2.3 Nature of Crystals

For the ideal case, where the x-rays are reflected from an infinitely thick ideal crystal and the absorption can be ignored, the incident x-ray is reflected only when Bragg's condition is strictly met. For actual reflection, however, since only the finite number of atoms can contribute to the reflection, the angle which can reflect x-rays has a finite width near the Bragg angle θ_B (e.g., see References [5,6]). The angular dependence of the reflected light intensity is called the rocking curve. According to the width of the rocking curve, crystals can be classified roughly into two types, perfect or mosaic crystals. A perfect crystal has a sharp curve ($\delta\theta \sim 10''$) with a high peak intensity, whereas a mosaic crystal has a broad curve ($\delta\theta \sim 10'$) with a low peak intensity. Thus, in general, a perfect crystal is preferred for high-resolution studies. On the other hand, a mosaic crystal is preferred when the efficiency is important because its integrated reflectivity is higher than that of a perfect crystal. Typical perfect crystals are Si ($2d = 3.84\,\text{Å}$ for (220) and $6.27\,\text{Å}$ for (111)) and Ge ($2d = 4.00\,\text{Å}$ for (220), $6.53\,\text{Å}$ for (111)), whereas a typical mosaic crystal is (pyrolytic) graphite ($2d = 6.71\,\text{Å}$). In a sense, all crystals other than perfect crystal can be classified as a mosaic crystal, but they have individual differences so that some of them have properties close to those of a perfect crystal. A list of the crystals which are frequently used for x-ray spectroscopy is compiled in the book by Thompson and Vaughan [7].

One must consider that the reflection efficiency is also depends on the polarization of the incident x-rays. The reflection efficiency ratio of an x-ray linearly polarized within the dispersion plane to that perpendicular to the dispersion plane is 0 at $\theta_B = 45°$ and is 1 at $\theta_B = 0°$ and $90°$. When the angle increases from $\theta_B = 0°$ to $45°$, the ratio monotonically decreases from 1 to 0 although the curve depends on the crystal. Similarly for $\theta_B = 45°$ to $90°$, the ratio monotonically increases from 0 to 1. If $\theta_B = 45°$ is used, then the crystal can be used as a polarizer [8].

4.3 Classification by Crystal Shape

Depending on the application, several shapes of crystals are used. Figure 4.3 shows four typical spectrometers where different crystal shapes are used. The geometrical optics rays starting from a point source are also shown. As seen in Figure 4.3, a point source results in a point image on the detector for some spectrometer configurations, whereas a point source results in a curved image for other configurations. In general, the former has a high spectral resolution compared with the latter.

4.3.1 Flat Crystal Spectrometer

The simplest crystal spectrometer is the one utilizing a flat crystal. When the x-rays of interest are not from a parallel light beam, such spectrometers are usually used with an entrance aperture or a slit. In such cases, the incident angle depends on the position of the crystal as shown in Figure 4.2c, and as a result, the wavelength of the incident x-ray is converted

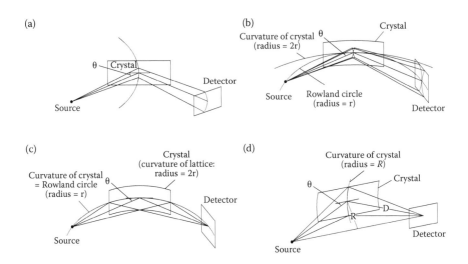

FIGURE 4.3
Various Bragg spectrometers. (a) Flat crystal spectrometer, (b) Johann spectrometer, (c) Johansson spectrometer, and (d) von Hámos spectrometer.

into the position on the detector. Thus, usually a position-sensitive detector is used for flat crystal spectrometers. Since there is no focussing mechanism (focussing plane) for a flat crystal spectrometer, the arrangement of the elements has no restrictions, that is, one does not need to align the elements so precisely. This is the main merit of a flat crystal spectrometer compared to other spectrometers. In general, spectrometers using a curved crystal need precise adjustment of the position and the angle of the elements (crystal and detector) because of its focussing properties. For a flat crystal spectrometer, the distance between the source and the crystal can be determined independently from that between the crystal and the detector in principle. However, considering the intrinsic focussing property of a crystal, one must make these two distances the same for high-resolution studies with a high position resolution detector [9]. The dispersion and thus the resolution $\lambda/\delta\lambda$ becomes large as the total distance L between the source and the detector increases, whereas the efficiency decreases as L^{-2}. The total efficiency is generally low compared to that of curved crystal spectrometers although it can be overcome by using a detector with a large effective size in the direction vertical to the dispersion plane [10].

Figure 4.4 shows the flat crystal spectrometer used for spectroscopic studies at the electron beam ion trap (EBIT) in Tokyo [10]. Since the source in an EBIT is a line with a width of 100-μm or less, it acts as an entrance slit for a dispersive spectrometer such as a flat crystal spectrometer. Thus, normally no entrance slit is used when a flat crystal spectrometer is used at an EBIT light source. For this spectrometer, a position-sensitive proportional counter with a large effective area is used to overcome the low efficiency of the flat crystal.

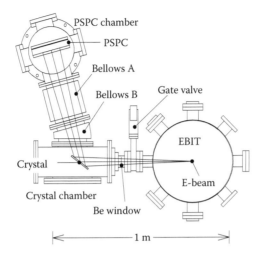

FIGURE 4.4
Flat crystal spectrometer at the Tokyo EBIT. (From N. Nakamura, *Rev. Sci. Instrum.* 71, 4065, 2000. With permission.)

When a slit-less flat crystal spectrometer is used for spectroscopy with an EBIT, it is extremely important to fix the position of the source. Thus, the reference line source should also be produced in the EBIT with the same operation conditions. This is because different EBIT operational conditions could lead to a displacement of the electron beam, that is, source position. Transitions in hydrogen-like light ions are often used as wavelength references because they are well known both theoretically and experimentally, whereas several clever methods are applied to obtain absolute wavelengths [11,12]. Since many hours of data accumulation are often needed, the stability of the system (source position, electronics, etc.) should also be carefully checked.

4.3.2 von Hámos Arrangement

As described in the previous section, one of the ways to overcome the low efficiency of a crystal spectrometer is to use a detector with a large effective area. Another way is to bend the crystal in the direction vertical to the dispersion plane in order to focus the x-rays onto the detector. This type of spectrometer is called von Hámos [11] and is shown in Figure 4.3b. A crystal with a radius of curvature R should be placed in such a way that the relation $D = R/\sin\theta_B$ is established, where D is the distance between the source and the crystal. The distance between the crystal and the detector should also be D. Thus, the crystal and the detector must be precisely positioned with respect to the source, unlike for a flat crystal spectrometer. Owing to the focussing property in the direction perpendicular to the dispersive plane, it is not needed to use a

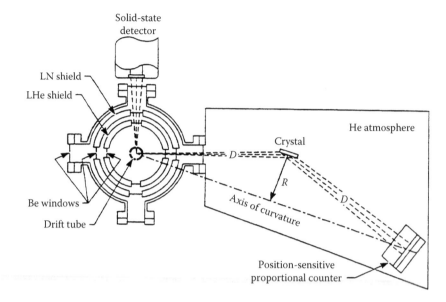

FIGURE 4.5
von Hámos crystal spectrometer at the LLNL EBIT. (From P. Beiersdorfer et al., *Rev. Sci. Instrum.* 61, 2338, 1990. With permission.)

detector with a large height. The detector plane should be placed on the focal line, which is the axis of curvature of the crystal.

Figure 4.5 shows an example of a von Hámos spectrometer. This instrument is used for spectroscopic studies with an EBIT in Livermore [13]. Again, the source in the EBIT acts as an entrance slit. In this case, a position-sensitive proportional counter with an effective depth is placed perpendicularly to the reflected x-rays to avoid any oblique x-ray incidence. Such oblique incidence would worsen the spectral resolution even more than the defocussing effect introduced by the geometry. By using a similar type of spectrometer but with an extended path length, the group at the Livermore EBIT has succeeded in obtaining very high resolution spectra, limited only by the natural line profile in the case of neon-like cesium [14].

4.3.3 Johansson

As described in Section 4.3.1, for a flat crystal spectrometer, the incident angle of the x-ray emitted from a point source depends on the position of the crystal. On the other hand, a crystal which is configured such that the incident angle does not depend on the position is used in a Johansson spectrometer. A Johansson crystal can be produced by polishing the surface of the crystal with a radius of curvature R after bending it with a radius of curvature $2R$. In the Johansson arrangement, the Rowland circle (a focal curve) has a radius R and comes in contact with the crystal surface. In this arrangement, since the x-rays emitted from a point on the Rowland circle have the same incident angle with respect to the lattice plane, if the incident angle fulfill Bragg's condition, all the x-rays are reflected and focused on the symmetric point on the Rowland circle. Thus, if an entrance slit or a point source is placed on the Rowland circle, then the Johansson spectrometer works as a monochromator. On the other hand, a slit-less Johansson spectrometer works as a polychromator for general sources which have a finite size and placed inside or outside the Rowland circle. The x-rays having the same wavelength focus on a point on the Rowland circle regardless of the position of the source.

Johansson-type spectrometers have almost ideal focussing properties, that is, there is no astigmatism unlike Johann-type spectrometers which are described in the following section. However, Johansson spectrometers are not widely used due to the difficulty in the procedure to produce the correct crystal shape, as described above. In fact, errors in the production process can easily result in a decline in the imaging properties and the focussing can be similar or worse than for Johann-type instruments.

4.3.4 Johann

A Johann spectrometer uses a crystal just bent with a radius of curvature $2R$. Similar to a Johansson crystal, the Rowland circle has a radius R and comes in contact with the crystal surface. Since the crystal for a Johann-type spectrometer is easily prepared and the optical efficiency is quite good, such spectrometers are widely used. However, the spectral resolution is not as good as for Johansson-type instruments due to the existence of astigmatism.

One of the merits of the Johann (and also the Johansson) geometry is that the x-rays of same wavelength will be focused on a point on the Rowland circle regardless of the position of the source. It means that a Johann spectrometer does not need an entrance slit, which is well met with plasma sources with a large volume. This merit is also useful for spectroscopy with an EBIT whose plasma size is relatively small, because in this case a reference light source can be positioned outside the EBIT. Figure 4.6 shows the Johann spectrometer designed at the Oxford EBIT [15]. In this arrangement, the EBIT source is placed inside the Rowland circle and the reference source can be placed near the Rowland circle, that is, outside the EBIT. By using this method, absolute wavelength measurements have been performed [16,17].

4.3.5 Spherical

All the curved crystals introduced above have a radius of curvature in only one direction. On the other hand, crystals with a radius of curvature in two directions can also be used as a spectrometer. For example, a spherically bent crystal is often used. The characteristics in spectral dispersion of a spherically

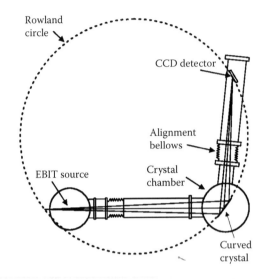

FIGURE 4.6
Johann spectrometer at the Oxford EBIT. (From M. R. Tarbutt, *Spectroscopy of Few-Electron Highly Charged Ions*, Thesis. Oxford: Oxford University, 2000. With permission.)

bent crystal are practically the same with those of a Johann crystal, but it is possible to observe the spatial distribution as well as the wavelength of radiation. Figure 4.7 shows an example of a spherically bent crystal, used for spectroscopy with an EBIT, where the source and the detector are placed outside the Rowland circle. In this arrangement, the spatial distribution along the direction which is normal to the dispersive plane can be observed on the detector. Nakamura et al. [18] measured the ion temperature in the EBIT plasma by observing the spatial distribution in a harmonic trap potential.

4.3.6 Transmission Spectrometer

As the x-ray energy of interest becomes high, the reflection efficiency becomes small. Thus, transmission-type spectrometers are often used for high-resolution spectroscopy for x-ray energies higher than 15 keV. Figure 4.8 shows a transmission-type spectrometer [19] based on the DuMond geometry [20]. The principle of a DuMond spectrometer is similar to that of a Johann spectrometer. However, in the DuMond geometry, a crystal cut perpendicular to a lattice plane is bent with a radius of curvature $2R$. The Rowland circle is defined as a circle with a radius R in contact with the crystal surface (i.e., the lattice surface is perpendicular to the arc of the Rowland circle at the contact position). In an arrangement such as Figure 4.8 where the source is placed on the Rowland circle, the crystal acts like a monochromator, that is, only x-rays within a narrow band width can be diffracted, whereas almost all other x-rays penetrate through the crystal without diffraction. Thus, a detector (no

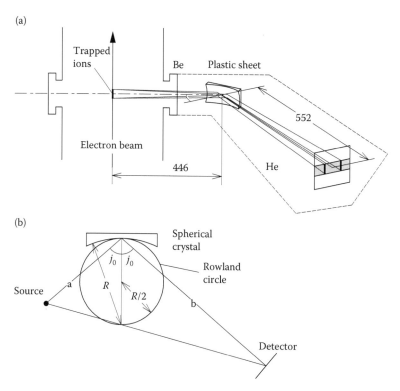

FIGURE 4.7
Spherical spectrometer used at the Tokyo EBIT. (a) Side view and (b) top view of the arrangement. (From N. Nakamura et al., *Rev. Sci. Instrum.* 70, 1658, 1999. With permission.)

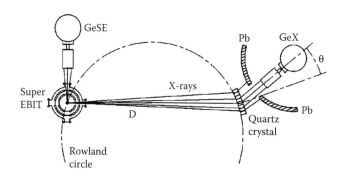

FIGURE 4.8
Transmission-type spectrometer used at the LLNL EBIT. (From K. Widmann et al., *Rev. Sci. Instrum.* 68, 1087, 1997. With permission.)

need for position sensitivity) is placed beyond the crystal and shielded to prevent it seeing the directly transmitted photons. Generally, the efficiency of this type of spectrometer is not so high, even if the transmission itself is high, because the system must be scanned by moving the crystal and the detector with respect to the source to obtain the spectrum. Thus, high-resolution spectroscopic studies for high-energy x-rays (>20 keV) have not, at least so far, received much attention.

4.4 Detector

For most crystal spectrometers, the diffracted x-rays are detected with a position-sensitive detector. The position sensitivity is needed at least along the direction of the spectral dispersion. In many cases, however, the position sensitivity is needed also in the direction perpendicular to the dispersion plane. For example, for measuring both the spatial distribution and the wavelength of radiation simultaneously with a spherically bent crystal, one must use two-dimensional position-sensitive detectors. Another example is found in the case of a large detector. As seen in Figure 4.3, diffracted x-rays make a curved image on the detector when a flat crystal is used. To obtain high-resolution spectra by correcting the curved image, two-dimensional position-sensitive detection is needed [10]. Table 4.1 lists typical position-sensitive detectors and some of their properties. In the following sections, three types of position-sensitive detectors in common use with crystal spectrometers are described.

4.4.1 Film

When the x-ray flux is high enough, an x-ray film is one of the most simple and useful detection devices. It utilizes the darkening of photographic emulsion. The photoelectronic and Compton effects of silver bromide (AgBr)

TABLE 4.1

Position-Sensitive X-ray Detectors

Detector Type	Position Sensitivity	$E/\Delta E$	Δx (mm)	Mode
Proportional counter	1D or 2D	~10	~0.1	Pulse
X-ray film	2D	None	~0.01	Integration
CCD	2D	~100	~0.01	Integration
Imaging plate	2D	None	~0.1	Integration
Arrayed Ge or Si	1D	~100	~1	Pulse

included in the emulsion are responsible for the darkening effect by x-rays. The image resolution and the efficiency thus depend on the size of the AgBr crystals. In general, as the size becomes larger, the resolution becomes worse, whereas the efficiency becomes better. The characteristics of a film can be represented by the x-ray exposure dependence of the photographic density. In the region where the density shows linear dependence on the logarithm of the exposure, the γ-value is defined as $\gamma = \tan \alpha$, where α is the slope of the linear dependence. A higher γ-value allows ones to obtain an image with higher contrast.

One of the advantages of an x-ray film is the cost. Since it does not need any precision electronics and it is widely used for medical applications, the price is much cheaper than other instruments such as a proportional counter and a charge-coupled device (CCD). On the other hand, one of the disadvantages is that it cannot be used for weak flux because the noise level is rather high. In addition, it is practically impossible to make time-resolved measurements.

4.4.2 Proportional Counter

It is often needed to detect a single x-ray photon, that is, pulse-counting mode, especially when the x-ray flux is very weak. One of the most common pulse-counting devices for the x-ray region is a proportional counter. A proportional counter consists of a chamber filled with gas and a thin anode wire going through the chamber. A positive high voltage is applied to the anode wire with respect to the chamber so that the electrons produced during the interaction of the gas molecules with an incident x-ray photon will be accelerated toward the wire. The accelerated electrons ionize other gas molecules and so on forming an electron avalanche and hence the electron number is multiplied. Due to the fact that the electric field is most intense near the anode wire, the electron avalanche is confined to this region and many electron–ion pairs are formed. As the electrons move faster than the molecular ions, they are absorbed quickly by the wire leaving an ion cloud in the region of the anode. By detecting the electrons with a charge-sensitive amplifier connected to the anode wire, single x-ray incident event can be detected. There are several methods to use a proportional counter as a position-sensitive detector. The simplest method is to use a resistive wire for the anode. In this method, two charge-sensitive amplifiers are connected to both ends of the wire. The amount of charges delivered to each amplifier is inversely proportional to the distance between the position where the electron cloud is absorbed and the end of the wire. Thus, the position X can be obtained from $X = Q_A/(Q_A + Q_B)$, where Q_A and Q_B are the amount of charges delivered to the each end of the anode wire.

Not only the anode, but also the cathode can be used for position detection. Figure 4.9 shows an example of a position-sensitive proportional counter whose cathode is used for position detection. In this counter, a specially made

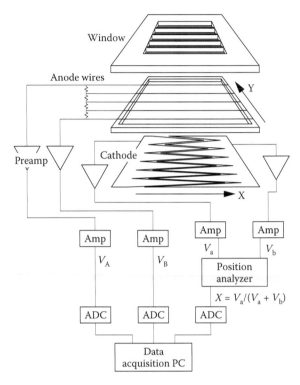

FIGURE 4.9
Position-sensitive proportional counter used at the Tokyo EBIT. (From N. Nakamura, *Rev. Sci. Instrum.* 71, 4065, 2000. With permission.)

cathode called a back-gammon type is used. As seen in Figure 4.9, a back-gammon cathode has two wedge-shaped electrodes facing each other. The ion cloud generated near the wire produces induced charges which then spread out over the cathode electrode. The induced charges are divided between the two electrodes depending on the position of the ion clouds, that is, the position of the incident x-ray. By detecting the divided charges Q_A and Q_B by charge-sensitive amplifiers and again by taking the ratio $Q_A/(Q_A + Q_B)$, the position of the incident x-ray can be obtained. The proportional counter shown in Figure 4.9 has position sensitivity also for the direction (Y) perpendicular to the position-sensitive direction (X) measured by the cathode. Here several anode wires are connected with resistors so that the absorbed electron charges are divided between the two ends. By taking the ratio of the amount of charges, the particular wire nearest to the electron avalanche can be determined, that is, the nearest wire to the incident x-ray. Obviously, the position resolution for this direction is determined by the interval between the wires.

Other than the back-gammon cathode, several types of cathode are used for position detection. For example, cathodes based on delay line techniques have been used [21].

4.4.3 Charge-Coupled Device

A CCD, which is probably the most popular device for visible light detection, can also be used as an imaging device in the x-ray range. There are two ways to detect an x-ray photon with a CCD: one is the indirect method and another is the direct method. In the indirect method, a scintillator is placed before the CCD, and the visible photons emitted from the interaction of the incident x-ray with the scintillator are detected. In the direct method, the incident x-ray is detected directly with a CCD. The principle for detection of a photon is the same both for the visible and x-ray ranges; however, several things should be considered depending on the wavelength range of interest. For visible light, the electrode and the insulation layer on the silicon substrate are almost transparent. However, for x-rays, absorption by such layers is so large that they should be made as thin as possible. For soft x-rays, extreme and vacuum ultraviolet light, the absorption is rather serious even for thin layers. In such a case, photons are injected from the opposite side of the electrodes so that they interact with the silicon substrate directly. This "back-illuminated" CCD is often used for the VUV to soft x-ray region. For hard x-rays, the absorption at the electrode is not serious so that a "front-illuminated" CCD can be used. However, since the transmission in silicon also becomes large, the depletion layer in which x-ray should be absorbed is made as thick as possible to increase the detection efficiency.

Similar to a Si detector, the number of electron–hole pairs produced in a CCD is proportional to the energy of the incident x-ray photon. Thus, each pixel can act as a energy-dispersive x-ray detector. For this to be meaningful, the photon count rate, or expose time, should be limited to ensure that no more than one photon is registered on any pixel. It is reported elsewhere in this volume that the energy resolution of around 140–150 eV can be obtained (see Figure 5.24). This energy resolution is often very useful when a CCD is used as the detector for a crystal spectrometer. In a crystal spectrometer, quasi-monochromatic x-rays should reach the detector. Thus, by excluding the x-rays with energies far from those of interest, by pulse height discrimination, the spectrometer background can be reduced and hence a higher signal-to-noise ratio can be obtained. Although the proportional counter also has energy resolution, the resolution is much worth compared with that of a CCD. CCD detectors are discussed in detail in Chapter 5.

4.5 Summary

Although several new techniques, such as a micro-calorimeter, are being developed, crystal spectrometers still play a leading role in high-resolution x-ray spectroscopy. Since it is rather old and well established, one can learn its basic properties from the existing papers and books. Thus, in this chapter, several actual examples of crystal spectrometers used in the spectroscopy of highly charged ions have been shown. These examples will help one to select the most suitable spectrometer depending on the purpose and the source type. In particular, since an EBIT is a special source which is a thin line with a width of less than 0.1 mm, the arrangement should be carefully considered to make good use of its characteristics. Good examples are efficient slit-less spectrometers such as ones shown in Sections 4.3.1 and 4.3.2 where the source is regarded as the "entrance slit" of the spectrometer.

Generally, the efficiency of wavelength-dispersive spectrometers is not so high regardless of the wavelength range. For the x-ray range, the fact that there are neither mirrors nor lenses generally available makes the efficiency even lower. Thus, further development of highly charged ion sources can be expected to increase the number of stored ions and hence lead to better spectroscopic results in future.

References

1. H. F. Beyer, H. J. Kluge, and V. P. Shevelko, *X-ray Radiation of Highly Charged Ions*, Springer Series on Atoms and Plasmas. Berlin: Springer, 1997.
2. W. R. Johnson and G. Soff, *At. Data Nucl. Data Tables* 33, 405, 1985.
3. M. H. Chen and K. J. Reed, *Phys. Rev. A* 47, 1874, 1993.
4. F. S. Porter, G. V. Brown, K. R. Boyce, R. L. Kelley, C. A. Kilbourne, P. Beiersdorfer, H. Chen, S. Terracol, and S. M. Kahn, *Rev. Sci. Instrum* 75, 3772, 2004.
5. B. Cullity, *Elements of X-Ray Diffraction*. Reading, MA: Addison-Wesley, 1956.
6. W. H. Zachariasen, *Theory of X-ray Diffraction in Crystals*. New York: Dover Publications, Inc., 1945.
7. A. C. Thompson and D. Vaughan, *X-ray Data Booklet* (2nd edition). Berkeley, CA: Lawrence Berkeley National Laboratory, 2001.
8. J. R. Henderson, P. Beiersdorfer, C. L. Bennett, S. Chantrenne, D. A. Knapp, R. E. Marrs, M. B. Schneider et al., *Phys. Rev. Lett.* 65, 705, 1990.
9. T. Mizogawa, *Phys. Scripta T* 73, 403, 1997.
10. N. Nakamura, *Rev. Sci. Instrum.* 71, 4065, 2000.
11. H. Bruhns, J. Braun, K. Kubicek, J. R. C. López-Urrutia, and J. Ullrich, *Phys. Rev. Lett.* 99, 113001, 2007.
12. D. Klöpfel, G. Hölzer, E. Förster, and P. Beiersdorfer, *Rev. Sci. Instrum.* 68, 3669, 1997.
13. P. Beiersdorfer, R. E. Marrs, J. R. Henderson, D. A. Knapp, M. A. Levine, D. B. Platt, M. B. Schneider, D. A. Vogel, and K. L. Wong, *Rev. Sci. Instrum.* 61, 2338, 1990.
14. P. Beiersdorfer, A. L. Osterheld, V. Decaux, and K. Widmann, *Phys. Rev. Lett.* 77, 5353, 1996.
15. M. R. Tarbutt, *Spectroscopy of Few-Electron Highly Charged Ions*, Thesis. Oxford: Oxford University, 2000.

16. T. E. Cowan, C. L. Bennett, D. D. Dietrich, J. V. Bixler, C. J. Hailey, J. R. Henderson, D. A. Knapp, M. A. Levin, R. E. Marrs, and M. B. Schneider, *Phys. Rev. Lett.* 66, 1150, 1991.
17. M. R. Tarbutt and J. D. Silver, *J. Phys. B* 35, 1467, 2002.
18. N. Nakamura, A. Y. Faenov, T. A. Pikuz, E. Nojikawa, H. Shiraishi, F. J. Currell, and S. Ohtani, *Rev. Sci. Instrum.* 70, 1658, 1999.
19. K. Widmann, P. Beiersdorfer, G. V. Brown, J. R. C. López-Urrutia, V. Decaux, and D. W. Savin, *Rev. Sci. Instrum.* 68, 1087, 1997.
20. J. W. M. DuMond, *Rev. Sci. Instrum.* 18, 626, 1947.
21. D. Vogel, P. Beiersdorfer, V. Decaux, and K. Widmann, *Rev. Sci. Instrum.* 66, 776, 1995.

5

CCD Detectors

Nick Nelms

CONTENTS

5.1 Introduction

A charge-coupled device, or CCD as it is more commonly known, is a one- or two-dimensional semiconductor (almost always silicon) detector. A CCD comprises a number of sensing elements or pixels, arranged in a regular array. The number of pixels is in the hundreds or thousands for linear (1D) devices and tens of thousands to millions for area (2D) devices. Pixel sizes vary from a few microns to a few tens of microns resulting in 2D devices with areas up to several square centimeters (Figure 5.1).

CCDs have been available for nearly four decades and can be found in a wide variety of commercial and scientific products and applications. The history of CCD development from the first production devices at Bell Laboratories in 1970 [1] has been well documented in many publications, including references [2,3]. Although not originally developed as such, their use as imaging detectors was realized very early on [4]. Since then, extensive CCD development has resulted in detectors with excellent performance across the waveband from near-infrared, through visible, ultraviolet, and into the soft x-ray region.

FIGURE 5.1
Various CCD detectors. (Courtesy of e2v Technologies Plc, Chelmsford, Essex, UK.)

Although complementary metal-oxide semiconductor (CMOS)-active pixel sensors are beginning to replace CCDs in a number of different areas (most notably in consumer products such as digital cameras), CCDs are a mature, well-understood technology with high performance and a long future in demanding scientific applications. The following sections present a description of the design and operation of a CCD detector followed by an introduction to their application in spectroscopic measurements.

5.2 CCD Design

5.2.1 Overview

A CCD is a semiconductor photon detector, usually fabricated from p-type silicon. It comprises an array (either one- or two-dimensional) of elements or pixels capable of storing local, photon-generated charge. A typical two-dimensional CCD will have an image section, possibly a store section, and a line readout section (Figure 5.2), whereas a one-dimensional device is essentially just a line readout section.

The charge in the image region can be quickly shifted (by frame transfer) into the store section. The stored charge is then shifted one row at a time into the line readout section where each pixel is moved sequentially toward an output node and measured. In this way, a charge map or image of the whole CCD can be constructed.

5.2.2 Charge Generation

Charge is generated in the silicon lattice whenever a photon is absorbed due to the photoelectric effect. In this case, a single electron–hole pair is generated for each photon absorbed. Silicon has a bandgap energy of ~1.1 eV corresponding

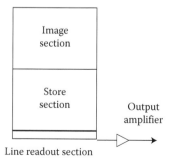

FIGURE 5.2
Typical CCD arrangement.

to a wavelength of ~1.1 μm and consequently is transparent to photons with a wavelength longer than this. Photons with a wavelength shorter than this, however, have a finite chance of being absorbed.

For higher-energy photons (x-rays), the situation is not quite the same. Absorption of an x-ray photon (<100 keV), in the silicon lattice of a CCD, results most probably (~92% [5]) in the ejection of a photoelectron from the K-shell of a silicon atom. The photoelectron is created with energy E, which is given by

$$E = E_x - E_b, \tag{5.1}$$

where E_x is the energy of the incident x-ray photon and E_b the silicon K-shell binding energy, ~1.85 keV. This reaction will only occur if the incident photon has energy greater than the binding energy of the K-shell. Lower-energy photons may interact with the L-shell, M-shell, and so on. The excited silicon atom may de-excite either by the Auger process or by fluorescent emission of a photon. In the case of photon emission, the photon will have an energy equal to the binding energy of the K-shell minus the binding energy of the L-shell, that is, $1.85 - 0.1 = 1.75$ keV. Secondary ionization by the photoelectron and reabsorption of the fluorescent photon can then occur until there is insufficient energy remaining to promote electrons from the valence band into the conduction band. This is a statistical process that results in the creation of an average number (N) of electron–hole pairs and is given by

$$N = \frac{E_x}{\omega}, \tag{5.2}$$

where E_x is the energy of the interacting photon and ω the mean ionization energy. Since some of the ionization energy is lost to the crystal lattice (phonons), ω is somewhat larger than the bandgap energy of silicon and is found by experiment to be ~3.65 eV. N is a statistical quantity but since the creation of electron–hole pairs is not mutually exclusive, the variation is less than that given by purely random statistics. The usual Poissonian variance is modified by an empirical quantity known as the Fano factor [5], F (~0.15 for silicon [6]) and for N is given by

$$\sigma_N^2 = FN. \tag{5.3}$$

5.2.3 CCD Structure and Charge Storage

The storage elements of a CCD are essentially metal-oxide semiconductor (MOS) capacitors. The capacitors comprise a conducting electrode deposited onto the silicon substrate with a thin insulating layer of silicon dioxide in-between. Each pixel actually contains a coupled number of MOS capacitors, commonly two, three (Figure 5.3), or four resulting in the so-called 2-, 3-, or 4-phase device, respectively.

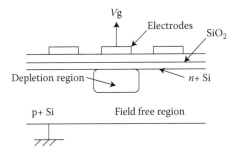

FIGURE 5.3
Simplified buried channel CCD pixel structure.

Application of a positive voltage to one of the conducting electrodes (gates) will create a region depleted of holes beneath the electrode. The $n+$ silicon layer modifies the shape of the electrostatic potential profile of this depletion region, creating a "buried channel," which allows charge collection away from the surface, where trapping sites would otherwise reduce the charge collection and transfer efficiency. The electrostatic potential profile, in terms of variation of electric field with depth in the silicon, can be derived using Poisson's equation. Photon charge (or thermal charge; see Section 5.2.5.2) generated within or close to this depletion region will be captured and held (or stored)—away from the surface—while the gate potential is applied.

In a 3-phase device, the three electrodes are actually common to one whole row of CCD pixels. In the horizontal direction, it is the potential applied to the gate of one of the electrodes which holds the charge in position whilst in the vertical direction "channel stops" are implanted to prevent migration of the electrons.* The channel stops are heavily doped, electrically inactive, p-type regions. The dimensions of the electrode groups and the spacing between the channel stops define the size of a pixel, typically between 10 and 40 μm^2 (although smaller, larger, and rectangular pixels are possible). This array of pixels collectively makes up the imaging area of the CCD. A number of factors determine the charge storage capacity, including pixel dimensions and gate voltage. Typical "full-well" capacities are in the range of 10^5 to 10^6 electrons.

5.2.4 CCD Characteristics and Quantum Efficiency

Independent of geometry (pixel size and array dimensions), CCDs are available in a number of different forms, for example, front-illuminated, back-illuminated (or back-thinned), and deep depletion (or high resistivity). These options are related to the material the CCD is made from and the way the CCD is fabricated. A standard CCD (if there is such a thing) is front-illuminated and manufactured using low-resistivity (of order 50 Ω cm) epitaxial silicon ($\sim$25 μm thick) on a bulk silicon substrate (see Figure 5.4).

---

* See Figure 5.8 for a better understanding.

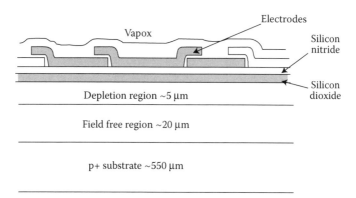

FIGURE 5.4
Standard 3-phase front-illuminated CCD (not to scale).

When the gate potential is applied, the epitaxial silicon is typically depleted to around 5 μm. Efficient charge collection only occurs when the charge is generated in this depletion region. Charge generated in the field-free region may be lost or spread into neighboring pixels.

When a photon is incident onto a detector, there is a finite chance that the photon will be absorbed and consequently "detected." The ratio of incident photons to photons "detected" is known as quantum efficiency (QE) and in an ideal case it would be 100%. Detected is placed in quotes in this instance since a photon may be absorbed but still not be detected at the output of the CCD (e.g., if the photon is absorbed in the field-free region). Absorption is limited by two factors—at longer wavelengths, photon capture is limited by the absorption length of photons in silicon and ultimately by the bandgap (equivalent to around 1.1 μm wavelength). At shorter wavelengths (in the blue and UV parts of the spectrum), the electrode structures (see Figure 5.5) start to absorb the photons before they reach the sensitive part of the detector. However, the standard CCD described above provides good QE when the incident photons are in the visible waveband (Figure 5.5) with performance reducing as expected toward the blue (shorter wavelengths).

An alternative CCD structure is available that provides improved performance at shorter wavelengths which is the back-thinned or back-illuminated CCD. A back-thinned CCD has the same structure as a front-illuminated device and is manufactured in the same way but with some additional steps. Once the standard structure is complete, the substrate and field-free regions are removed from the underside of the detector (by mechanical and chemical means) and then mounted effectively upside down in the package. In this case, the incident light now falls on the back face of the detector, hence the names back-thinned and back-illuminated, where there is no electrode structure in the way to absorb the shorter wavelength light. In this way, much higher QEs are achieved in the blue as can be seen in Figure 5.5. QEs can also

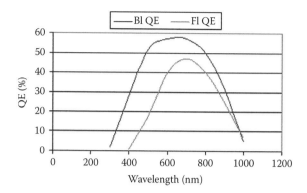

FIGURE 5.5
QE curves for front- and back-illuminated CCD detectors (no AR coating).

be improved for both front- and back-illuminated CCD types by the application of a suitable antireflection (AR) coating. Silicon has a high refractive index which results in high surface reflectivity when the detector is operated in air or vacuum. Approximately 30% of incident light is reflected which immediately lowers the prospective QE. With appropriate AR coating, the reflectivity can be reduced to the few percent level across a wideband. In demanding applications, specific AR coating can be applied to provide high QE in specific parts of the waveband (e.g., UV).

In the soft x-ray region of the spectrum, photon detection at low energies (of order <500 eV) is again influenced by the electrode structure (in the front-illuminated case) and by the depletion depth at higher energies. With a standard CCD, significant numbers of x-ray photons in the keV range are not absorbed within the 5 μm depletion depth. Consequently, CCDs for direct x-ray detection are manufactured using high-resistivity epitaxial silicon, since the depletion depth is related to the resistivity. With epitaxial layers of order 80 μm and a resistivity of 4000 Ω cm, a depletion depth of 30–40 μm can be achieved with the same electrode potential. This results in much higher QE figures as can be seen in Figure 5.6.

5.2.5 Charge Transport and Measurement

Charge generation and storage is usually allowed to continue for a certain "integration" time after which it is desired to measure the charge stored in each pixel. During integration, generated charge is held in the potential well of each pixel by maintaining the voltage on the gate of one electrode. If the voltage on this electrode is reduced to zero and simultaneously the voltage on an adjacent electrode is raised, the potential well effectively shifts and consequently the charge stored in it is transferred, or coupled, from beneath the first electrode to the second.

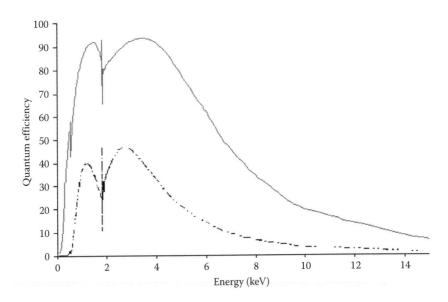

FIGURE 5.6
X-ray QE of standard (bottom curve) and deep-depletion CCD (top).

If the groups of three electrodes for each row of CCD pixels are arranged consecutively as in Figure 5.7 with common electrical connections as shown, and the potential of electrode 3 is lowered as electrode 1 is raised, the charge will be transferred from one CCD row to the next. This procedure can be repeated as many times as it is necessary to move the charge to the end of the CCD array, where the row of charges is finally shifted into a series of pixels known as the readout register. Each pixel of the readout register is aligned

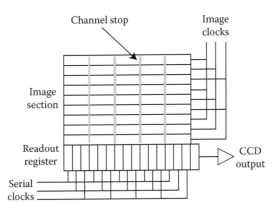

FIGURE 5.7
Typical CCD array schematic. The store section is omitted for clarity.

with a single column of the imaging area, and the gate electrodes for these pixels are arranged in a perpendicular direction to those of the image section. This allows a single row of CCD pixel charges to be transferred consecutively toward a charge-sensitive amplifier at the end of the line readout register.

5.2.6 Charge Read Out

Figure 5.8 shows a simple on-chip output charge amplifier and comprises three sections: the output node, which is essentially a diffused diode, a reset field effect transistor (FET), and an output FET.

The output node has an associated capacitance onto which the pixel charge is transferred from the line readout section (see Section 5.2.4). This causes a change in voltage at the gate of the output FET of

$$\Delta V_g = \frac{Q}{C_o}, \tag{5.4}$$

where Q is the charge transferred and C_o the output node capacitance. The load resistor R_L in Figure 5.8 is external to the CCD and is used to bias the output FET in the linear region of operation. With typical node capacitances of <0.1 pF and an output FET gain of ~0.7, a CCD will have a typical output responsivity of a few μV/electron.

To prevent saturation of the output amplifier, the DC level of the output FET is normally restored before each pixel charge is transferred onto the output node. This is achieved by briefly turning on the reset FET. It is often the case that both the readout register and the output node have charge storage capacity of two or more times that of the image section pixels. This allows for addition of the charge packets within single pixels which is known as on-chip binning. This is discussed in more detail in Section 5.3.4.4.

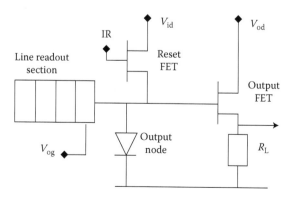

FIGURE 5.8
Typical CCD output amplifier.

5.2.7 Noise Sources

Before looking at methods to read the data from a CCD, it is necessary to understand the various contributions to the noise on this measurement. The various noise sources present in a CCD output measurement are either signal-related (shot noise) or device-related (dark current, charge transfer losses, reset noise, and output amplifier noise).

5.2.7.1 Signal Shot Noise

Signal shot noise refers to the statistical variation in the average number of photons from a source arriving at a detector in a fixed time. This is a Poissonian process with variance given by

$$\sigma_N^2 = N, \qquad (5.5)$$

where N is the average number of photons arriving at the detector in time t. This is the case when a large number of photons are emitted. In the soft x-ray region, CCD detectors are capable of single photon detection, and the signal shot noise as described above is obviously not relevant since the true number of photons arriving is known (although the Poissonian production of x-ray photons from the source still applies). In this situation, the signal shot noise can be interpreted as the statistical variation of charges produced when an x-ray photon is absorbed in the silicon lattice of the CCD. This process is described in detail in Section 5.2.2 and results in the generation of N electron–hole pairs (Equation 5.2) with a variance given by

$$\sigma_N^2 = \frac{FE_x}{\omega}, \qquad (5.6)$$

where F is the Fano factor, E_x the energy of the incoming photon, and ω the mean ionization energy in silicon.

5.2.7.2 Dark Current

Thermal generation of charge, occurring without photo-ionization (hence the name dark current), occurs when thermal excitation of the silicon lattice provides enough energy (greater than the bandgap of silicon, $\sim 1.1\,\mathrm{eV}$) to lift an electron from the valence band to the conduction band. At room temperature, dark currents of order 1–$2\,\mathrm{nA/cm^2}$ are typical, dominated by surface charge generation. This is equivalent to tens of thousands of electrons per pixel per second. With a full-well capacity of a few hundred thousand electrons, dark current will quickly saturate the device if it is not cleared (by clocking out the CCD). Dark current generation is a Poissonian process with the r.m.s. noise contribution (σ_{dc}) equal to the square root of the number of thermal charges. With the previously mentioned level of dark current, the noise is substantial, of the order of 100–200 electron r.m.s. for the number of charges generated in 1 s. Dark current generation in a buried channel CCD occurs in three areas, the depletion region, the field-free region, and the silicon–silicon oxide interface, and can be written as

$$I_d = \frac{en_i x_d}{2\tau} + \frac{eD_n n_i^2}{L_n N_a} + \frac{esn_i}{2}, \tag{5.7}$$

where n_i is the intrinsic carrier concentration, x_d the depletion depth, τ the effective lifetime of the minority carriers in the depletion region, D_n the diffusion constant, L_n the diffusion length for electrons, s the surface recombination velocity, and N_a the dopant concentration. In Equation 5.7, the first term is the contribution from the depletion region, the second term is from the field-free region, whereas the third term is due to the thermal generation of charges at the silicon–silicon oxide interface. It is the third term which typically dominates dark current generation and it has a temperature dependence of the form

$$\exp\left(-\frac{E_g}{kT}\right), \tag{5.8}$$

where E_g is the bandgap energy of silicon, k is Boltzmann's constant, and T the absolute temperature. Reducing the temperature of the CCD, therefore, will reduce the level of dark current generation. Operating at $-100°C$, a typical CCD operating temperature for scientific work, dark current will be of the order of a few electrons per pixel per hour, and makes no significant contribution to the overall noise performance of the CCD. As a rule of thumb, dark current is approximately halved for every 6–8° reduction in operating temperature.

A variation of CCD design which operates in "inverted mode" is also available (also known as multiphase pinned devices). As indicated above, the dominant dark current contribution is from electronic states that exist within the surface interfaces. In normal operation, the surface region is depleted and so thermal electrons are able to jump from the valence band into the surface states where they appear as dark current. In inverted mode operation, the potential across the image (and store) section clock gates is reversed, typically by raising the substrate voltage and turning off the clock voltage (during integration only). This has the effect of transferring holes from the p+ channel stops to the surface states and so greatly reducing thermal charge generation in this region. An overall reduction in dark current generation of two orders of magnitude is possible compared to standard devices. There is a disadvantage with inverted mode devices. Since the clock gate voltage is reduced during integration, the potential well, where the charge collection in a pixel occurs, is not created. This is compensated for by introducing an implant into the silicon beneath the pixel that generates a potential difference allowing charge to be accumulated. The storage capability (full-well capacity) is severely reduced when compared to noninverted mode devices.

5.2.7.3 Charge Transfer Losses

In an ideal CCD, transfer of electrons or charge from one electrode to another would be a loss-less process, that is, if n electrons are stored under the first electrode, after the completion of the transfer there would be n electrons under the second electrode. In a real CCD, however, this is not the case but buried channel devices come very close to achieving this ideal. Charge transfer efficiency (CTE) is usually of the order of 99.9995% or better for a single transfer. If the charge loss per transfer is small, then the total charge lost per pixel, Q_1, can be approximated by

$$Q_1 \approx N(1 - \text{CTE})n, \tag{5.9}$$

where N is the number of transfers and n the number of electrons in the charge packet. The charge loss is Poissonian and the r.m.s. noise associated with a read-out charge value is given by

$$\sigma_{CT} = \sqrt{Q_1}. \tag{5.10}$$

5.2.7.4 Reset Noise

Resetting the output diode to the reference potential (Section 5.2.6) is subject to thermal variation, with r.m.s. noise (in volts), and is given by

$$\sigma_{reset} = \sqrt{\frac{kT}{C_o}}, \tag{5.11}$$

where k is Boltzmann's constant, T the absolute temperature, and C_o the output node capacitance. At $-100°C$, with a typical output node capacitance of 0.1 pF, this is equivalent to ~100 electron r.m.s. This noise source is one of the largest contributions to overall cooled CCD read noise. X-ray spectroscopic performance, where signal charges are of the order of 500–3000 electrons, would be very poor with this level of noise. Fortunately, the reset noise can be effectively eliminated using a technique known as correlated double sampling (CDS), where the reset level is sampled and subtracted from the measured charge level in each pixel. This operation is discussed in more detail in Section 5.3.5.

5.2.7.5 Transistor Noise

The output FET has two main noise components, flicker or $1/f$ noise and Johnson noise. The $1/f$ noise, so called because it is inversely proportional to frequency, is thought to be due to trapping and release of signal charge carriers in the conduction channel. The Johnson noise component arises from the random thermal motion of the electrons and has a flat band response. Due to its temperature dependence, it is also known as thermal noise. Postprocessing of the CCD output signal is usually employed to reduce the impact of both $1/f$ and white noise from the transistor.

5.2.7.6 Total Noise

The noise sources discussed above contribute to the total read-noise of the CCD output signal. Since the noise sources are uncorrelated, the total read-noise is equal to the quadratic sum of the individual values and is given by

$$\sigma_T = \sigma_N^2 + \sigma_{dc}^2 + \sigma_{CT}^2 + \sigma_{reset}^2. \tag{5.12}$$

Methods of noise component reduction are discussed in detail in Section 5.3.5.

5.2.8 CCD Performance

The performance of a CCD detector can be expressed in different ways depending on the measurement being performed. A CCD datasheet will often give the read-noise of the detector output stage only, assuming that CDS (or other noise reduction technique) will be applied externally. Typical noise figures for a scientific device are in the region of a few electron r.m.s. for read-out speeds of a few tens kilopixels/s to a few megapixels/s. In this case, the dynamic range of the CCD is often expressed as the full-well capacity divided by the quoted read-noise. Although this gives some idea of CCD performance, it is sometimes more instructive to examine the signal-to-noise ratio (SNR). Figure 5.9 plots the SNR for a CCD with a total read-noise of 10 electron r.m.s. (after removal of the reset noise).

From the graph, it is clear that for a low-noise CCD, as the signal level increases, the SNR quickly becomes dominated by the shot noise from the

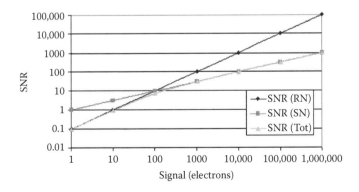

FIGURE 5.9
Plot of SNR for a CCD with a total read-noise of 10 electron r.m.s. SNR for signal with just read noise (RN), signal with just shot noise (SN), and overall SNR (Tot) are plotted.

signal itself. However, for low-level signals, the read noise can have a detrimental effect on the CCD performance. This is often the case when x-ray signals are being measured and/or when applications are "photon starved."

5.3 CCD Detector Operation

This section presents the operating requirements and methods of a CCD detector with a discussion of various output signal processing and noise reduction techniques.

5.3.1 Introduction

The operation of CCDs, in contrast with many other detectors, requires a large number of control signals. These include clock signals for the movement of the charge in various sections of the device and a number of different voltage sources for correct biasing of the detector. Use in scientific photon counting applications also requires low noise operation, which in turn requires low-noise electronics to both drive the CCD and process the output signal and to remove inherent detector noise. The noise contribution from thermally generated "dark current" has already been discussed in Section 5.2.5.2 and can easily swamp the signal that requires measurement. By cooling the detector to temperatures of the order of $-100°C$, it is possible to reduce the dark current noise to negligible levels and this is now a standard technique in scientific CCD detector operation. The remainder of this section assumes cooled CCD conditions as the starting point for achieving low-noise operation.

5.3.2 Detector Operation Overview

Figure 5.10 shows a typical arrangement used for low-noise operation of a cooled CCD. Electronic circuits provide the bias voltages, and a clock-sequencing device (often programmable) generates the clocking signals. The clock signals are then level-shifted to adjust the amplitude necessary for charge transfer. The CCD output is amplified and then passed through a signal processor, usually employing a method of CDS, for noise reduction. Finally, the output signal from the CDS processor is digitized, using an analog-to-digital converter (ADC) and stored on a computer. The following sections describe each of these requirements in detail. Specific CCD descriptions are based upon an e2v 3-phase, frame-transfer CCD. CCDs are also available as full-frame devices, or with 2- or 4-phase operation. Although not discussed here in detail, the basic principles are the same and it is a relatively simple matter to understand the operation of these devices from a description of 3-phase operation.

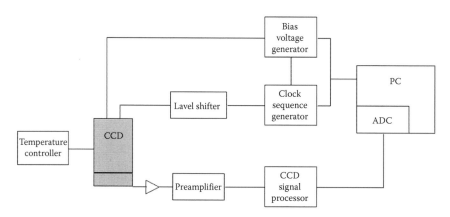

FIGURE 5.10
Typical arrangement for low-noise operation-cooled CCD system.

A CCD is a CMOS integrated circuit, comprising a large number of MOS capacitors and an output amplifier. The capacitor-type structures facilitate charge storage and transfer, whereas the output amplifier provides charge-to-voltage conversion. Figure 5.11 shows typical clock and bias voltage requirements for a 3-phase, frame-transfer CCD.

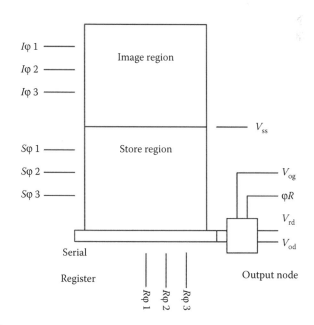

FIGURE 5.11
Typical clock and bias voltage requirements for a frame-transfer CCD.

5.3.3 CCD Biasing

The substrate of a CCD is usually connected to ground or some potential above this. e2v CCDs typically require a substrate voltage in the range 0–6 V. All other voltages (clocks and bias) are referenced on-chip to substrate.

5.3.3.1 CCD Output Amplifier

Since the CCD substrate material is n-type, the on-chip output amplifier of a CCD is typically a p-channel depletion metal-oxide semiconductor field effect transistor (MOSFET) arranged as a source-follower (Figure 5.12). R_L is either a load resistor or a constant current source and is external to the CCD.

The output transistor needs to be operated in the saturated part of its characteristic for linear operation. A low drain current allows the transistor to function in "buried-channel" mode with current conduction taking place away from possible charge trapping sites near the silicon surface. This can result in greatly reduced low-frequency noise, especially at low temperature. The low drain current typically means that the transistor is operating in depletion (a depletion MOSFET is capable of operating in either depletion or enhancement parts of the characteristic). This in turn means that the output source tends to sit at a higher potential than the FET gate voltage.

The exact operating point of the output transistor is determined by the gate–source voltage and the drain current i_d. These parameters are related by the transconductance, g_m, of the FET as follows:

$$i_d = g_m v_{gs} = g_m(v_g - v_s), \qquad (5.13)$$

where v_g is the gate voltage and v_s the source voltage. In practice, the drain current and gate–source voltage are set with a combination of load resistor R_L and the reset transistor voltage V_{rd}.

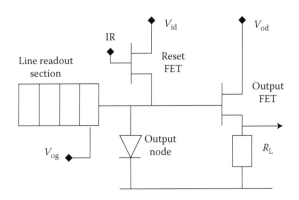

FIGURE 5.12
CCD output amplifier—FET source-follower.

The output source voltage can be described by

$$v_s = i_d R_L \qquad (5.14)$$

and combining this with Equation 5.13, we obtain a relationship for the output voltage in terms of the input voltage as follows:

$$v_s = \frac{R_L g_m}{1 + R_L g_m} v_g. \qquad (5.15)$$

With a typical load resistor value of $5\,k\Omega$ and an FET transconductance of $500\,\mu\text{Mho}$, we obtain $v_s \sim 0.7 v_g$, that is, a reasonably good "follower."

The output gate voltage, V_{og}, is used to bias the end of the serial register to prevent charge from flowing back under the final electrode. Transfer of charge Q onto the capacitive (C_o) output node causes a voltage change

$$\Delta v_g = \frac{Q}{C_o} \qquad (5.16)$$

at the gate of the output transistor. The corresponding change in the output source voltage Δv_s can be calculated using Equation 5.15. Typical node capacitance is of the order of 0.1 pF, giving output responsivities of $1.6\,\mu\text{v/electron}$ for the simple amplifier discussed. Multiple FET output amplifiers are now common which provide higher responsivities (of order of a few $\mu\text{V/electron}$) and greater noise immunity for the output signal.

5.3.4 CCD Clock Sequencing

In a CCD, the pixels are defined in the horizontal direction by the electrode structures and in the vertical direction by the channel stops (Figure 5.8). In a frame-transfer device, this format is used to create an image region and a store region of pixels (Figure 5.11), where the pixels in any row are linked by a common electrode structure but divided by the perpendicular channel stops. At the base of the store region, a line readout section or serial register is created. The electrodes in this register are arranged at right angles to, and to coincide with, the pixel structure of the rows in the store region. A single row of pixels may be transferred from the store region into the serial register and then clocked, one pixel at a time, into the output node of the CCD. The operation of a frame-transfer CCD can be divided into three parts: image integration, frame transfer, and store region read out.

5.3.4.1 Image Integration

Image integration is simply the exposure time of the image section to the source, prior to transfer of the image section charge into the store section. Integration usually occurs with a single electrode biased, typically phase 2 in a 3-phase device.

5.3.4.2 Frame Transfer

Once the desired integration time is achieved, a frame-transfer operation is performed to transfer charge from the image section to the store section. The image section and store section clocks are operated simultaneously in the order as shown in Figure 5.13. The crossover point typically occurs at 50% amplitude, with rise times of the order of 500 ns. Once this operation is complete (typically a few tens of ms), integration in the image section can begin again. Whether imaging optical or x-ray photons, it is usual for the store section of the device to be covered with an opaque material to prevent photon detection during readout corrupting the image. This effectively provides an electronic shutter.

5.3.4.3 Store Region Read Out

After frame transfer, the charge in the store section is transferred one row at a time into the serial register by performing a single cycle of the store section clocks only. The charge in the serial register is then transferred one pixel at a time onto the output node of the CCD. The clock pattern to perform this operation is shown in Figure 5.14. While read-out of the store section is occurring, the image section is available for integrating the next image. With frame-transfer times of several milliseconds, very high duty cycles can be achieved for modest integration times.

The rise and fall times of the reset and serial clocks are typically 100 ns. The crossover point is shown at 50% but can vary with CCD type. In some cases, 100% clock overlap may be required to ensure complete charge transfer.

The read-out of charge from a single row of the store section of the CCD proceeds as follows:

1. A single cycle of the three store section clocks transfers the charge from the final row of the store section into the serial register (Figure 5.13).

2. The output node of the CCD is reset by pulsing the reset clock input high for typically 150 ns.

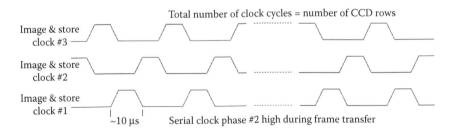

FIGURE 5.13
Frame-transfer clock operation.

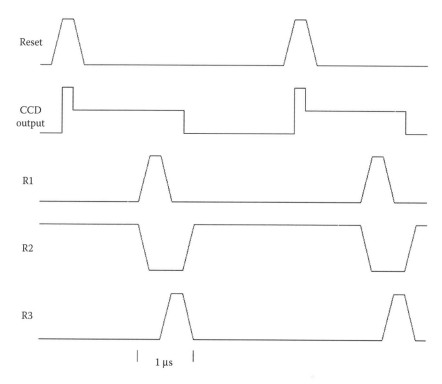

FIGURE 5.14
CCD output clock pattern.

3. The CCD output reset level is then available for sampling (see Section 5.3.5).

4. The serial register clocks are cycled. On the falling edge of R3, the charge is transferred from the end of the serial register onto the output node of the CCD.

5. The CCD pixel charge level is then available for sampling.

6. Repeat from step 2 for the number of pixels in a row.

7. Repeat from step 1 for the number of rows in the store section.

5.3.4.4 On-Chip Binning

An extremely useful side effect of the clocked nature of CCDs is binning. This is where the signal from two or more rows (when clocking the store section or image section) or from two or more pixels in the read-out register are combined by cycling the relevant clocks extra times without reading out the data in between. As an example, two-by-two binning in a frame-transfer device is achieved by cycling the store clocks twice to transfer two rows of charge into the read-out register pixels. The read-out register pixels are then read out by cycling the register clocks twice before measuring the signal at the output node. In this way, the CCD appears to be made up of an array of super pixels, each one two-by-two actual pixels in size. This has a number of advantages—CCD read-out time is reduced (by approximately a factor of 4 in this case), since the binning is performed on chip the read-noise associated with a super pixel is the same as for a single pixel. Binning can of course be performed in any number of ways, three-by-three or more (if charge levels permit), row binning in spectroscopic applications (see Section 5.4.1), windowing to avoid reading out unilluminated areas of the CCD image, and so on. Many CCDs also include a charge drain which allows unwanted image rows with charge to be disposed of without having to read them out through the output node. This is typically achieved by activating a bias voltage that sinks the charge from any row clocked into the serial register when activated.

5.3.5 CCD Signal Processing

5.3.5.1 Preamplifier

Since the CCD is typically operated in a vacuum chamber (for cooled operation), it is usually located some distance from the electronics. Also, the output signal of a CCD is quite small (~1–5 μV/electron). Consequently, a pre-amplifier is normally situated as close to the output of the CCD as possible to reduce the effects of noise pick-up before the output signal is processed.

A typical CCD output transistor noise spectrum is shown in Figure 5.15. At slow scan read-out rates of 20 kpixels/second, the CCD noise contribution is <5 electrons r.m.s. With an output responsivity of 1.6 μV/electron this is equivalent to ~8 μV r.m.s. noise at the input of the preamplifier. It is important in amplifier selection and design that the CCD noise dominates at this point if low noise operation is to be achieved.

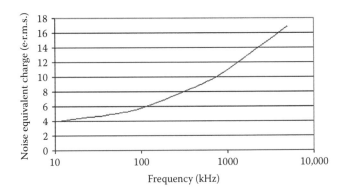

FIGURE 5.15
Typical CCD output transistor noise spectrum.

5.3.5.2 Correlated Double Sampling

In slow-scan, low-noise operation, the greatest contribution to the read-noise from a cooled CCD is generated by uncertainty in the DC level achieved after resetting the output node of the device (Section 5.3.4.3). This variation, typically of the order of a few hundred electrons, can be effectively removed by a technique known as CDS [7,8]. This method involves sampling the output of the CCD twice for each pixel. First, the CCD output node is reset, allowed to settle, and then sampled. This value (the reset level) is stored. The pixel charge is then clocked onto the output node, allowed to settle again and then sampled. This new level is equal to the charge level plus reset level. If the previously stored reset level is now subtracted from (correlated with) the new level, a value for the pixel charge alone is obtained independent of the variation in the previous reset level. This technique can be performed in various ways, either in the analog or the digital domain. Three common methods are described below.

5.3.5.3 Clamp and Sample

The basic circuit is illustrated in Figure 5.16.

The simplest way to understand the operation of the circuit is to study signal waveforms at different points. These are illustrated in Figure 5.17. The CCD output waveform is as described in Section 5.3.4.3, and Figure 5.17 can be correlated with Figure 5.14 to link the sequences together. Once the reset level has settled, the clamp switch is momentarily closed (clamp signal) which clamps the voltage at point C to ground. The clamp is then released, and the pixel charge clocked onto the CCD output. The change in voltage is transferred across the capacitor, and the voltage at point C now represents the pixel charge signal only, since the variation due to the reset level has been removed. The sample switch is closed, and the charge level is stored on the final capacitor.

The noise performance of the clamp and sample technique is limited by the need to low-pass filter the CCD output; otherwise, the clamped signal will be influenced by high-frequency noise. Since the filtering slows the rise and

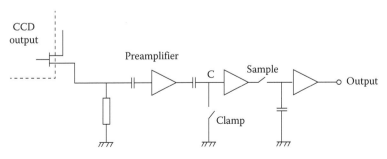

FIGURE 5.16
Basic clamp and sample circuit.

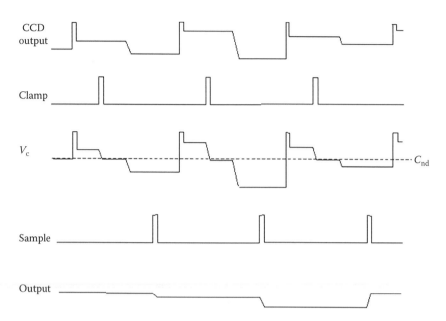

FIGURE 5.17
Clamp and sample circuit waveforms.

fall times of the CCD signal, the minimum time between the clamp (reset) and sample operations is limited, reducing the ability to discriminate against low-frequency noise.

5.3.5.4 Dual-Slope Integration

There are various schemes using dual-slope integration. A typical arrangement is shown in Figure 5.18.

Again, the operation of the circuit is best illustrated by studying the output waveform, illustrated in Figure 5.19.

After allowing the reset level to settle, switch A is closed for a time period t. During this period, the reset level signal is integrated onto the feedback capacitor of the integrator. Integrator hold is then selected, by opening switch A (switch B is already open) and the pixel charge clocked onto the output node. Switch B is now closed, the output of which has a reverse polarity to the output of switch A, and this "negative" signal integrated with the previously held reset level, for the same period t. The reversal of polarity between the reset and charge level integrations produces the desired subtraction (correlation) of the noisy reset level. Integrator hold is selected again while the output value is digitized (ADC Strobe) and then switch C is closed to reset the integrator ready for the next pixel.

The dual-slope integrator generally produces the best noise performance of the CDS circuits discussed here. The main reason for this is the inherent

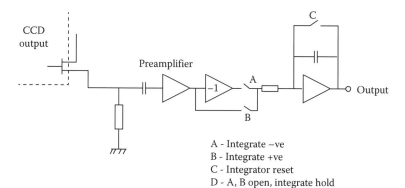

FIGURE 5.18
Dual-slope integrator CDS circuit.

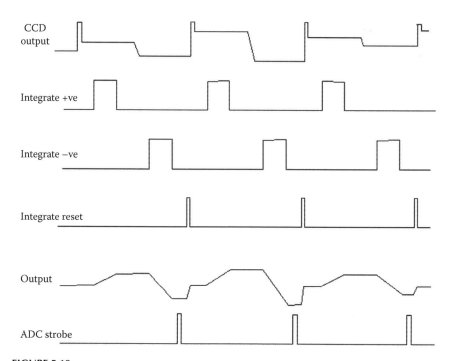

FIGURE 5.19
Dual-slope integrator output waveform.

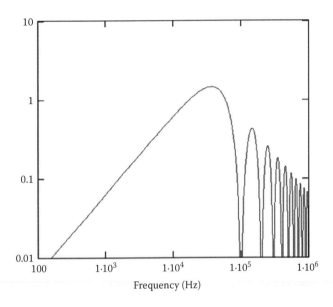

FIGURE 5.20
Noise spectrum of dual-slope integrator ($t = 10\,\mu s$).

low-pass filtering of the CCD signal during the integration phases, which essentially averages the input level over the integration period t (typically 5–20 μs). The noise spectrum for a dual-slope integrator with an integration period of 10 μs is shown in Figure 5.20.

If we now correlate this curve with the noise spectrum of the CCD output transistor, a noise spectrum of the sampled CCD signal is obtained (Figure 5.21). Not only is reset noise eliminated, but also the noise contribution from the CCD output transistor is greatly reduced especially at lower

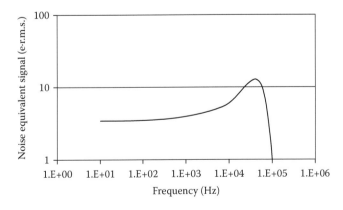

FIGURE 5.21
Noise spectrum of CCD signal after dual-slope integration ($\tau = 10\,\mu s$).

frequencies. Noise levels of $<10\,\mu V$ r.m.s. are achievable, equivalent to ~6 electron r.m.s. with a typical node capacitance of 0.1 pF.

5.3.5.5 Digital Double Sampling

In this method, the subtraction is performed in the digital domain. Amplification and some form of low-pass filtering are usually applied to the CCD output signal. The reset level is digitized and then the pixel charge clocked onto the CCD output node. This new level, representing the variable reset level plus the pixel charge level, is also digitized and the two values correlated. This method can be combined with a variation of the dual-slope integration process, where the reset and charge levels are integrated and stored (in the analog domain) separately and then digitized and subtracted in the digital domain.

5.3.5.6 Intermittent Reset and Reset-on-Demand

In low-signal applications (e.g., x-ray spectroscopy), it may not be necessary to reset the output node of the CCD after every pixel since output node saturation is not going to occur. In this situation, a modified read-out pattern can be adopted to reduce the read-out time of the image. A typical approach is as follows:

1. Transfer store section row into the serial register.
2. Reset the CCD output node.
3. Sample the output signal and store it.
4. Transfer the pixel charge onto the output node.
5. Sample the output signal and store it.
6. Correlate output signal with previously stored value.
7. Repeat from step 4 for the number of pixels in a row.
8. Repeat from step 1 for the number of rows in store section.

In this way, the current pixel charge level is determined by subtracting the output value (reset level + charge n + charge $n-1$ + charge $n-2+\cdots$) from the previously stored output value (reset level + charge $n-1$ + charge $n-2+\cdots$). Instead of resetting just once per row, it is of course possible to monitor the output level and reset the output node when it gets close to saturation (reset-on-demand).

5.4 CCD Spectroscopy

Spectroscopic applications can be divided into two broad categories, energy (or wavelength)-dispersive and energy-resolving. In the first case, a

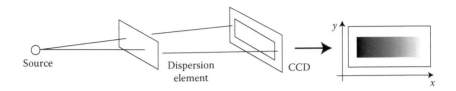

FIGURE 5.22
Typical dispersive spectroscopy arrangement.

position-sensitive detector (either area or moving) provides the spectroscopic information at the output of a dispersive element, while in the second case the intrinsic energy-resolving capability of the detector provides the spectroscopic information. In many ways, CCDs are the ideal detector for such applications. They are available in position-sensitive linear or area arrays, with spectral response across a wide range of wavelengths and energies. They are very linear, offer a high SNR, and can even provide background rejection to improve performance further.

5.4.1 Dispersive Spectroscopy

In dispersive spectroscopy, the inherent position sensitivity of the CCD provides the ability to measure a wavelength-resolved signal. In a typical set-up, the CCD is placed at the focus of a dispersing element or system, as shown in Figure 5.22. This arrangement is applicable across the waveband from IR to x-rays by using the appropriate dispersion element (in transmission or reflection) such as grating or monochromators in the IR–UV or a curved crystal in the x-ray region.

From Figure 5.2, two observations can immediately be made:

1. The source spectrum is dispersed across the CCD columns (x-axis).
2. Each row of the CCD that is illuminated contains a separate but identical spectrum (y-axis).

In its most basic form, this arrangement allows the spectrum to be captured simply by reading out the CCD image. The signal from each row can be added together and then plotted as a histogram of position along the CCD, resulting in a plot similar to that shown in Figure 5.23.

This is analogous to the results that may be obtained with a photographic plate located at the focus. However, a CCD offers a number of distinct advantages inherent in its design and operational capabilities. The regular pixel lattice of the CCD provides an inherent wavelength resolution for a spectrum dispersed across the array. For example, in a system where the measured

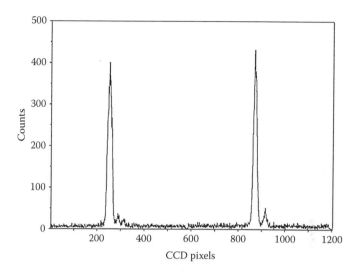

FIGURE 5.23
Spectral histogram obtained from a CCD image.

dispersed waveband, λ_r, covers n CCD pixels, the spectral resolution is given by

$$\Delta\lambda = \frac{\lambda_r}{n}(\text{nm/pixel}). \qquad (5.17)$$

Consequently, calibration is a relatively simple matter. The system is illuminated by a source with known spectral features. The location of these features can then be used to determine the waveband across a number of CCD columns. Substituting these values into Equation 5.17 will then provide the calibration in nm/pixel. From observation 2 above, it can be seen that the individual rows of the image can be added together to provide an improvement in the SNR of the resulting spectrum. Since each row can be treated as a separate measurement, the signal will increase in line with the number of samples, whereas the noise will increase with the square root of the number of samples (assuming similar signal levels in each row). Consequently, adding four rows together will provide a factor of 2 improvement in the SNR. In a normal CCD image, the noise contributions to the pixel signals in each row would include both the signal shot noise and the CCD read noise. However, the inherent binning capability of the CCD (see Section 5.3.4.4) allows any number of rows to be added together on chip before reading out just a single row. Not only is this faster than reading out all the rows, but in low-signal applications where the CCD read noise makes a significant contribution, a great improvement in the SNR can be obtained over the traditional postimage row addition method.

5.4.2 Energy Resolving

In the UV–visible–NIR waveband, each absorbed incident photon creates a single electron–hole pair (see Section 5.2.2) and so there is no way to distinguish the wavelength of the incoming photon from the measured signal. However, as the energy of the photon increases and we move into the x-ray waveband, each absorbed photon creates an average number of electron–hole pairs as discussed in Section 5.2.2. For example, an incident photon with an energy of 1.49 keV will create a charge cloud of around 400 electrons. If the CCD is operated in such a way that the number of incident x-ray photons per frame is small, so that there is little chance of pile-up, then x-ray spectroscopy can be performed using the CCD's inherent energy resolution. Each photon will generate a number of electrons directly proportional to the incident energy and these electrons will be captured in one or more (adjacent) pixels. Assuming in the first instance that the charge is captured into single pixels, then a histogram of CCD signal versus number of pixels will provide an energy spectrum of the x-ray source (Figure 5.24).

In Figure 5.24, x-rays from an Fe-55 radioactive source (with energies of 5.9 and 6.4 keV) have been used to illuminate a CCD. In the histogram, three peaks can be seen. These are the two x-ray lines (Kα and Kβ) and on the left is the "zero energy" peak, that is, all pixels that did not absorb a photon during integration. This shows how such a system is often calibrated, using

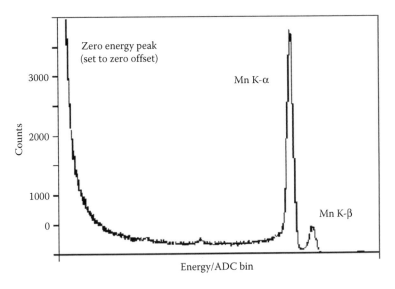

FIGURE 5.24
CCD histogram of Fe-55 x-rays.

a radioactive source which emits x-rays at known energies. In this case, the energy scale corresponds to the signal amplitude measured for each pixel and it is a simple matter to calculate the energy represented by each histogram bin since the zero energy point and a fixed energy point are known. The overall CCD noise (including dark current and other CCD/electronics-related noise sources) is obtained by measuring the standard deviation of the zero energy peak in ADC units and converted to eV or electrons r.m.s. (noise equivalent charge) by using the gain calibration.

5.4.3 Caveats

There are a number of precautions that must be taken when using CCDs to ensure not only their safe operation, but also that the results are as expected. CCDs are CMOS integrated circuits and as such are sensitive to damage from static electricity discharge. Care must be taken when handling devices and open connectors. CCDs (even those for x-ray detection) by their nature are very sensitive to visible light. Usually, this is a good thing but it can also be the cause of problems. Care must be taken to ensure that unwanted light does not enter the system and corrupt the measurement. Vacuum feed-through connectors with glass-to-metal seals and Penning gage filaments are just two examples of causes of poor performance in systems that aim for photon detection capability. In dispersive systems, CCD alignment is also important. Although systems can be calibrated, data processing is far easier with a properly aligned detector. Successful on-chip binning may also not be possible if the detector is not oriented correctly with the spectrum. As we have seen, dark current production is a function of temperature. Since the majority of scientific CCD detectors are cooled for operation temperature, stability is important to avoid variations in dark current accumulated during different integration and read-out periods. CCDs are typically controlled to within $\pm0.5°C$.

References

1. W. Boyle and G. Smith, *Bell Syst. Tech. J.* 49, 587, 1970.
2. G. W. Fraser, *X-ray Detectors in Astronomy* (1st edition). Cambridge University Press, Cambridge, UK, 208, 1989.
3. I. S. McLean, *Electronic and Computer-aided Astronomy* (1st edition). Springer-Praxis, Chichester, UK, 81, 1989.
4. E. I. Gordon, *IEEE Trans. Nucl. Sci.* NS-19, 190, 1972.
5. U. Fano, *Phys. Rev.* 72, 26, 1947.
6. D. H. Lumb and A. D. Holland, *IEEE Trans. Nucl. Sci.* NS-35, 534, 1988.
7. G. R. Hopkinson and D. H. Lumb, *J. Phys. E. Sci. Inst.* 15, 1214, 1982.
8. R. Kansy, *IEEE J. Solid State Circuits* SC-15, 373, 1980.

Further Reading

This chapter is only meant as a brief introduction to CCDs, what they are, how they operate, and why they are useful in spectroscopic applications. The following books provide a much more detailed study of these extremely useful detectors:

Scientific Charge Coupled Devices by J. R. Janesick, 1st edition, SPIE Publications, 2001.

Electronic Imaging in Astronomy: Detectors and Instrumentation by I. S. McLean, 2nd edition, Springer-Praxis, 2008.

6

Microchannel-Plate Detectors in Atomic Physics Applications

Ottmar Jagutzki

CONTENTS

6.1 Introduction

Microchannel plates (MCPs) are the most important detection devices in modern atomic physics. MCPs are especially suited for detecting various atomic particle species (including photons) produced in atomic collisions not only due to their high quantum efficiency (QE) in the relevant energy range, but also because of the high temporal and spatial resolution over a fairly large sensitive surface with high particle throughput. MCPs are ultra-high vacuum compatible and fairly insensitive to external magnetic and electric fields, which makes them easy to use in a typical scattering experiment often involving electrostatic and/or magnetic guiding fields for charged particles [1].

An MCP detector for single particle detection usually contains a stack of at least two MCPs and a dedicated "read-out anode" placed on the output side (see Figure 6.1). Sometimes (i.e., for photon detection) the front MCP surface is coated or a transmission photo-cathode is placed in front of the MCP stack.

The disadvantage of an MCP-detector comes from the fact that it cannot directly determine energy or type of the particle. Therefore, a dedicated spectrometer setup or filter is required for discriminating or translating energy, charge, mass, and/or other particle properties into the measurable parameter position (in two dimensions) and time, that is, time-of-flight (TOF).

Extracting this MCP-inherent "triple-differential" (3d) information is technically very difficult. Up to date there is no all-purpose read-out concept retrieving the 3d information with optimal precision on time and position

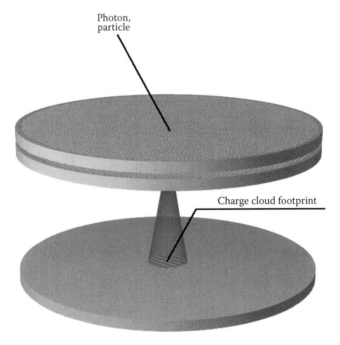

FIGURE 6.1
Sketch of an MCP detector for single particle counting. The incoming particle triggers an electron avalanche in the channels of the MCP stack. The charge cloud is extracted toward the read-out anode. The distance between the anode and the MCP stack and the extraction field determines the footprint size of the charge cloud on the anode.

at highest particle throughput. Therefore, many different read-out anode concepts have been developed over the years, each optimizing only one or a few of the desirable MCP features, that is, those most relevant for the specific application, compromising on others. This chapter shall describe the general function of MCPs and give an overview about the different concepts relevant to atomic/molecular physics, which is probably the most demanding field for MCP use.

6.2 Operation of MCPs

6.2.1 The MCP as a Secondary Electron Multiplier

The operation principle of the MCP is the same as of a single secondary electron multiplier (SEM), for example, the so-called channeltron: the inside surface of a lead–glass capillary tube with high aspect ratio is chemically treated to yield a high-emission probability for electrons if hit by an energetic particle, for example, by a primary electron from a scattering process. An SEM tube has an electrical resistance of several tens of $M\Omega$ to $G\Omega$. If a voltage of a few 100 V or more is applied, then electrons drifting in the tube will gain enough energy in the electric field that they can generate one or more secondary electrons in the tube wall, which will also be accelerated in the field and hit the wall repeatedly, generating a large electron avalanche. SEMs must be operated in high vacuum conditions of $<10^{-4}$ h Pa to avoid continuous glow discharge, eventually destroying the device.

An MCP is a dense package of millions of individual SEM pores (see Figure 6.2), stacked side by side with a distance of only a fraction of the pore diameter, which gives an MCP macroscopically the shape of a thin plate without visible structure by naked eye [2]. MCPs with diameters up to 120 mm or even more can be purchased. The MCP's conductive surface layers are used to bias all pores with the same high voltage. But each pore will "fire" individually (locally), when a primary electron starts an avalanche, that is, when an energetic particle hits the MCP at some pore's position or next to it on the MCP front face. The pore size (more precisely: the center-to-center pore distance) limits the localization of a detected particle (spatial resolution limit). Nowadays MCPs with pore sizes as small as 2 μm are commercially available although dual-use regulations restrict the availability. Since small pores also have practical disadvantages especially for large-sized MCPs, a typical pore size for MCPs used in atomic physics is 12 μm with 15 μm center-to-center distance between neighboring pores, yielding a fill factor of 50% or more, the so-called open-area ratio (OAR). The pores are tilted by an angle of typically 10° with respect to the surface normal and thus to the electric field vector. This limits the electrons' free propagation path between wall encounters. The tilt angle also prevents the unwanted "ion feedback" effect: Free ions emitted

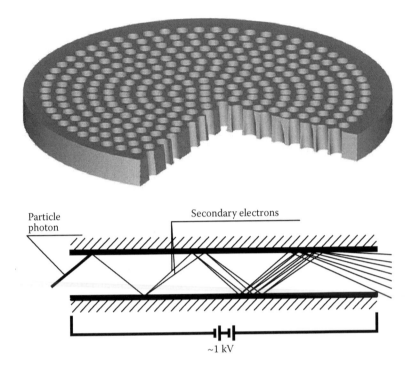

FIGURE 6.2
Operation principle of a MCP array as a dense pack of secondary electron multipliers: an incoming particle creates a primary electron, gaining enough energy in the field to produce at least one additional (secondary) electron as it hits the channel wall. This happens several times for each created electron and eventually an electron avalanche exits the MCP pore as a localized charge cloud. The cloud is collected on a read-out anode yielding a "footprint" which depends mostly on distance and electrostatic field between anode and MCP stack.

from the surface wall or generated in the residual gas must be stopped by a wall encounter before they can gain enough energy in the field to trigger a second (delayed) avalanche in the channel. The typical aspect ratio (L/D) of the pores is between 40:1 and 80:1, resulting in a typical plate thickness of 1 mm.

6.2.2 MCP Gain and Count Rate

While a single SEM tube such as a Channeltron, with its at least 10 times larger diameter and higher aspect ratio, can generate an electron avalanche of $>10^7$ electrons per primary particle at low particle flux, an MCP with straight channels can only reach a gain of few 10^5 before field-charge effects and a critical field gradient in the pores (>1000 V/mm) will limit further gain increase.* By

---

* Special MCPs with curved pores and large aspect ratio (e.g., 120:1) can yield a gain $>10^6$ also as a single stack.

stacking two (chevron stack, V-stack) or three MCPs (Z-stack), a total gain of up to 10^8 or even higher can be obtained, especially if spacers between MCPs in a stack allow a distribution of the charge output into several pores of the next MCP. High gain values $>10^6$ are required to "count" individual particles by means of a time-correlating electronic signal pickup chain, otherwise the signal is lost in the electronic noise.

Due to the statistical nature of secondary electron production in the channel wall the MCP gain for individual particles has a considerable variation. Typically, the pulse height distribution (PHD) has a full-width at half-maximum (FWHM) as large as the mean of the gain distribution (see Figure 6.3). Therefore, it is hardly possible to accurately distinguish, for example, between single particle hits on the MCP surface and multiple simultaneous hits, or between particles of different species or those carrying different energies. Although the mean gain of MCPs varies with these parameters, the gain variation for a subset of particles with same characteristics can be large as well.

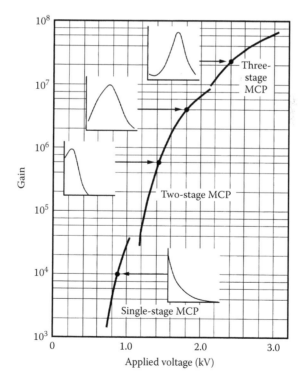

FIGURE 6.3
Typical gain and PHD for single particle counting with MCP stacks as function of the total bias and the number of MCPs in the stack. (From Hamamatsu Photonics K.K., *MCP Assembly, Technical Information Manual*, 2006. With permission.)

It is to note, however, that an MCP can only emit a maximum output current which is, as a rule of thumb, about as high as the so-called strip current: The strip current is the current across the MCP at a maximum bias voltage applied (i.e., just below the critical field for discharge) in the absence of particle input. The high MCP resistance prevents a higher current output because the charge drain cannot be replenished fast enough. Typical MCP resistance is a few tens to hundreds of $M\Omega$ which allows a maximum current output of $10–100\,\mu A$. Therefore, there is a trade-off between the achievable gain and the maximum count rate. The typical rate limit in single pulse counting mode of MCPs is a few million particles per second. If an MCP with high strip current is selected as the rear MCP, higher count rates maintaining high gain may be achieved, but there is a risk of "thermal runaway" due to the negative thermal coefficient of MCPs: A high current will heat up the MCP and this temperature rise will increase the strip current further until the MCP becomes thermally destroyed.

If an MCP is exposed to a rather focused particle flux, local count rate saturation effects can kick in even before the "global" count rate limit is reached: It takes a single MCP pore several tens of milliseconds to fully recover from a discharge.

It is to note that these considerations on rate limitations are only valid for the single particle counting mode of MCPs, that is, if a certain minimum gain is required. Physically, SEMs and MCPs are current amplifiers. If a certain flow of charged particles, for example, an electron current, enters the SEM pore, then an amplified output current will exit until the maximum output of few $10\,\mu A$ is reached. Even billion electrons per second or more can thus be "detected" and imaged by an appropriate charge-coupled device (CCD) (see next section) as long as the gain is kept low. But the low gain at such high fluxes is not sufficient for registering each electron individually and for attributing a precise time tag to it, as required in many atomic physics experiments.

Finally, it should be noted that MCPs are fairly immune to moderate electrostatic and magnetic fields. However, fields $>0.1\,T$ will affect the gain, depending on pore diameter and orientation with respect to the field [3]. It is important to note that the effect of even smaller magnetic fields and that of stray E-fields to the functionality of a read-out anode in an MCP detector must also be considered (see below).

6.2.3 Quantum Detection Efficiency and Lifetime of MCPs

One of the most important properties of any detector is its QE. Ideally the QE, expressed as a probability that a particle hitting the detector is registered, should be near 100%. This is especially desirable if coincidence measurements between many particles are of interest and the total QE is the product of each particle's QE. SEMs in general reach up to 90% QE for specific particles and energies. But although SEMs can detect a wide spectrum of particles and energies the QE is not always high [4]. In the case of MCPs, this "optimal" QE is further reduced by the OAR to about 75%, that is, particles hitting the MCP between pores may not create a primary electron triggering an avalanche because the primary electron can escape the field gradient penetrating from a nearby pore or gets buried in the bulk material without re-emission of a secondary electron. However, particles can have a QE near unity (i.e., higher than the geometrical single particle maximum) if they create on average more than one primary electron on impact, or a large number of primary electrons per particle is produced by a converter foil (e.g., CsI) in front of the MCP. It is also possible to mount a biased mesh in front of the MCP, producing a field gradient to turn back escaping primary electrons toward the MCP surface. However, there is a trade-off between mesh transmission for incoming particles and QE increase. Also the spatial or temporal resolution may be affected from such QE-increasing measures by rescattering and micro lens effects.

The QE is also dependent on the incident angle of the particle [5,6]. Highest QE is typically achieved for particles hitting the MCP at an angle between 10° and 20° (as in the case of electrons) with respect to the pore orientation which makes normal incidence to the MCP surface a favorable angle for particle impact (given the typical pore tilt angle). The reason for a decrease toward larger angles is that the primary electron produced in a pore has a decent probability to "escape" without starting an avalanche because it has not been created deep enough in the pore (in the potential well). On the other hand, if the incoming particle comes more or less aligned with the pore, then it will hit the wall at a grazing angle and probably very deep in the pore, which can reduce the emission probability of a primary electron or shorten the effective channel length (smaller avalanche). The latter is especially important if a minimum charge is required for registering the charge cloud at all, that is, in pulse counting mode. Such "anode loss" can also happen whenever a low gain and/or broad PHD are combined with high noise background: Small avalanches cannot be discriminated from the noise. Anode loss is independent from the QE of the primary electron and must be considered as another factor which reduces the total detection probability. But if the MCP stack provides a good PHD over the whole area, then the anode loss is a negligible contribution to the total detector efficiency, which then becomes equal to the QE of the primary interaction.

Finally and most importantly, the QE depends strongly on the particle species and energy. It is very beneficial in atomic physics applications that a QE > 60% can be obtained for electrons in the energy range between 200 eV and 2 keV and for light ions or neutral atoms/molecules (<40 amu) between 2 and 100 keV.* Slower electrons/ions born on ground potential can be accelerated to favorable energies by biasing the MCP front face to +200 or ±2000 V, respectively. Particles with more than 100 keV ("projectiles") or electrons above 2 keV energy will still produce a decent QE. This widely covers the

---

* Neutral meta-stable atoms with negligible kinetic but high potential energy can also trigger an electron avalanche when autoionizing at the MCP surface.

particle type and energy range of most interest in atomic physics. Ions with very high masses (clusters, bio-molecules) need a kinetic energy of 10 keV or more on impact to achieve a QE of 50% (also obtainable by an adequate MCP bias). Additional coating (e.g., MgO) on the MCP front surface can further improve the QE of such ions (and also for electrons).

The situation is less favorable for photon detection. MCPs have zero QE for photon wavelengths >200 nm and only a few percent to 15% in the VUV (<200 nm) and x-ray regime [7,8]. Coating of the MCP surface with KBr or CsI increases the QE up to 30% at VUV energies but these coatings are unstable in ambient air, which makes such detectors difficult to handle [9].

By the use of sealed detector tubes the optical and near-UV wavelengths becomes accessible with a QE <20%: A photo-cathode on an optical window, an MCP stack, and a read-out anode (usually a phosphor screen) are sealed vacuum-tight as a photomultiplier tube and form a so-called image intensifier. Nowadays, such devices can also be equipped with a variety of pulse-counting read-out anodes which turns them into useful tools in atomic physics as open-face MCP detectors [10].

The life-time of MCPs is usually expressed as the accumulated charge output which an MCP stack can deliver before the gain drops significantly. This "life-time" charge output of an MCP is about 1 C, which numerically corresponds to 600 billion particles "counted" at a gain of 10^7. But this number is misleading not only because in daily lab use the MCP "life" is usually limited by the occurrence of vacuum or mechanical accidences. The gain drop as a function of accumulated charge is only low (and consequently the life time long), when the MCPs have been artificially aged by an elaborate process, the so-called scrubbing. Scrubbing removes all volatile impurities (i.e., gases) from the MCP bulk material. This is achieved by exposing the biased MCP stack to a high and uniform dose of electrons or VUV-light until about 10% of the targeted 1 C life-time charge output is exhausted. During this scrubbing, the MCP gain drops by a factor of 10–100. This is not necessarily a problem, if the MCP bias can later be raised to compensate this gain loss.

Without scrubbing the gain drops 10–100 times more rapidly and the life time is accordingly lower. Prescrubbing of MCPs before use is not an option if they are exposed to ambient air because the MCPs will reabsorb gases. The presence of these gas atoms in the bulk material not only leads to an enhanced gain, but will also bring back the much faster gain drop characteristics of nonscrubbed MCP.

Exposing a biased MCP to particles in an experiment produces a "natural" scrubbing process. If this does not happen uniformly (in case of nonuniform MCP irradiation), then the MCP will eventually experience an unequal gain response as a function of position which is a significant problem for many MCP applications. Eventually an MCP will have to be exchanged because this affects the local detection efficiency, long before the nominal charge output life time is reached. In this case, usually only the last MCP in the stack needs to be replaced.

There are also other effects than gain fatigue which can limit the life time of MCP. It turns out, for example, that certain particle species such as Helium ions with about 2 keV kinetic energy cause a not yet fully understood premature aging process on the MCP input side. While gain fatigue is mainly a problem of the second- or third-stage MCP, here the MCP front surface, that is, the first few 100 μm of pore depth seems to be affected, probably by a sputtering process which alters the chemistry of the emissive layer in the pores and reduces the electron emission yield.

6.3 Particle Imaging and Timing with MCPs

6.3.1 Particle Imaging

Originally MCPs have been developed to boost the sensitivity of night vision devices. In a standard night vision scope, a transmission photo-cathode converts photons into electrons. These electrons are accelerated in an electric field and projected onto a phosphor screen (read-out anode) which emits several photons per incident electron, thus effectively amplifying the incoming light. Introducing MCPs between the photo-cathode and the phosphor allows an increase of the gain so that even single photons hitting the photo-cathode can be visualized on the phosphor screen. For scientific purposes the human eye is replaced by a CCD camera or the CCD is coupled to the phosphor by a fiberglass taper. Since this method is not limited to photomultiplier tubes, open-face MCP detectors with a phosphor screen can be used to detect particles in the same way. Since CCD devices are usually not vacuum compatible, the CCD device is usually placed in air observing the phosphor screen through a vacuum window.

This well-matured technique allows imaging of atomic particles even at high fluxes and at low system costs. It is robust, very tolerant to electronic noise, and can give a high spatial resolution of 1000×1000 pixels or more. However, there are three disadvantages:

(1) The imaging quality is affected by CCD-inherent noise background (especially problematic at low count rates): Dark (not illuminated) areas turn "gray" over time.
(2) The dynamic range is low. Bright image areas get saturated and blur the image (pixel overflow).
(3) Nonuniform gain response across the MCP area will proportionally affect the imaging and thus requires a choice of premium MCP quality ("imaging grade" or better).

Although some of these problems can be addressed by cooling and elaborated read-out schemes, the most challenging disadvantage is the absence of precise time information on the detected particles. Standard CCD cameras operate at a "frame rate" of typically 50 Hz, such as standard video cameras, corresponding to a temporal resolution of 20 ms (1/50 Hz). Even high-speed cameras based on CMOS-technique can only reach a few 1000 frames/s, which is more than seven orders of magnitude slower than the MCP-inherent temporal resolution. Therefore, CCD/CMOS cameras are not ideal detection devices in atomic physics, which often requires identifying individual particles with high temporal resolution.

By gating the MCP bias or, in case of visible photon detection, gating the photo-cathode or a grid between photo-cathode and the MCP stack a high temporal resolution can be achieved. This "stroboscopic" method has a rather limited application range.

However, it is possible to combine the CCD/CMOS techniques with other read-out methods and thus accomplish a more general time- and position-sensitive read-out concept for MCPs will be discussed in one of the next sections.

Nevertheless, there are a number of applications in atomic physics where the missing time information and the nonideal imaging characteristics of CCD devices are not crucial. An example is the technique of velocity imaging of molecular fragments where even the three-dimensional momentum distribution of molecular break-up can be deconvoluted from its two-dimensional projection onto an MCP detector with a CCD read-out [11]. As long as there is a symmetry in the break-up pattern, an inverse Abel-transformation can reconstruct the three-dimensional break-up without the need for a third dimension information, for example, the TOF.

6.3.2 Timing Measurements with MCP Detectors

Due to their compact pores MCP detectors can provide a better temporal resolution than other SEM devices like Channeltrons and are therefore often used in place of SEMs even if spatial resolution capability is not required. A temporal resolution on the order of a few 10 ps can be obtained with adequate timing electronics (see below). The temporal resolution is usually defined as the FWHM of a "time peak," for example, as experimentally obtained from a measurement on a process with vanishing temporal width.

Coincident detection of particles or correlating particles to outer trigger events (e.g., a laser shot) also requires time determination with at least a few nanoseconds' precision, a task easily achievable since the MCP pulse duration is on the order of a nanosecond.

Even if the time information is not at all desired and only the particle position on the MCP detector is of interest, coincidence operation can very effectively remove a disturbing background from a measurement. Standard CCD read-out without additional electronic circuits cannot be used here unless a millisecond temporal precision is sufficient for the experimental task.

Determining the time information of a particle requires an MCP operation in a single pulse counting mode, bringing on a high gain demand (double or triple MCP stack) and thus count rate limitations. On the other hand, this counting mode has advantages since it does not require premium MCP quality with high gain uniformity across the active area (unless highest temporal precision is aimed for): As long as each charge cloud from the MCP stack contains a certain minimum number of electrons above the electronic noise level equivalent, each particle will be registered as "one" and the temporal resolution will be equal to or smaller than the typical MCP pulse width.

Picking up the signal and "counting" the particle is achieved by a simple RC-circuit followed by a fast amplifier and discriminator unit, which provides a logic signal for a counter and/or time digitizer device (e.g., a time-to-digital converter (TDC) or a time-to-amplitude converter (TAC) followed by an analog-to-digital converter (ADC)). A "timing anode" is in its simplest form—just a metal plate placed behind the MCP stack. For achieving optimal temporal resolution, impedance-matched cone-shaped anodes are commercially available (no position sensitive anode can be used in this case). But it is also possible to simply pick up the timing signal off an MCP stack bias contact. Typically the rear MCP stack contact is used for this, but a pick-up from MCP front end contact or even from an intermediate MCP bias contact can work as well (see Figure 6.4). This allows placing a position sensitive anode behind the MCP stack without compromising much on the achievable temporal resolution which has been reported to be about 40 ps FWHM [12].

While the signals on a read-out anode have a negative polarity, signals picked up from an MCP contact are positive (the MCP stack as a whole experiences an electron loss during charge drain) and have several 10 mV mean pulse height (on 50 Ohm impedance termination) under typical MCP gain conditions for counting mode.

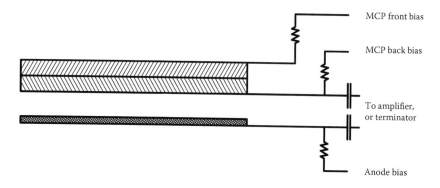

FIGURE 6.4
Pick up of timing signals from an MCP contact or from a timing anode. MCP and anode bias are supplied via resistors (typically $>10\,\text{k}\Omega$), preferably placed close to the respective contacts. Capacitors of few nF are used to decouple the signal for amplification and further electronic processing or for terminating a signal with proper impedance (to avoid ringing).

Aspect ratio, number of MCPs in the stack and pore size do not only determine the maximum gain (i.e., pulse height), but also the pulse width and thus the achievable temporal resolution [3]. Basically, the thinner the MCP stack, the smaller is the pulse and the better is the achievable temporal resolution. Other details like introducing spacers between the MCPs will also determine pulse shape. But a short avalanche path will generally keep the longitudinal spread of the charge cloud along the channels small and thus produce a short-time avalanche. Since small pore diameter usually means short pores (if the typical aspect ratio is maintained), such small-pore MCPs not only have the potential for optimal spatial resolution, but also have the potential for highest temporal resolution if used with adequate follow-up electronic, for example, fast amplifier, precise timing discriminator, and a high resolution time digitizer.

The common discriminator type for precise timing is the so-called constant fraction discriminator (CFD) which can produce a digital output signal which (in theory) does not jitter as a function of the input pulse height (see Figure 6.5). In practice, CFDs can obtain a temporal precision of $1/100$ of the input signal's rise time. Since also today's commercial time digitizing circuits have a temporal resolution of few 10 ps or even better this high temporal precision of MCPs can actually be used for highly demanding TOF experiments.

6.3.3 MCP Detectors for Simultaneous Time and Position Read-Out

6.3.3.1 Charge-Coupled Devices

Since phosphor/CCD systems are readily available and the additional pick up of a timing signal from an MCP contact is technically easy, it is straightforward to combine these two methods. Unfortunately, this limits the count rate to one particle per frame (typically <1000 Hz) since otherwise the particle counts cannot be correlated. Only if the rate is low one can indeed measure position and time of a particle with highest precisions by means of CCD, especially if a spot-centroiding method corrects for blur effects and cooling allows virtually background-free and pin-sharp imaging.

If the desired count rate is higher, then other methods than the classical CCD approach must be used, which requires replacing the phosphor screen by an anode structure which is able to electronically pick-up the charge cloud and encode its footprint position. These "electronic anodes" bring about their own limitations on spatial resolution and particle throughput which will be discussed in the following sections.

6.3.3.2 Pixel Anodes

The simplest way of turning an anode for timing read-out into a position- and time-sensitive anode is subdividing it into many separate anode elements, each equipped with an individual timing electronics chain for a

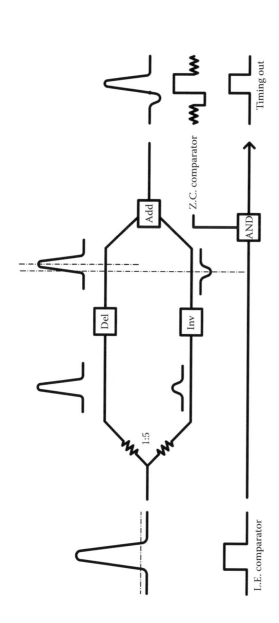

FIGURE 6.5

Operation principle of a CFD for enhanced timing performance. The input signal is divided by resistors into two signals with different pulse heights (the "CFD-fraction," here: 0.2). While one of the signals is delayed for a certain time ("CFD delay"), which is a little smaller than the rise time of the signal, the other becomes inverted. Superimposing these two signals produces a bipolar signal with a zero crossing that does (in theory) not jitter as function of different pulse heights. Two comparators are used to determine the time of the zero crossing; While the "leading-edge" (L.E.) comparator sets a trigger level ("CFD threshold") just above the electronic noise, the "zero-crossing" (Z.C.) comparator has a trigger level ("CFD walk") near zero, that is, it triggers in the electronics noise and is only in defined states "high" or "low" while the bipolar signal is present. A logic "AND" gate between the two comparator outputs serves as the pulse height independent logical timing signal ("CFD output").

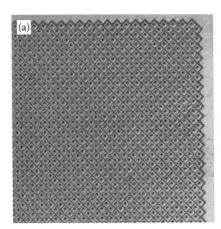

FIGURE 6.6
Photographs of the front (a) and back sides (b) of a crossed-strip pixel anode. The diamond-shaped pads on the front side collect the charge cloud and are interconnected in rows via the back side, forming an array of parallel tracks for each dimension. Each strip is connected to an individual timing electronics chain with amplifier, CFD, and time digitizer. (From Werner, U., unpublished, Universität Bielefeld, 2008. With permission.)

parallel read-out [13]. Using a so-called pixel anode* or multianode would be the ideal read-out concept for MCP if it had enough pixels to allow for a high spatial resolution. Ideally, there should be one pixel per MCP pore, or at least the equivalent of the typical CCD chip resolution (1000 × 1000). To date, this cannot be achieved due to the corresponding electronic complexity. If n is the number of position bins in one direction, then the number of necessary electronic channels is n^2. In practice, it is already quite difficult to operate pixel anodes with more than about a 100 electronic channels, which corresponds to only 10–15 position bins per dimension. The problem is not only the price of the electronics or the space/power consumption required for it but also the need for setting the electronic thresholds for each channel individually. This is necessary to account for nonperfect MCP gain uniformity and tolerances in the electronic devices. Furthermore, cross-talk and nonuniform pixel response make this anode type difficult to use. Providing one individual vacuum feed-through per pixel poses another serious mechanical challenge. However, if the demand on spatial resolution is low, a pixel anode is the ideal tool for applications with high count rates and/or for particle burst (multihit) detection with high temporal resolution.

A variation of the "true" pixel anode is the use of crossed-strip (crossed-wire) anode structures (see Figure 6.6) for improving the spatial resolution without increasing of the number of the electronic channels. One electronic channel is used for each strip of interconnected pixel elements or each wire

---

* Although CCD cameras also employ pixel structures, those are not considered to be pixel anodes in this context.

[14–16]. Crossed-strip pixel anodes need only $2n$ electronic channels. With a practical upper limit on electronic channels the crossed-strip anode approach thus allows up to 100 position elements per dimension, that is, 100×100 effective image pixels. The disadvantage compared to "true" pixel anodes is the less parallel read-out: The position measurement of two particles arriving within the electronic dead-time (typically 10 ns) is ambiguous. However, this ambiguity can partially be resolved by introducing a third strip array [17].

If multihit tolerance is not required at all, then one can further improve the ratio of the required electronic channels to the number of effective image pixels by introducing a binary coding scheme of the strips known from the MAMA (multianode microarray) detectors [18] used for space telescope instrumentation and from the CODACON design [19]. By using two separate sets of strips for each dimension it is not necessary to have one separate electronic channel for every strip. Several strips can be connected to the same read-out channel as long as it is ensured that the charge cloud footprint covers two neighbor strips, for example, one of each set (and for each dimension, see Figure 6.7). Then the position in each dimension can be determined by correlating those electronic channels which have "fired." For a given number n of targeted effective image pixels per dimension (e.g., 1000) the standard crossed-strip anode would require $2n$ electronic channels (2000, respectively). The MAMA concept reduces the channel number to approximately 4 times the square root of n (128, respectively).

Still, due to the limited spatial resolution (or loss of multihit ability in case of the MAMA concept) and the overall detector complexity common to all pixel anode approaches this technique has not yet reached the stage of an

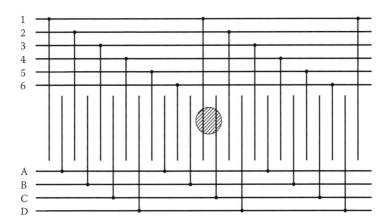

FIGURE 6.7
Principle of the MAMA encoding for one dimension. If the charge cloud footprint covers two neighboring tracks, the electronic channels connected to each horizontal line detect a signal in coincidence, here: channel 1 and channel C. By this method, only 10 electronic channels can uniquely encode 24 different position pixels.

ideal all-purpose read-out concept. Recently, a novel type of pixel anode, the MEDIPIX/TIMEPIX sensor [20], based on modern semiconductor technology has been introduced for MCP readout. Although resembling a CCD-device the MEDIPIX chips feature noise-free single photon/particle counting at GHz rates with high position resolution and even moderate temporal resolution (see Section 6.3.4.3).

6.3.3.3 Charge Dividing Anodes

Reducing electronic complexity of pixel anodes can be "bought" by introducing a more elaborate electronic read-out. While the individual read-out circuits for pixel anodes simply register the presence of a charge pulse above a minimum detection threshold, it is also possible to measure the amount of this charge accurately. A read-out anode can be structured in such a way that information on the accumulated (relative) charge from the anode elements encodes the centroid position of the charge cloud footprint on the anode.

The simplest version of such a charge dividing anode is the quadrant anode [21], which looks like a pixel anode with only four elements (see Figure 6.8). But whereas pixel anodes are placed close to the MCP stack and require a small charge cloud footprint (i.e., about as big as the pixel), here, a drift gap

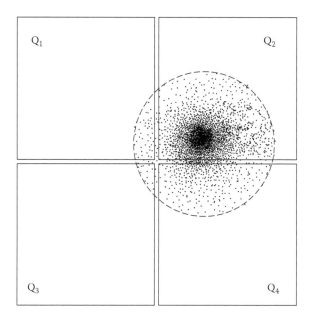

FIGURE 6.8

Sketch of a quadrant anode. If the charge cloud footprint from the MCP is enlarged, so that for all particle positions on the MCP each of the four quadrants receives a charge share, it is possible to determine the centroid of the footprint (e.g., the particle position) by measuring and comparing the relative charge portions on the quadrants.

between the MCP and the anode is introduced to significantly enlarge the charge footprint [22] such that each quadrant receives a share. If the charge on each quadrant is precisely determined for each particle, then it is possible to compute the centroid of the charge distribution, corresponding to the particle position on the MCP. Introducing such a centroiding method decouples the achievable spatial resolution from the size of the anode structures, which is the limiting factor for pixel anodes. If a centroiding method is employed, then only the accuracy of the relative charge measurement determines spatial resolution.

The timing signal is picked up from an MCP contact as was described before and correlated with the anode signals. Therefore, such an anode concept allows both a high spatial and temporal resolution with a small number of anode elements and electronic channels (<10). The achievable particle rate is a function of the charge integrating electronics' dead-time, that is, much lower than imposed by a CCD's frame rate.

The quadrant anode is a very simple structure, but has some disadvantages: Although its systematic image nonlinearity can be corrected for, the nonlinear charge/position response affects the spatial resolution which becomes position-dependent, for example, poorer in the outer imaging area. The very large footprint is also quite susceptible to electrostatic and magnetic fringing fields.

To overcome these problems, different anode structures have been developed, for example, the wedge&strip (W&S) anode (see Figure 6.9). Due to its quasi-repetitive pattern it can operate with footprints of only a few millimeters size and a linear charge/position correspondence can be achieved [23].

As long as the structures are precisely machined and the charge footprint size covers several pitches of the structure elements, the spatial resolution is indeed only limited by the accuracy of the charge measurement. In practice, the signal-to-noise ratio limit allows about 1000 position bins across the active anode dimension, which is comparable to the typical spatial resolution achieved by CCD. The "price" for such a high precision is the use of "slow" read-out electronics (charge amplifier) with several microseconds integration time. This limits the achievable count rate to about 100.000 particles per second. Beyond that the spatial resolution deteriorates until pile-up effects will finally make this method unsuitable for even higher rates. All signals from all particles have to be processed through the same electronics chains, therefore multihits can not be analyzed at all.

This deadlock between resolution and rate limitations cannot be resolved, but partially relaxed by introducing more sophisticated anode structures. Examples are the tera-wedge anode [10] and the vernier anode [24,25], the latter using nine electrodes shaped in a way that a "coarse" position determination is refined by "phase" coordinates, which in combination allow a much more precise position determination. This is equivalent to the Vernier scale on a caliper rule which can improve length measurement by a factor of 10. Accordingly, the vernier anode allows imaging with almost 10,000 pixels.

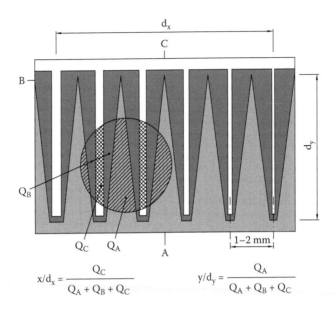

$$x/d_x = \frac{Q_C}{Q_A + Q_B + Q_C} \qquad\qquad y/d_y = \frac{Q_A}{Q_A + Q_B + Q_C}$$

FIGURE 6.9
Sketch of a 3-jaw wedge&strip (W&S) anode. If the charge cloud covers several (at least two) pitches of this quasirepetitive structuring of three electrodes ("wedge," "strip," and "meander"), measuring the relative charge portions allows for determining the footprint centroid. The footprint can be much smaller compared to the quadrant anode and the charge/position response over the active area is linear. Unlike in this sketch, a real anode contains 20 to over 50 very narrow and precisely machined strips and wedges across the active area. See also [31] for a common 4-jaw W&S design.

6.3.3.4 Spatial Resolution

At this point, it is necessary to define the term spatial resolution more accurately because for different anode concepts different definitions exist which can differ by a factor 2 at least. For CCD and pixel anodes the number of pixels or the pixel size are often used for defining the spatial resolution (in relative or absolute terms).

In optics, especially in astronomy, the image definition is often expressed as the distance between two point-like objects which are just distinguishable in their image. It turns out that this corresponds roughly to the FWHM of a line scan through the image of point-like spots, that is, broadened only by the nonperfect resolution response.

This is an established spatial resolution definition for MCP read-out anodes and independent from the concept of pixels. This "FWHM-resolution" is straightforward to measure with the help of pin-hole masks in front of the MCP. Alternatively, the edge blur of any shadow mask allows an estimation of the resolution expressed in FWHM [26].

Sometimes, it is not the FWHM that is used to specify resolution, but the RMS (root mean square or standard deviation) of a Gaussian fit through a spot profile, which is a little less than half (0.43) of the Gauss-fitted FWHM value. This is also a useful definition because it turns out that structures with size of the RMS are just "apparent" in the image. Furthermore, for a given CCD or pixel anode the obtainable optical resolution expressed in RMS corresponds numerically quite well to its nominal pixel size.

Even for continuous anodes like the W&S the image information encoded as analog signal heights will at some point undergo a digitization for position computation and further data treatment. This electronic binning should be smaller than the RMS resolution value or it must be properly weighted with other resolution limits, for example, imposed by electronic noise. On the other hand, a high number of pixels on a chip or of the digitization circuit does not guarantee an equally precise resolution performance, if noise or blur effects are present. It is therefore important to specify the "valid" resolution in pixels or RMS, which takes into account all resolution-limiting effects.

6.3.3.5 The Resistive Anode (Encoder)

A disadvantage of the structured charge dividing anodes is that the structures need to be machined with micrometer precision. Therefore, the most common commercially available charge dividing anode uses a different approach: The charge cloud from the MCP is collected on a continuous layer of certain surface resistivity (typically $10\,k\Omega/\square$). If there are several low-impedance contacts on such a resistive sheet, then an MCP-induced charge will drain off to all contacts according to Kirchhoff's law: The distance to each contact is proportional to the resistance. This concept, the so-called resistive anode encoder (RAE) thus allows position determination by comparing different charge portions draining off the contacts. Boundary effects and the presence of the pickup contacts lead to a nonlinear position/resistance correspondence. But this can be corrected for by specially shaped anodes with border resistors (see Figure 6.10) [27] or software. The RAE uses similar read-out electronics like the W&S anodes with the same dead-lock between resolution and rate. However, as long as the count rates are moderate and multihit detection is not required, the RAE or other charge dividing anodes are very powerful alternatives to CCD or pixel anodes and feature high spatial and temporal resolution, especially for small-sized MCPs of 40 mm or smaller active area: A high relative spatial resolution (expressed in number of valid pixels) corresponds to a low absolute spatial resolution (measured in micrometer).

6.3.3.6 Image Charge Method

In recent years, the method of image charge coupling through an insulator to a read-out anode [28] was adopted also for MCP detectors: The charge output from the MCP stack is collected on a resistive layer deposited on a substrate of few mm thickness made of an insulating material. The sheet resistivity is

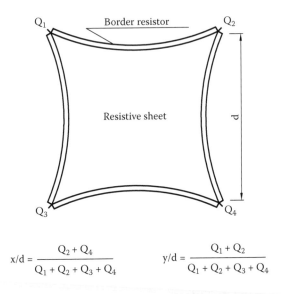

$$x/d = \frac{Q_2 + Q_4}{Q_1 + Q_2 + Q_3 + Q_4} \qquad y/d = \frac{Q_1 + Q_2}{Q_1 + Q_2 + Q_3 + Q_4}$$

FIGURE 6.10
Sketch of a resistive anode (encoder). If the charge is locally collected on the resistive sheet (only small footprint required), the amounts of charge measured on the four corner contacts allow for a determination of the position. The special shape and properly matched border resistors ensure a linear charge/position response over the active area. With such an anode design, it is also possible to embed a central hole without affecting the image linearity.

typically $10\,\text{M}\Omega/\square$ and thus about three orders of magnitude higher than on the RAE. As the charge cloud is collected on this "resistive screen" an image charge is created on any pickup electrode placed behind the insulating material (see Figure 6.11) [29,30]. Although the charge cloud footprint on the resistive screen can be very small, the image charge footprint size at the

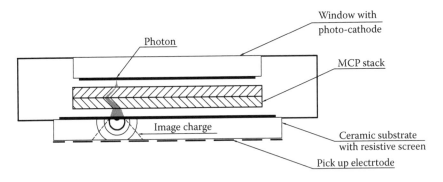

FIGURE 6.11
Sketch of an MCP photomultiplier (image intensifier) with resistive screen on ceramic substrate and reconfigurable read-out electrode (anode) placed outside the sealed tube. Although the charge cloud footprint can be small, the projection of the image charge on the read-out electrode mediated by capacitive coupling is broadened. An open-face MCP detector for particle counting can be built likewise, for example, mounted on a flange with the read-out electrode on the air side.

position of the pickup electrode is enlarged to a few mm as a function of thickness and permittivity of the insulating substrate, the desired value for most centroiding anode types (charge dividing and others). This has several advantages, including:

(1) No drift region between the MCP and the anode for enlarging the footprint is required and the footprint is not affected by fringing fields.

(2) Image distortions produced by (secondary electron mediated) charge redistribution on the anode and a certain noise component due to quantized charge collection are eliminated.

(3) The pick-up structures connected to sensitive electronics can be operated at ground potential.

(4) The resistive screen can be used as a mechanical vacuum wall separating the read-out structure with nearby electronic circuits from the MCP stack in vacuum, thus saving a number of vacuum feed-throughs.

Especially the latter is very beneficial for novel single photon counting multiplier tubes [31,32].

6.3.3.7 Delay-Line Anodes

In order to reduce the rate-limiting electronic dead-time effects a different anode concept than charge weighting was developed which uses only fast timing electronics circuits: the delay-line anode. As the name implies, the delay-line anode employs an electronic delay-line to slow down signal propagation speed across the anode. An example is a crossed-strip anode structure (as used for pixel anodes) but with the strips interconnected by LC circuits or by meander-shaped tracks (see Figure 6.12) [33,34]. Each circuit or meander loop introduces a delay (typical 1 ns) for a traversing signal so that an anode with 50 mm active area has a single path delay of several 10 ns. Typical strip distance is 1–2 mm. Other layouts like the serpentine delay-line anode [35] and the helical wire delay-line (HWDL) anode [36] use meander-tracks on a substrate or helically wound wires (see also below) which act simultaneously as pickup and delay-line elements.

In all those designs, the arrival time of signals induced by an MCP charge avalanche is proportional to the relative signal propagation time toward the two delay contacts (independently for each dimension). Since only timing electronics as used for genuine TOF-applications are employed, the dead-time between detected particles can, in principle, be on the order of the MCP signal width, that is, <10 ns, which principally allows operating at the highest achievable MCP count rates and is also fairly suitable for multihit detection.

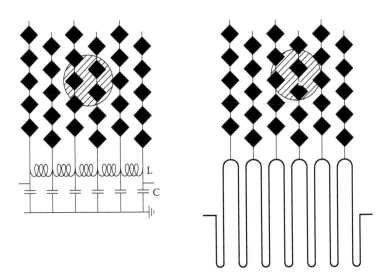

FIGURE 6.12
External delay-line elements placed on a crossed-strip anode structure. Only one dimension is shown. The signal is picked up by the strips and split into two signals traveling along the delay-line toward the terminals. The delay-line is either formed by an LC-circuit chain or a meander-shaped track. The delay per pitch not only ensures that the charge portions picked up by the strips superimpose to a properly shaped single signal which propagates along the delay-line toward the terminals, but also allows for a decent spatial precision as a function of the delay per pitch and the given electronic temporal resolution.

It is important to note that the delay-line anode also operates via charge cloud centroiding with the charge cloud footprint covering several pitches. Therefore, the pitch distance does not limit the spatial resolution as long as signal rise time and delay between pitches are fairly matched so that individual charge signals from each strip are merged into one (broader) signal propagating along the delay-line chain. The achievable spatial resolution is thus only limited by the temporal precision in determining the relative signal arrival times on the anode contacts. Since modern timing electronic circuits allow a precision of <50 ps, a spatial resolution of typically <50 micron can be obtained easily, corresponding to 1000 pixels or more for a 50 mm active anode.

Even for large anode formats of >50 mm a high absolute spatial resolution can be maintained so that large anodes can have an especially high relative spatial resolution (many pixels). Since signal damping introduced by LC or meander-shaped tracks limits the useful anode size, large detectors usually employ HWDL anode arrays. In case of the HWDL, even the absolute spatial resolution may improve as the anode size is increased. The HWDL consists of wire pairs which are continuously wound around a support frame for both dimensions (like a double-helix, remotely similar to the microscopic DNA-structure). The wires perform simultaneously as charge collectors and as signal propagation (delay-) lines. Adequate wire biasing ensures fairly equal charge sharing between both dimension's layers and that one wire of a pair receives an excess charge relative to the other. Thus, a differential signal is induced which has very low damping on such a parallel transmission line. Although the signals travel at speed of light, the loop length ensures a proper delay per pitch (typically 1 mm) as required for the charge centroiding.

The HWDL is a powerful anode concept for large formats from around 50 mm active diameter [37] to the largest MCP sizes of 120 mm or more. The disadvantage is the delicate three-dimensional wire structure. But by using adequate materials commercial HWDL anodes have been demonstrated to possess sufficient thermal and mechanical robustness against baking demands and vibration as is found in a typical lab environment.

Since recent developments of commercial high-throughput time digitizing electronics with low dead-time, the delay-line method became almost the desired "all-purpose" read-out concept for MCPs in the counting mode:

- Spatial and temporal resolutions are nearly as high as theoretically achievable at all with MCPs, even for large formats.
- The electronics' complexity is comparably low and the circuits can cope with the highest possible particle throughput in counting mode.
- The anode is robust and not very complicated in production.
- Operating several delay-line detectors (DLDs) in coincidence is straightforward simply by adding more synchronized read-out timing channels to the digitizing electronics.

The DLD only suffers from the nonparallel read-out scheme [38]. In a multi-hit experiment with many simultaneously arriving particles, the electronic chains of the delay-line anode will not yield useful information due to the small but nonvanishing electronic dead-time.

6.3.4 Multihit Detection with MCP Detectors

6.3.4.1 Pixel Anodes and the Hexanode

Nowadays atomic physics experiments demand not only for the operation of several detectors in coincidence at maximum rate with high spatial and temporal resolution, but is often also required to detect burst of particles, the so-called multi hits. The term multihit has likewise been defined before but it is necessary to refine this by introducing the terms multiplicity of a multihit, which is the number of particles in the burst, and pulse pair resolution, which is the minimum time between two consecutive particles (that a certain anode concept has to cope with).

For MCP detectors it is not so obvious discriminating a multihit from a series of "single hits" by definition because this depends on the dead-time of the electronics and/or of the anode concept. In atomic physics, it became common to draw the border line somewhere around a few microseconds time between particles. This is partially due to the typical range of commercial "multihit TDCs" opposed to slower analog circuits with a few microseconds' conversion time per particle, as in the case of charge integrating amplifiers or TAC/ADCs (which are not considered multihit capable devices). It also turns out that a few microseconds' period is the upper limit on pulse-pair resolution required for detecting ionic fragments from a molecular break-up in a reaction microscope [1], opposed to the time between "single" (uncorrelated) particles even at high count rates (which is $>1\,\mu s$).

For read-out anodes using only fast timing circuits (i.e., pixel and delay-line anode) the electronic dead-time is on the order of 10 ns. Modern multihit TDCs have a pulse-pair resolution of <10 ns and can collect a large number of hits per channel in the given range. These numbers determine the practical multihit limitations of pixel and delay-line anodes.

If the temporal distance between two particles is smaller than the electronic dead-time, defined by the TDC or the circuits before, one can speak of "quasisimultaneous" particles. If one wants to detect such multihit events, then it is necessary to have a parallel read-out chain like "true" pixel anodes (one electronic channel per image pixel) or employ a redundant read-out scheme (see below). Pixel anodes are ideally suited for multihit detection and are successfully employed for such applications.

But sometimes the demand for spatial resolution in multihit detection tasks is too high for using a pixel anode, given the nowadays manageable electronic complexity. For multihits with low multiplicity and/or modest time between two consecutive particles efforts of using the delay-line anode have been made. Although signals propagate for several 10 ns on a delay-line anode before they are processed by the electronics this does not mean that the effective dead-time for the particle detection is high as well. Signals from several particle hits can simultaneously travel along the delay-line without interference. The particle dead-time is only determined by the electronics circuits' dead-time of about 10 ns, as in the case of crossed-strip pixel anodes. By introducing the so-called Hexanode (see Figure 6.13) with its third layer it became possible to reconstruct the relative times and positions for a simultaneous particle pair with high resolution as long as the two particles are spatially separated by about 10 mm [39]. If the typical time between particles in a burst is several 10 ns with only an occasional quasi-simultaneous pair, the Hexanode can measure position and time of each particle very accurately even for bursts of high multiplicity.

A reduction of the electronic dead-time and thus a further improvement of the multihit tolerance of delay-line anodes was achieved by bypassing the CFD circuit and digitizing the analog signals from the delay-line anode terminals with fast sampling ADCs. Thus even signals with slight overlap ("camel pulses") can be properly timed after pulse-shape digitization and "soft(ware)-CFD" algorithms [40]. Even the pulse height of signals can be used for the event reconstruction because all signals induced by the same particle on the

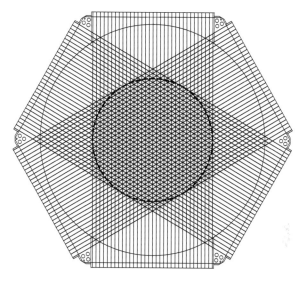

FIGURE 6.13
Sketch of a hexanode formed by an HWDL array. Charge footprints within the circle are registered by all three delay lines, ensuring redundant information on each particle's position. Thus, it is possible to recover position and time information for several particles beyond the electronic dead-time limit: even simultaneously arriving particle pairs can be detected as long as they have a minimum spatial distance of at least 10 mm.

detector elements have a certain height, which might be different for another particle in the multihit event. The latest commercially available generation of such fast sampling ADCs allows a high throughput and it seems only a matter of developing fast pulse-shape algorithms until the "traditional" CFD/TDC timing read-out chains with their dead-time limitation will be replaced by this new read-out method. However, the software development is a very complex task and due to the small number of parallel electronic chains and the given signal width there will still remain a practical limit on the number of quasi-simultaneous particles in a burst which can be analyzed.

6.3.4.2 CCD and Hybrid Detectors

For very multiplex and/or quasi-simultaneous multihit events (e.g., explosion of a large cluster or molecule) with high spatial resolution demand methods with CCD cameras have been developed which are redundant to the pixel anode and hexanode approaches. The "price" for using a CCD is of course the low event rate (only one multihit per CCD camera frame as it is also the case for single particle counting with CCD). As long as this is not a limiting factor one can, for example, combine a pixel read-out with a CCD camera by weaving a crossed-strip array into the phosphor screen [41]. Due to the large number of parallel read-out channels this hybrid method can analyze a higher number of quasi-simultaneous hits than the hexanode: While the pixel anode provides a precise temporal resolution and a coarse spatial resolution for each particle, the CCD refines the spatial resolution for each counted particle. However, limitations for truly simultaneous particles remain. Also, the complexity of the electronic circuits and of correlating the hits to the phosphor spots is high.

To overcome these disadvantages a different technique working entirely with "optical" read-out has been developed: A pair of CCD cameras is imaging the same phosphor screen behind an MCP stack [42]. Electronic shutters for both cameras are triggered by the pulsed ionization source (e.g., a laser shot). While one of the cameras has a long exposure time the other camera is gated with a time constant of about the decay time of the phosphor's light emission characteristics. This modulates the brightness of a particle "spot" in the CCD image with the relative particle arrival time for this second camera. Comparing the brightness of each spot in the two exposures allows a fairly precise time determination for each of the quasi-simultaneous particles arriving in a burst.

This method cannot detect "late" hits, for example, the burst duration can be only a few times the phosphor's light emission decay constant (which can be selected from few ns to several 10 ns), but within this time range there is virtually no limitation on the number of particles in the burst.

6.3.4.3 MEDIPIX/TIMEPIX Sensors

As mentioned earlier, there is a very recent development of pixel detectors from the CERN Electronics Department based on most advanced semiconductor technique, the so-called MEDIPIX chips. These can also be applied for MCP read-out and find first applications in atomic physics [43]. Although initially designed as an advanced-performance photo-diode array for noise-free x-ray counting at GHz rates this detector type can also be used as a sensor for electron detection and consequently as an anode for MCP read-out.

A pixelated array of 256×256 diodes with $55\,\mu$m pitch (Medipix2, active area about $14 \times 14\,\text{mm}^2$) is placed on an ASIC read-out chip, backing each individual pixel with an amplifier, a twofold discriminator (separately tuneable), and a 14 bit counter. The dead-time per pixel is $10\,\mu$s and the counters are read out with about 1 kHz frame rate. The latter is similar to CMOS/CCD frame rates but since the MEDIPIX operates fully digital there is no read-out noise, dark current or over-spill effect as common even to the most advanced CCD chips. On top of being a true counting device, a modification of the MEDIPIX (the so-called TIMEPIX) sensor now even allows for a time tagging of each individual count with 10 ns precision and/or a charge measurement.

Since the detection threshold for photons and electrons is 4 keV the MEDIPIX/TIMEPIX sensors are already suitable as standalone detectors for some atomic physics applications. If placed behind a MCP the whole range of particles/photons can be detected. By butting four chips together, the active area can be increased to $28 \times 28\,\text{mm}^2$ with 512×512 pixel and the effective position resolution can further be improved by charge cloud centroiding below the physical pixel size [44], even at comparably low MCP gain (thus allowing for counting rates well above 1 MHz).

Although the temporal resolution of 10 ns may not be sufficient for a typical TOF application one can refine the precision by correlating these time tags from the TIMEPIX sensor to signals picked up with <100 ps temporal resolution from the MCP stack (which must then be operated in saturation). This principally allows multihit detection with 10 ns dead-time at about 1 MHz detection rate and high temporal/spatial resolution over ∼30 mm active diameter.

6.4 Outlook on Future MCP Detector Concepts

These developments on semiconductor detection techniques will sooner or later revolutionize MCP read-out. To date, these sensor developments are extremely expensive and exclusive to larger collaborations. Therefore, other inventive concepts will be needed in the mean time for the next step toward the flexible, affordable all-purpose read-out anode for MCP detectors.

As soon as fast sampling ADC systems with high data throughput and fairly large channel number will become affordable, the next logical step is introducing centroiding methods to pixel anodes. It has been shown that crossed-strip anodes equipped with precise charge amplifiers and digitizers can reach a spatial resolution by centroiding (similar to the W&S anode) which is a factor of 100 more precise than the pitch distance, for example, sufficient to image even individual MCP pores [45].

Using about 100 sampling ADC channels on a 50×50 crossed-strip anode, each operated as a fast charge digitizers with "poor" resolution (integration time only on the order of the natural signal width), position centroiding of a large footprint may at least allow for a 20-fold resolution refinement compared to the pitch distance. Thus 1000×1000 pixels spatial resolution could be achieved with high temporal resolution.

Such a detector would be fairly multihit capable (especially in three-layer version similar to the Hexanode) due to the high number of parallel channels. It would also be easier to operate than classical pixel anodes because no detection thresholds need to be set and "cross-talk" between strips due to a broad charge cloud footprint is not a problem but a feature. The concept would bring together almost all currently pursued read-out methods. But even before this anode/read-out concept may become available, steady improvements of electronic circuits in general will help optimizing each of the present approaches for MCP read-out and give atomic physicist over time even better tools to master the next experimental challenges.

References

1. J. Ullrich, R. Moshammer, A. Dorn, R. Dörner, L. Schmidt, and H. Schmidt-Böcking, Recoil-ion and electron momentum spectroscopy: Reaction-microscopes. *Rep. Prog. Phys.* 66, 1463–1545, 2003.
2. J. L. Wiza, Microchannel plate detectors. *Nucl. Instrum. Meth.* 162, 587–601, 1979.
3. G. W. Fraser, The gain, temporal resolution and magnetic-field immunity of microchannel plates. *Nucl. Instrum. Meth.* A 291, 595–606, 1990.
4. M. Krems, J. Zirbel, M. Thomason, and R. D. DuBois, Channel electron multiplier and channelplate efficiencies for detecting positive ions. *Rev. Sci. Instrum.* 76, 093305:1–7, 2005.
5. R. S. Gao, P. S. Gibner, J. H. Newman, K. A. Smith, and R. F. Stebbings, Absolute and angular efficiencies of a microchannel-plate position-sensitive detector. *Rev. Sci. Instrum.* 55, 1756–1759, 1984.
6. G. W. Fraser, The ion detection efficiency of microchannel plates (MCPs). *Intern. J. Mass Spectrom.* 215, 13–30, 2002.
7. O. Siegmund, J. Vallerga, and A. Tremsin, Characterization of Microchannel Plate Efficiency. In O. H. W. Siegmund (Ed.), *Proceedings of the SPIE Vol. 5898: UV, X-ray, and Gamma-Ray Space Instrumentation for Astronomy XIV*, 2005, pp. 58980H1–11.
8. G. W. Fraser and J. F. Pearson, Soft X-ray energy resolution with microchannel plate detectors of high quantum detection efficiency. *Nucl. Instrum. Meth.* 219, 199–212, 1984.
9. M. J. Whiteley, J. F. Pearson, G. W. Fraser, and M. A. Barstow, The stability of CsI-coated microchannel plate X-ray detectors. *Nucl. Instrum. Meth.* 224, 287–297, 1984.
10. J. S. Lapington, J. Howorth, and J. Milnes, A reconfigurable image tube using an external electronic image readout. In R. E. Longshore (Ed.), *Proceedings of the SPIE 5881: Infrared and Photoelectronic Imagers and Detector Devices.* 2005, pp. 588109:1–10.
11. A. J. R. Heck and D. W. Chandler, Imaging techniques for the study of chemical reaction dynamics. *Annu. Rev. Phys. Chem.* 46, 335–372, 1995.
12. A. Vredenborg, W. G. Roeterdink, and M. H. M. Janssen, A photoelectron–photoion coincidence imaging apparatus for femtosecond time-resolved molecular dynamics with electron time-of-flight resolution of $\sigma = 18\,\mathrm{ps}$ and energy resolution $\Delta E/E = 3.5\%$. *Rev. Sci. Instrum.* 79, 063108:1–9, 2008.
13. S. Bouneau, P. Cohen, S. D. Negra, D. Jacquet, Y. Le Beyec, J. Le bris, M. Pautrat, and R. Sellem, 256-anode channel plate device for simultaneous ion detection in time of flight measurements. *Rev. Sci. Instrum.* 74, 57–67, 2003.
14. J. Becker, K. Beckord, U. Werner, and H. O. Lutz, A system for correlated fragment detection in dissociation experiments. *Nucl. Instrum. Meth.* A 337, 409–415, 1994.
15. M. Lavollée, A new detector for measuring three-dimensional momenta of charged particles in coincidence. *Rev. Sci. Instrum.* 70, 2968–2974, 1999.
16. U. Becker, Angle-resolved electron–electron and electron–ion coincidence spectroscopy: new tools for photoionization studies. *J. Electron Spectrosc. Relat. Phenom.* 112, 47–65, 2000.

17. W. Koenig, A. Faibis, E. P. Kanter, Z. Vager, and B. J. Zabransky, A multiparticle 3D imaging technique to study the structure of molecular ions. *Nucl. Instrum. Meth.* B 10–11, 259–265, 1985.

18. J. G. Timothy, G. H. Mount, and R. L. Bybee, Multi-anode microchannel arrays. *Trans. Nucl. Sci.* 28, 689–697, 1981.

19. J. F. Pearson, G. W. Fraser, C. H. Whitford, M. R. F. Siggel-King, F. M. Quinn, and G. Thornton, The development of a fast imaging electron detector based on the CODACON concept. *Nucl. Instrum. Meth.* A 513, 183–186, 2003.

20. X. Llopart, R. Ballabriga, M. Campbell, L. Tlustos, and W. Wong, Timepix, a 65k programmable pixel readout chip for arrival time, energy and/or photon counting measurements *Nucl. Instrum. Meth.* A 581, 485–494, 2007.

21. M. Lampton and R. F. Malina, Quadrant anode image sensor. *Rev. Sci. Instrum.* 47, 1360–1362, 1976.

22. A. S. Tremsin and O. H. W. Siegmund, Spatial distribution of electron cloud footprints from microchannel plates: Measurements and modelling. *Rev. Sci. Instrum.* 70, 3282–3288, 1999.

23. C. Martin, P. Jelinsky, M. Lampton, R. F. Malina, and H. O. Anger, Wedge-and-strip anodes for centroid-finding position-sensitive photon and particle detectors. *Rev. Sci. Instrum.* 52, 1067–1074, 1981.

24. J. S. Lapington, A. A. Breeveld, M. L. Edgar, and M. W. Trow, A novel imaging readout with improved speed and resolution. *Nucl. Instrum. Meth.* A 310, 299–304, 1991.

25. J. S. Lapington, B. Sanderson, L. B. C. Worth, and J. A. Tandy, Imaging achievements with the Vernier readout. *Nucl. Instrum. Meth.* A 477, 250–255, 2002.

26. G. W. Fraser, M. A. Barstow, and J. F. Pearson, Imaging microchannel plate detectors for X-ray and XUV astronomy. *Nucl. Instrum. Meth.* A 273, 667–672, 1988.

27. M. Lampton and C. W. Carlson, Low-distortion resistive anodes for two-dimensional position-sensitive MCP systems. *Rev. Sci. Instrum.* 50, 1093–1097, 1979.

28. G. Battistoni, P. Campana, V. Chiarella, U. Denni, E. Iarocci, and G. Nicoletti, Resistive cathode transparency. *Nucl. Instrum. Meth.* 202, 459–464, 1982.

29. O. Jagutzki, J. S. Lapington, L. B. C. Worth, U. Spillmann, V. Mergel, and H. Schmidt-Böcking, Position sensitive anodes for MCP read-out using induced charge measurement. *Nucl. Instrum. Meth.* A 477, 256–261, 2002.

30. D. Céolin, G. Chaplier, and M. Lemonnier, High spatial resolution two-dimensional position sensitive detector for the performance of coincidence experiments. *Rev. Sci. Instrum.* 76, 043302:1–7, 2005.

31. J. Barnstedt and M. Grewing, Development and characterization of a visible light photon counting imaging detector system. *Nucl. Instrum. Meth.* A 477, 268–272, 2002.

32. A. Czasch, V. Dangendorf, J. Milnes, S. Schössler, R. Lauck, U. Spillmann, J. Howorth, and O. Jagutzki, Position and time sensitive photon counting detector with image charge delay-line readout. In W. Becker (Ed.), *Proceedings of the SPIE on Advanced Photon Counting Techniques II*, Vol. 6771, 2007, pp. 67710w:1–12.

33. O. H. W. Siegmund, J. Zaninovich, A. Tremsin, and J. Hull, In O.H.W. Siegmund and M.A. Gummin (Eds.), *Proceedings of the SPIE on EUV, X-Ray, and Gamma-Ray Instrumentation for Astronomy IX*, Vol. 3445, 1998, pp. 397–406.

34. J. H. D. Eland, Simple two-dimensional position-sensitive detector with short dead-time for coincidence experiments. *Meas. Sci. Technol.* 5, 1501–1504, 1994.
35. P. G. Friedman, R. A. Cuza, J. R. Fleischman, C. Martin, D. Schiminovich, and D. J. Doyle, Multilayer anode with Crossed serpentine delay lines for high spatial resolution readout of microchannel plate detectors. *Rev. Sci. Instrum.* 67, 596–608, 1996.
36. S. E. Sobottka and M. B. Williams, Delay line readout of microchannel plates. *IEEE Trans. Nucl. Sci.* 35, 348–351, 1988.
37. E. Liénard, M. Herbane, G. Ban, G. Darius, P. Delahaye, D. Durand, X. Fléchard et al. Performance of a micro-channel plates position sensitive detector. *Nucl. Instrum. Meth.* A 551, 375–386, 2005.
38. O. Jagutzki, V. Mergel, K. Ullmann-Pfleger, L. Spielberger, U. Spillmann, R. Dörner, and H. Schmidt-Böcking, A broad-application MCP detector system for advanced particle or photon detection tasks: Large area imaging, precise multi-hit timing information and high detection rate. *Nucl. Instrum. Meth.* A 477, 244–249, 2002.
39. O. Jagutzki, A. Czasch, R. Dörner, M. Hattaß, V. Mergel, U. Spillmann, K. Ullmann-Pfleger et al. Multiple hit read-out of a microchannel plate detector with a three-layer delay-line anode. *IEEE Trans. Nucl. Sci.* 49, 2477–2483, 2002.
40. G. Da Costa, F. Vurpillot, A. Bostel, M. Bouet, and B. Deconihout, Design of a delay-line position-sensitive detector with improved performance. *Rev. Sci. Instrum.* 76:013304:1–8, 2005.
41. R. Wester, F. Albrecht, M. Grieser, L. Knoll, R. Repnow, D. Schwalm, A. Wolf et al. Coulomb explosion imaging at the heavy ion storage ring TSR. *Nucl. Instrum. Meth.* A 413, 379–396, 1998.
42. D. Strasser, X. Urbain, H. B. Pedersen, N. Altstein, O. Heber, R. Wester, K. G. Bhushan, and D. Zajfman, An innovative approach to multiparticle three-dimensional imaging. *Rev. Sci. Instrum.* 71, 3092–3098, 2000.
43. G. Gademann, Y. Huismans, A. Gijsbertsen, J. Jungmann, J. Visschers, and M. J. J. Vrakking, Velocity map imaging using an in-vacuum pixel detector. *Rev. Sci. Instrum.* 80, 103105:1–7, 2009.
44. B. Mikulec, A. Clark, D. Ferrere, D. L. Marra, J. McPhate, O. Siegmund, A. Tremsin et al., A noiseless kilohertz frame rate imaging detector based on microchannel plates read out with the Medipix2 CMOS pixel chip. *Nucl. Instrum. Meth.* A567:110–113, 2006.
45. O. H. W. Siegmund, A. S. Tremsin, J. V. Vallerga, R. Abiad, and J. Hull, High resolution Cross-strip anodes for photon counting detectors. *Nucl. Instrum. Meth.* A 504, 177–181, 2003.

7

Coincidence Techniques in Atomic Collisions

John A. Tanis

CONTENTS

7.1 Introduction

7.1.1 Overview

In atomic collisions, the products that are observed following a reaction are photons, electrons, and ions. In addition to simply recording the number of events, the energy and angular distributions of the emitted photons or particles are often of interest. Although much can be learned from spectral distributions (energy or angle), frequently it is necessary to obtain more detailed information by isolating and identifying specific outcomes of a collision reaction. Such detailed information can be gained through the use of *coincidence techniques*, in which two or more reaction products from the same collision event are associated (or correlated) in time (see, e.g., [1–5]). This association is accomplished by electronically recording the time difference Δt between detection of the products of interest and necessarily assumes that the detected reaction products resulted from a single event. It is noted that if the detection rate for one or both of the individual reaction products is very high, then *accidental coincidences* from two or more separate collision events occur. To the extent that most of the recorded coincidences are due to single-collision events, coincidence techniques provide a powerful tool that permits detailed insights into collision reactions and consequently are widely used in all areas of atomic collision physics.

While the specific instrumentation used in different coincidence setups can vary widely, from a collection of individual electronic modules to rather large-scale arrays integrated onto a single "board," it remains true that the underlying features of all coincidence techniques share a common structure and logic. It is these common aspects that will be presented in this chapter. The discussion is aimed at the graduate students just beginning their research and who are called upon to set up a coincidence circuit for the first time, or for the more advanced scientist who has not previously used coincidence techniques but finds the need to employ them. Hence, much of what is presented here will already be familiar to the experienced user of coincidence circuits. The viewpoint of the novice user is taken because it is unlikely that the scientist already experienced in coincidence techniques will benefit from any overview of coincidence techniques, but will instead require more detailed and specialized aspects relevant to a given circuit. Moreover, no attempt is made to describe individual modules from particular manufacturers because the precise specifications and controls of individual units can vary widely, although the intended use or application might be the same.

7.1.2 Using Coincidence Techniques to Study Reactions: An Example

As an example, consider the following collision reaction (see Refs. [3,4] for similar collision systems):

$$O^{7+}(1s) + He \rightarrow \begin{cases} O^{6+}(1snl) + He^{+} + h\nu & \text{(single capture)} & (7.1a) \\ O^{6+}(1snl) & \text{(transfer ionization)} & (7.1b) \\ \quad + He^{2+} + e^{-} + h\nu & & \\ O^{7+}(1s) + He^{+} + e^{-} & \text{(single ionization)} & (7.1c) \\ O^{7+}(1s) + He^{2+} + 2e^{-} & \text{(double ionization)} & (7.1d) \\ O^{8+} + He^{+} + 2e^{-} & \text{(electron loss + single ioniz.)} & (7.1e) \\ O^{8+} + He^{2+} + 3e^{-} & \text{(electron loss + double ioniz.)} & (7.1f) \end{cases}$$

Here, the He target atom undergoes single or double ionization, whereas the incident projectile captures an electron, remains unchanged, or loses an electron. Additionally, the projectile, the target, or both could be excited in the collision and subsequently emit a photon as part of the de-excitation process (Equations 7.1a and 7.1b). Each of the six possible outcomes represents a specific and unique collision reaction. Thus, to properly characterize the collision reactions listed in Equations 7.1a through 7.1f and to provide detailed information on theoretical calculations, it is necessary to isolate the individual processes by associating each of the target ionization states with a specific outgoing projectile charge state. It should be noted that the other reaction products might be detected as well, that is, the photons [1,5] or electrons [2] that are emitted as a result of the collision process. While the detectors required for these latter products would be different, in general, coincidences between any pair of reaction products can be detected and the coincidence techniques used to isolate individual collision outcomes remain the same.

A simple schematic of an apparatus that could be used to study the reactions listed in Equations 7.1a through 7.1f is shown in Figure 7.1. An incoming ion A^{q+} with speed v interacts with the target in the collision region which may consist of a collision cell (shown) or a gas jet. As a consequence of the collision, the target may become ionized, whereas the projectile may or may not change charge. The target ions are typically born at very low energies and may be extracted from the collision region with a transverse electric field of a few volts per centimeter to several hundred volts per centimeter depending on what information is to be obtained from the reaction. Meanwhile, the projectile ions can be passed through a magnetic (shown) or electrostatic analyzer to separate the outgoing beam into its individual charge-state components. The individual components can then be recorded with various types of solid-state or gas-filled particle detectors. Individual solid-state detectors may be used for each beam component or a single position-sensitive solid-state detector may be used to collect all the charge state components simultaneously.

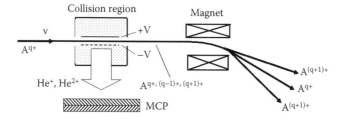

FIGURE 7.1
Schematic of apparatus that can be used to study reactions such as those listed in Equations 7.1a through 7.1f. In this setup, recoiling target ions are detected in coincidence with projectile charge-changing events. Electrons and photons may also be ejected in the collision but they are not detected in the apparatus shown here.

Consideration of the various types of available detectors is outside the scope of the present discussion.

To determine specific collision outcomes, or reaction *channels*, target ionization (He^+ or He^{2+}) must be associated with the outgoing projectile ion that caused the ionization [4]. Since a given target ion and the associated projectile ion resulted from a single collision event, that is, they occurred in *coincidence* with each other, there must be a definite time correlation between them. If this time correlation can be recorded, then a given collision process is uniquely identified.

7.2 Essentials of a Coincidence Circuit

In this section, the basic elements of a coincidence circuit are presented, along with a discussion of the general operating logic of the circuit. Following this, the individual electronic modules utilized in the coincidence circuit are considered in more detail to give an overview of their basic operation. A discussion of the basic elements of a coincidence circuit, the modules used in it, and applications can also be found in Ref. [6]. Prior to discussing a coincidence circuit, however, it is desirable to briefly describe the types of signals (linear and logic) employed at the various stages of the circuit.

7.2.1 Linear and Logic Signals

In a coincidence circuit, both *linear (analog)* and *logic (digital)* pulses are utilized at various stages and so it is useful to distinguish between them and to describe their basic features. In Figures 7.2 and 7.3, respectively, are shown the typical characteristics of linear and logic pulses, as well as a listing of some of the electronic modules used in processing each of these signals. Both linear and logic signals can be "slow" or "fast" depending on the application.

Linear pulses, so called because the *relative amplitude* is preserved when the signal is processed by a given module, are classified as slow or fast depending

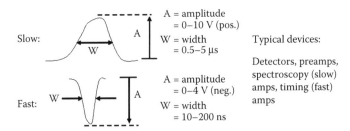

FIGURE 7.2
Examples of slow and fast linear pulses, their characteristics, and devices that produce or process these pulses.

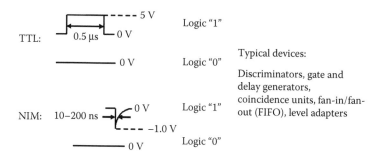

FIGURE 7.3
Examples of slow (TTL) and fast (NIM) logic pulses, their characteristics, and devices that produce or process these pulses.

on the width of the pulse: 0.5–5 μs for slow pulses and 10–200 ns for fast pulses as shown in Figure 7.2. Typically, slow pulses have amplitudes in the range 0–10 V, whereas fast pulses have amplitudes of 0–4 V. Both slow and fast signals can be positive, negative, or bipolar. In the latter case, the pulse has both a positive and a negative half cycle.

On the other hand, logic pulses have fixed amplitude and also have long and short pulse widths classified according to the type of logic. In Figure 7.3, both transistor-transistor-logic (TTL) and Nuclear Instrumentation Module (NIM) logic pulses are shown. While other types of logic pulses are sometimes used, the two shown are common and illustrate the necessary principles. The essential feature of logic pulses is that they have two possible states: Logic "1" and Logic "0," allowing electronic decision-making based on the particular state. TTL Logic 1 signals have a fixed amplitude, usually +5 V, and a width of 0.5 μs, whereas Logic 0 has amplitude of 0 V, constituting slow logic. On the other hand, fast NIM logic signals have a negative Logic 1 amplitude of about −1.0 V and a width that can range from 10 to 200 ns, and a quiescent state of 0 V for Logic 0.

7.2.2 Block Diagram of a Coincidence Circuit

Figure 7.4 presents a simplified block diagram of a coincidence circuit [6] showing the basic electronic functions required to process signals from the individual detectors used to record the reaction products from the collision processes listed in Equations 7.1a through 7.1f and to correlate these signals in time. A variety of electronics modules with specific functions are required, involving a combination of both linear and logic input and output pulses. Again it is noted that instead of detecting target ions and outgoing projectile ions, electrons or photons produced in the collision could also be detected and correlated in time.

Referring to Figure 7.4, the detector signals generated from the separate products of a collision event are passed to a preamplifier located in close physical proximity to the detector to prevent attenuation of the signal by

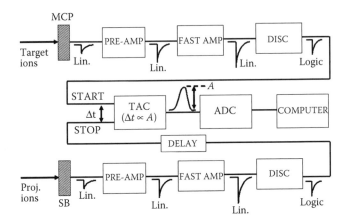

MCP = microchannel plate
SB = surface-barrier detector
Lin. = linear signal

FIGURE 7.4
Block diagram showing the typical components of a coincidence circuit. The diagram shows the basic electronic modules required to process signals and the types of input and output signals associated with each module.

boosting the signal amplitude. Preamplifiers are usually designed (a) to preserve the fast rise time (typically 10–50 ns) of the signal to permit counting rates as high as possible, and (b) to provide suitable amplification for the linear amplifier that follows it. The fast linear amplifier, which might be located quite some distance away from the preamplifier (several tens of meters) further amplifies the signal (typically in the range 1–4 V) and provides some pulse shaping for subsequent time or energy (or both) analysis. Both the preamplifier and the fast amplifier are linear devices, that is, their output amplitudes are proportional to the input amplitude. The primary function of the amplifier is to provide a suitable signal that can be time or energy analyzed by the discriminator or ADC that follows it. In addition to signal amplification, the amplifier typically provides for shaping of the pulse by changing the rise and fall times, that is, the pulse width can be changed, which helps to distinguish true event signals from background electronic noise. There is a trade-off because larger pulse widths permit better signal-to-noise discrimination, but at the same time reduce the count rate capabilities and may also reduce the time resolution of the instrument.

Following the amplifier is a discriminator (DISC) that is used to generate a logic pulse that will be used as the START or STOP signal for a reaction event of interest, that is, a detected target ion, projectile ion, electron, or photon. The DISC analyzes the incoming signal amplitude (voltage) from the amplifier and determines if the amplitude exceeds the threshold voltage setting of the DISC. If it does, then a logic "1" pulse is generated; if not, then no output signal (a logic "0") is generated.

Two logic signals representing different reaction products, with a time difference Δt between them, are then used as the START and STOP inputs to a time-to-amplitude (TAC) converter, which converts the time difference into an amplitude A (voltage) proportional to Δt as indicated in Figure 7.4. The time difference includes intrinsic differences due to particle flight times and individual response times of the detectors, but may also include additional delay (DELAY) that is introduced into the circuit to obtain time differences in a range required by the TAC. Typical TAC time ranges vary from a few picoseconds to several microseconds, while the linear output signal of the TAC typically lies in the range 0–10 V. Because any desired delay can be introduced into the inputs to the TAC, either signal can be chosen as the START or the STOP. Typically, the signal from the detector with the lowest count rate (detectors can have vastly different count rates due to geometry and detector efficiency) is used as the START to reduce the number of events that are processed by the TAC. Since an output from the TAC occurs only when there is both a START and STOP signal present within the selected time range of the TAC, reducing the number of START signals reduces significantly the "busy" time of the TAC, while causing no loss in capability for detecting correlated START and STOP events by the TAC.

The output signal generated by the TAC is then sent to an ADC in which the amplitude of the incoming signal is converted to an integer number called a *channel number* with typical maximum values of 2^{10} (1024) to 2^{13} (8096) and proportional to the output amplitude A. These numbers are then stored by a COMPUTER event-by-event so that different *gating conditions* can be applied to the collected data, and the data can later be replayed for more detailed analysis. This mode of data collection is referred to as *event mode* data acquisition. Gating conditions are used to apply various constraints on the collected data to isolate particular reaction outcomes. From the collected data, a *histogram spectrum* of events (counts) versus channel number can be derived. For events that are correlated in time, for example, a target ion that results from a collision with a projectile ion, a large number of events will occur in a small range of channel numbers because these events have a definite time relationship Δt between them. Hence, a peak corresponding to these *coincidence events* is seen in the histogram spectrum in the vicinity of the channel number corresponding to this time difference. A typical histogram spectrum for target ionization coincident with electron capture by the projectile is shown in Figure 7.5 [7]. It is noted that *random coincidences* also occur in which the START and STOP pulses did not originate with the same collision event. However, these latter coincidences are expected to appear approximately uniformly over the time range of the TAC and hence appear as a nearly constant underlying background in the histogram spectrum.

As a final point, we note that frequently the functions of the TAC and the ADC are combined into a single unit forming a time-to-digital converter (TDC) as shown in Figure 7.6. In this case, START and STOP signals with a time difference Δt are provided to the TDC, the same as for the TAC, but instead

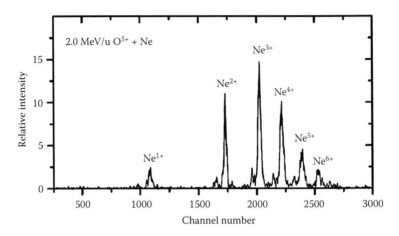

FIGURE 7.5
Histogram spectrum showing Ne target ion charge states coincident with single electron capture by the projectile resulting from collisions of 2.0 MeV/u O^{5+} with Ne, that is, $O^{5+} + Ne \rightarrow O^{4+} + Ne^{n+} + (n-1)e^-$.

TDC = time-to-amplitude converter

FIGURE 7.6
Schematic showing the TAC and ADC functions combined into a single module referred to as a TDC. See Figure 7.21 for more details.

of giving a linear output signal of amplitude A, Δt is converted directly into an integer number proportional to Δt.

Also, the functions of the ADC and the computer can be combined into a single unit to form a multichannel analyzer (MCA) as shown in Figure 7.7. In this case, the computer stores the data as one or more histogram spectra and has appropriate software for data manipulation and analysis. Usually, an MCA is used when only histogram spectra are being recorded, or when coincidence events are not being detected. When data are stored using an MCA, individual events are not recorded, that is, there is no event-mode data acquisition, so there is no possibility to later replay the data and apply different gating conditions to the physical parameters.

7.3 Individual Electronics Modules

In this section, the individual electronics modules required in a coincidence setup are described. The modules are divided into three categories: (1) *linear*, including detectors (slow and fast), preamps, spectroscopy (slow) amps, and

MCA = multi-channel analyzer

FIGURE 7.7
Schematic showing the ADC and COMPUTER functions combined into a single module referred to as an MCA. See Figure 7.22 for more details.

timing (fast) amps; (2) *logic*, including discriminators, gate and delay generators, multicoincidence units, fan-in/fan-out (FIFO) units, and level adapters; and (3) *other*, including TACs, ADCs, TDCs, and MCAs. Because several different modules of each type are commercially available with varying specifications and features, only the general operating characteristics, which are similar from module to module, are described.

7.3.1 Linear Devices

7.3.1.1 Detectors

There are many different types of detectors but the common feature is that most of them are designed to detect the individual products resulting from a reaction event. Specifically, these products can be positive or negative ions, neutral atoms, electrons, or photons. Typically, detectors produce small positive or negative output pulses when a particle or photon is incident on the detector as shown in Figure 7.8. Detectors are usually sensitive to the amount of energy deposited within them and hence the output signal is closely proportional to this energy with a degree of accuracy dependent on the type of detector and the specific application. The amplitude of the output signal is typically in the range 10–50 mV, with a width that is dependent on whether the detector has a slow or fast response time. For silicon and germanium detectors with relatively slow response times such as those used for photon and particle detection, the output signal can have widths up to $\sim0.5\,\mu$s. On the other hand, detectors with fast response times such as channel electron multipliers, microchannel plates, plastic scintillators, and phototubes

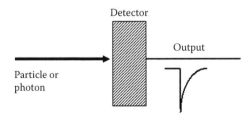

FIGURE 7.8
Schematic of a detector showing the incident particle or photon and the resulting output signal generated.

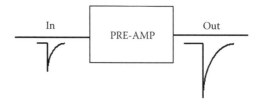

FIGURE 7.9
Schematic of a typical preamplifier. Usually the input signal is negative, whereas the output signal could be negative (shown) or positive.

used to detect photons, electrons, ions, and other particles have output signal widths in the range 10–50 ns. In the case of either slow or fast detectors, the output signal is usually directed to a preamplifier.

7.3.1.2 Preamplifiers

The main purpose of the preamplifier is to increase the amplitude of the detector output signal to a few tenths of a volt as indicated in Figure 7.9. Usually, the preamplifier is located in close physical proximity to the detector to prevent attenuation of the signal and the introduction of additional electronic noise. In addition to amplification, the preamplifier may do some shaping of the pulse to increase signal-to-noise discrimination, or to enhance its pulse height resolution for subsequent energy analysis of the pulse. Both the amplification and the shaping must make the signal suitable for input to the amplifier that follows. Sometimes preamplifiers also invert the polarity of the pulse (negative to positive), but this is usually done only for convenience in subsequent processing.

7.3.1.3 Spectroscopy (Slow) Amplifiers

A spectroscopy amplifier (SPEC AMP) is used to increase the pulse amplitude to a value in the range 0–10 V and also to modify the pulse shape for the best energy resolution. Both of these features are indicated in Figure 7.10. While the output pulse polarity is usually positive, there may be applications where it is negative, and so the output polarity can often be changed. Typically, for the best energy resolution, the pulse width should be in the range 1–10 μs. The output pulse amplitude must also be proportional to that of the input pulse in order to preserve linearity with the energy of the detected photon or particle. The output of a SPEC AMP is frequently used as the input to an ADC for direct energy analysis of the detected photon or particle.*

7.3.1.4 Timing (Fast) Amplifiers

As with a SPEC AMP, a fast timing amplifier (FTA), shown in Figure 7.11, is used to linearly increase the pulse amplitude but typically the output

---

* See Figure 7.18 for an explanation.

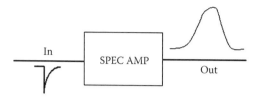

FIGURE 7.10
Schematic of an SPEC AMP. Typically the input signal is negative, whereas the output signal can be positive (shown) or negative. In addition to amplification, the output signal also undergoes shaping and may be inverted as indicated.

amplitude does not exceed 4 V. While the output is linear, the primary purpose of a fast amplifier is to produce a signal suitable for input to a DISC (see below) from which a logic signal can be derived for use in subsequent timing applications. Thus, any shaping done on the pulse must preserve the inherent time information while at the same time allowing true event signals to be discerned from electronic noise. As such, the widths of output pulses from an FTA usually do not exceed ~200 ns. The output from an FTA can be positive or negative (the latter being the usual case) and must match the input requirements of the DISC or other module that follows it.

7.3.2 Logic Devices

7.3.2.1 Discriminators

A DISC, shown in Figure 7.12, is used to select input linear pulses with amplitudes greater than a certain set value, or, equivalently, to reject pulses with an amplitude smaller than that value. To accomplish this, a voltage *level* (DISC LEVEL) is adjusted so that only pulses with amplitudes larger than this value will result in an output logic 1 signal. For timing applications, this output signal must have a fast rise time and a narrow width, that is, fast logic, and is typically a negative NIM logic signal (see Section 7.2.1 and Figure 7.3). There are two common types of discriminators: *leading edge* (LE) and *constant fraction* (CFD). For an LE DISC, the DISC LEVEL is set to a certain voltage above (or

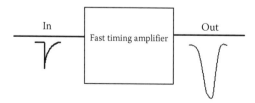

FIGURE 7.11
Schematic of an FTA. Usually the input and output signals are negative. In addition to amplification, the output signal also undergoes some shaping as indicated.

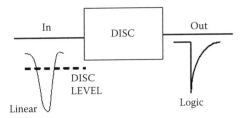

FIGURE 7.12
Schematic of a DISC. Usually the linear input signal is negative, whereas the logic output signal may be positive (TTL) or negative (NIM) (only the negative output is shown). DISC LEVEL refers to the voltage setting of the DISC (see text).

below) 0 V on the *leading (rising) edge* of the input pulse. On the other hand, in a CFD, the DISC LEVEL is set to a certain constant fraction of the input pulse height. Usually a CFD gives better time resolution than an LE DISC, especially when input pulses have significant variations in their amplitudes and relatively slow rise times.

7.3.2.2 Gate and Delay Generators

Gate and delay generators (GDGs) are used to regenerate a logic (GATE) signal and to introduce some delay prior to delivering the output logic pulse. The delay is typically adjustable in the range 10 ns to 1 ms, and a voltage output is usually provided for accurate setting of the desired delay with the width of the pulse determining the length of the delay. Figure 7.13 shows a schematic of a GDG and the timing sequences associated with the input and output logic pulses. Outputs from GDGs are often used to enable or to inhibit other modules and instrumentation in a data acquisition system such as TACs, ADCs, TDCs, scalars, and computers. In any coincidence circuit, the relative timing of the various pulses and the subsequent actions are critical and GDGs can provide the means to accurately adjust these times.

7.3.2.3 Multilevel Coincidence Units

Frequently, it is necessary to perform AND and OR functions on a multiple number of inputs in order to place "conditions" on the events of interest, for example, to accept or reject certain events depending on the outputs generated from various detectors. Both the inputs and outputs must be logic signals. The AND and OR functions and the placing of conditions can be accomplished with a multilevel coincidence unit as shown in Figure 7.14. As indicated in the figure, there is provision for several inputs, and the number of inputs to be analyzed for the desired function (AND or OR) can be selected with the COINC LEVEL. For example, there may be three logic inputs but an AND function might be required between any two of them, in which case a COINC

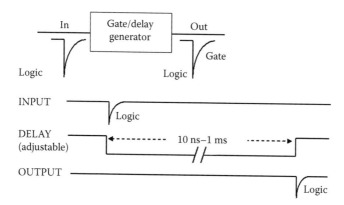

FIGURE 7.13

Schematic showing a GDG and the timing sequences associated with the INPUT, DELAY, and OUTPUT logic pulses. A GDG typically provides both negative (shown) and positive output logic signals.

LEVEL of 2 would be selected. Similarly, the input logic signals from any one of four detectors may be sufficient to "trigger" a subsequent action, and hence the OR function with a COINC LEVEL of 1 would be selected. Since multilevel coincidence units can vary quite widely, the operating manual for a given model must be consulted.

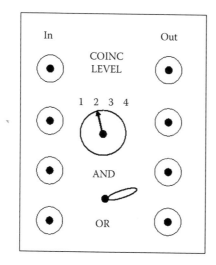

FIGURE 7.14

A typical multilevel coincidence unit. For multiple inputs, either AND or OR functions may be selected, as well as the number of inputs (COINC LEVEL) to be considered in the coincidence requirement.

7.3.2.4 Fan-In/Fan-Out

Sometimes, it is desired to mix a number of logic inputs and derive from them a single logic output, or, conversely, use a logic single input to produce a number of identical independent logic outputs. These functions can be accomplished with the so-called FIFO module as illustrated schematically in Figure 7.15. When used in the *fan-in* mode, multiple logic inputs, for example, from several discriminators, are mixed to provide a common output (actually, there may be several identical outputs for multiple subsequent uses). For example, the outputs from any one of several event channels might be used to initiate processing by an ADC or conversely to inhibit a scalar or other instrument. If all of the inputs occur within the signal processing time (a few nanoseconds) of the instrument, then the fan-in mode functions essentially as an OR module.

In other situations, a single input logic signal might be needed to simultaneously "trigger" several subsequent devices. This can be accomplished using the *fan-out* mode of the module. For example, the logic signal resulting from a particular detector channel is frequently used to provide simultaneous START signals to several TACs that receive their STOP signals from individual detector channels. In Equations 7.1a through 7.1f, this could correspond to detection of a He^+ or He^{2+} ion providing the common START signals, while the separate signals from O^{6+} (electron capture), O^{7+} (no charge change), and O^{8+} (electron loss) would provide the individual STOP signals.

7.3.2.5 Level Adapters

Frequently, it is necessary to translate one type of logic signal into another, for example, NIM to TTL or vice versa, and possibly the inverse of these logic

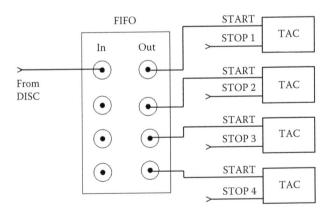

FIGURE 7.15
Schematic of an FIFO in which (a) a number of logic inputs can be mixed to provide a single logic output (not shown), or (b) a single input can be used to provide multiple identical logic outputs (shown). Mode (a) constitutes a fan-in, while mode (b) constitutes a fan-out.

signals might be needed as well (see Section 7.2.1). Such a situation can arise, for example, if the NIM output of a DISC is used as the input to a subsequent logic device that accepts only TTL signals. These logic translation functions are carried out using a level adapter as illustrated in Figure 7.16. If multiple outputs of the translated signal are required, then the level adapter can be followed by an FIFO to accomplish this task.

7.3.3 Other Devices

7.3.3.1 TAC Converter

To detect coincidences between two time-correlated events, a TAC converter can be used, in which the time difference Δt between two input logic signals, START and STOP, is represented as an analog output pulse with amplitude (voltage) A such that $\Delta t \propto A$. The functional operation of a TAC is illustrated schematically in Figure 7.17. The maximum time difference is selectable in ranges that vary typically from about 100 ps to 100 µs, while the maximum output voltage is typically 2, 4, or 10 V. For example, for a time range of 2 µs and a maximum output amplitude $A = 10$ V, a time difference $\Delta t_{min} = 0$ corresponds to $A = 0$ V, while $\Delta t_{max} = 2$ µs corresponds to $A = 10$ V. The analog output pulse from a TAC, which is proportional to the time difference Δt, can then be used as the input to an ADC (see below) for analysis of the pulse height (amplitude) and conversion to an integer number.

7.3.3.2 Analog-to-Digital Converter

An ADC is used to represent the amplitude (voltage) of an input analog signal as an integer number. Two common uses are to analyze the input signal corresponding to (1) the time difference Δt between two events, or (2) the energy

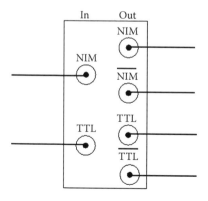

FIGURE 7.16
Schematic of a level adapter in which an input logic signal of one type can be translated to another type of logic signal or its inverse.

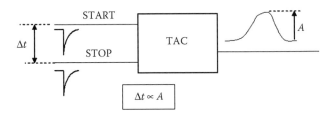

FIGURE 7.17
Schematic showing the functional operation of a TAC converter. The amplitude of the linear output signal is proportional to the time difference Δt between the input logic pulses.

of a detected reaction product. Each of these uses is illustrated in Figure 7.18. Functionally, an input voltage in a range such as 0–10 V is converted to an integer, called a *channel number*, in a range such as 0–1023 ($= 2^{10}$) such that 0 V corresponds to channel 0, and 10 V corresponds to channel 1023. In this example, each channel corresponds to about 10 mV. In an ADC, the number of conversion channels can typically be adjusted to give the desired resolution, for example, 2^9, 2^{10}, 2^{11}, 2^{12}, 2^{13}, and so on.

This resolution is chosen based on the time or energy resolution of the input analog signal to give a sufficient number of channels in the histogram peak corresponding to the time-correlated events or the energy of the detected particle or photon. In this regard, there is a trade-off between having too few channels such that time or energy information is lost and having so many channels where no additional time or energy information is gained at the expense of causing the ADC to be "busy" with unnecessary processing. Figures 7.5 and 7.19 [8] show typical time and energy (x-ray) histogram spectra, respectively, with an appropriate number of channels in the peak regions in each case.

Moreover, if one reaction outcome, for example, an electron, is associated with another reaction outcome, for example, a photon, using a TAC as in

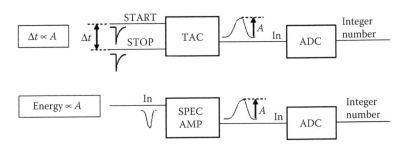

FIGURE 7.18
Schematic showing two common uses of an ADC: (upper) to analyze the linear signal from a TAC corresponding to the time difference between two events and (lower) to analyze the linear signal from an SPEC AMP corresponding to the energy of a detected particle or photon. In both cases, the linear input signal is converted to an integer number.

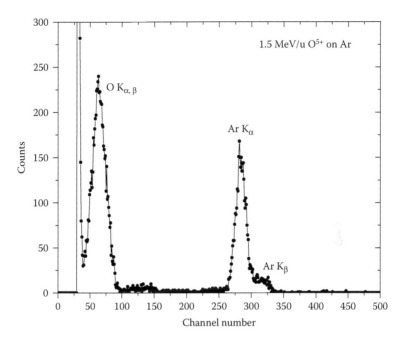

FIGURE 7.19
Typical x-ray energy spectrum obtained with a lithium-drifted silicon [Si(Li)] detector resulting from 1.5 MeV/u O^{5+} + Ar collisions [8]. The O K_α and K_β ($n = 2 \rightarrow 1$ and $n = 3 \rightarrow 1$, respectively) and Ar K_α and K_β x-ray transitions are indicated. The large spike at the low-energy end of the spectrum (near channel 40) is due to electronic noise.

Figure 7.18, then these coincidence events can be used to gate the spectrum obtained from a SPEC AMP to view only those events associated with coincidences recorded by the TAC. Such a case is shown in Figure 7.20 for x-ray events detected in coincidence with forward-going electrons traveling at the velocity of the projectile for 90 MeV/u U^{88+} + N_2 collisions [9]. In Figure 7.20 both the total noncoincident (singles) x-ray spectrum (gray) is shown as well as the contribution to the spectrum associated only with forward-going electrons (black). It is seen that the latter case leads to a very different spectrum and hence isolates a particular reaction channel resulting from the collision.

7.3.3.3 Time-to-Digital Converter

As already mentioned in Section 7.2.2, sometimes the functions of the TAC and the ADC are combined into a single module called a TDC, as shown in Figure 7.21. In a TDC, the time difference Δt between the input logic pulses is converted directly into an integer channel number such that Δt is proportional to the channel number. As with an ADC, the number of conversion channels can be adjusted to give the desired resolution, for example, 2^9, 2^{10}, 2^{11}, 2^{12}, 2^{13},

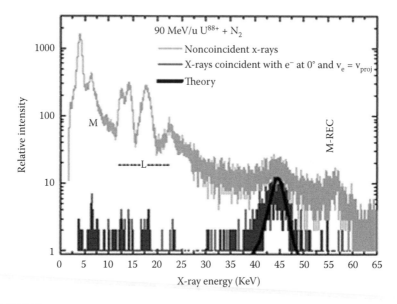

FIGURE 7.20
Spectra showing total x-ray emission (gray) in $90\,\text{MeV/u}$ $U^{88+} + N_2$ collisions and only those x-rays associated with forward ejected electrons (black) traveling with the velocity of the projectile. The notation REC stands for radiative electron capture. (Taken from M. Nofal et al., *Phys. Rev. Lett.* 99, 163201, 2007.)

in the peak(s) of interest. It should be noted that a TDC can be used to replace an ADC only when time differences between input logic signals are being analyzed. It cannot be used to analyze analog input signals carrying energy information, in which case an ADC must be used.

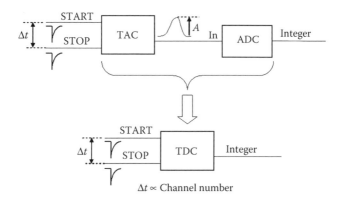

FIGURE 7.21
Schematic showing how the functions of a TAC and an ADC are combined into a single module forming a TDC. In a TDC, the time difference Δt between the input START and STOP logic pulses is converted directly into an integer (channel number).

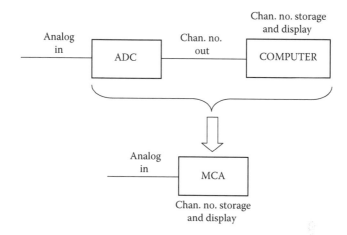

FIGURE 7.22

Schematic showing how the functions of an ADC are and the COMPUTER are combined into a single unit forming an MCA. The MCA usually allows for quick and simple manipulation of the stored spectra.

7.3.3.4 Multichannel Analyzer

Finally, we consider an MCA, which was also mentioned in Section 7.2.2. This module combines the functions of the ADC and COMPUTER into a single unit for storage, display, and manipulation of the histogram spectra formed by the output channel numbers from the ADC as shown schematically in Figure 7.22. The MCA is commonly used in experiments that do not detect coincidences, or for relatively quick and simple conversion, storage, and display of spectra. For more sophisticated data collection, such as the event-mode storage and multiple sorting and display options required in a coincidence experiment, the ADC and computer functions must be separated.

7.4 Conclusion

An overview of the essential features of a coincidence circuit have been presented, including a discussion of the general operation as well as descriptions of the various modules required to process signals at each stage of the circuit. The characteristics of typical linear and logic pulses have been summarized and their roles at the various stages were considered. Some typical histogram spectra resulting from specific components used in a coincidence circuit have also been presented.

Because the specifics of individual coincidence circuits can vary widely, only the basic features common to most such circuits have been considered here. The same approach has been taken in regard to the individual electronic modules that make up a coincidence circuit. No attempt has been made to discuss specific modules or components from particular manufacturers because the appearance, features, and operation of the available modules vary widely. Thus, when setting up a coincidence circuit, it is imperative to consult the specification and operating manuals for the actual modules being used.

The material presented in this chapter will be most useful to the researcher who encounters a coincidence circuit for the first time, either as a user of an existing circuit or as the person responsible for setting up a circuit. This viewpoint has been taken because, due to variations in the technical details associated with each individual coincidence circuit, it is unlikely that a general guide can be written to cover the detail required by the experienced user. Thus, it is hoped that the discussion presented in this chapter provides a useful starting point for the researcher who has not had previous experience with coincidence circuits.

References

1. J. A. Tanis, E. M. Bernstein, W. G. Graham, M. P. Stockli, M. Clark, R. H. McFarland, T. J. Morgan, K. H. Berkner, A. S. Schlachter, and J. W. Stearns, Resonant electron transfer and excitation in two, three, and four electron $_{20}Ca^{q+}$ and $_{23}V^{q+}$ ions colliding with helium. *Phys. Rev. Lett.* 53, 2551, 1984.
2. J. A. Tanis, G. Schiwietz, D. Schneider, N. Stolterfoht, W. G. Graham, H. Altevogt, R. Kowallik et al., Evidence for electron correlation during double capture in fast ($v \sim 10$ a.u.) collisions. *Phys. Rev. A* 39, 1571, 1989.
3. J. A. Tanis, E. M. Bernstein, M. W. Clark, S. M. Ferguson, and R. N. Price, Target ionization accompanied by projectile electron loss in fast $O^{6,7+}$ + He collisions. *Phys. Rev. A* 43, 4723, 1991.
4. O. Woitke, P. A. Závodszky, S. M. Ferguson, J. H. Houck, and J. A. Tanis, Target ionization and projectile charge changing in 0.5–8 MeV/q Li^{q+} + He ($q = 1,2,3$) collisions. *Phys. Rev. A* 57, 2692, 1998.
5. A. Simon, A. Warczak, T. ElKafrawy, and J. A. Tanis, Radiative double electron capture in collisions of O^{8+} ions with carbon. *Phys. Rev. Lett.* 104, 123001, 2010.
6. ORTEC Application Note, *AN 34 Experiments in Nuclear Science, Alpha, Beta, Gamma, X-ray, and Neutron Detectors and Associated Electronics: Introduction to Theory and Basic Applications* (2nd edition). ORTEC, Oak Ridge, Tennessee, 1976.
7. D. P. Cassidy, E. Y. Kamber, A. Kayani, and J. A. Tanis, Dissociative and non-dissociative charge-changing processes in 1.0–2.0 MeV/u O^{5+} + O_2 collisions. *J. Phys.: Conf. Ser.* 163, 012054, 2009.
8. T. Elkafrawy, A. Simon, and J. A. Tanis, X-ray processes associated with projectile charge changing in $\sim$1 MeV/u collisions of O^{5+} on Ar. *J. Phys.: Conf. Ser.* 194, 082033, 2009.
9. M. Nofal, S. Hagmann, Th. Stöhlker, D. H. Jakubassa-Amundsen, Ch. Kozhuharov, X. Wang, A. Gumberidze et al., Radiative electron capture to the continuum and the short-wavelength limit of electron-nucleus bremsstrahlung in 90 MeV U^{88+} ($1s^2 2s^2$) + N_2 collisions. *Phys. Rev. Lett.* 99, 163201, 2007.

8

Isoelectronic Trends in Atomic Properties

Alan Hibbert

CONTENTS

8.1 Introduction

In the literature, there are many instances of calculations being done at an equivalent level of accuracy for many ions of an isoelectronic sequence. It is then possible to establish patterns emerging along the sequence. But in some cases, marked deviations from smooth patterns occur, perhaps for a few neighboring ions, or even for a single ion in the sequence. It is important to know what the "normal" pattern should be, so that these deviations can be identified and explained, often in terms of configuration interaction or some other cancellation effect. Alternatively, and particularly, when calculations are done for just a single ion or a few neighboring ions, an isoelectronic pattern might be used to check the accuracy of such isolated results.

In this chapter, we will establish some trends that might be expected along isoelectronic sequences and use them to explain some results which might at first sight seem surprising or unexpected.

8.2 Perturbation Expansions in Inverse Powers of Z

In order to establish isoelectronic trends, it is useful to undertake a perturbation theory analysis of the Hamiltonian. We will give the results of calculations for two-electron ions, but the same theory applies to many-electron ions as well.

For a two-electron atomic system, the Hamiltonian is

$$H = \left(-\frac{1}{2}\nabla_1^2 - \frac{Z}{r_1} \right) + \left(-\frac{1}{2}\nabla_2^2 - \frac{Z}{r_2} \right) + \frac{1}{r_{12}}. \tag{8.1}$$

Now by substituting $r_i = s_i/Z$, then we obtain

$$H = Z^2 \left(-\frac{1}{2}\nabla_1^2 - \frac{1}{s_1} \right) + Z^2 \left(-\frac{1}{2}\nabla_2^2 - \frac{1}{s_2} \right) + \frac{Z}{s_{12}}, \tag{8.2}$$

where derivatives in ∇_i^2 are now with respect to s_i. Hence

$$H = Z^2\mathcal{H} = Z^2 \left[\left(-\frac{1}{2}\nabla_1^2 - \frac{1}{s_1} \right) + \left(-\frac{1}{2}\nabla_2^2 - \frac{1}{s_2} \right) + \frac{1}{Z}\frac{1}{s_{12}} \right]. \tag{8.3}$$

In general, for an N-electron atomic system, the Hamiltonian can be expressed as

$$H = Z^2\mathcal{H} = Z^2 \left[\sum_{i=1}^{N} \left(-\frac{1}{2}\nabla_i^2 - \frac{1}{s_i} \right) + \frac{1}{Z} \sum_{i<j} \frac{1}{s_{ij}} \right]. \tag{8.4}$$

The parameter $1/Z$ becomes a natural perturbation parameter.

8.2.1 Perturbation Theory

A possible way of solving Schrödinger's equation is by means of expressing the Hamiltonian as a sum of two terms. The first is a simplified Hamiltonian $\mathcal{H}_0$ which is easy to solve, and the second, $\lambda\mathcal{H}_1$, is a correction, where λ is a parameter included to denote smallness:

$$\mathcal{H} = \mathcal{H}_0 + \lambda\mathcal{H}_1. \tag{8.5}$$

If we now substitute the expansions

$$\varepsilon = \varepsilon_0 + \lambda\varepsilon_1 + \lambda^2\varepsilon_2 + \cdots \tag{8.6}$$

and

$$\Psi = \Psi_0 + \lambda\Psi_1 + \lambda^2\Psi_2 + \cdots \tag{8.7}$$

into Schrödinger's equation and equate powers of λ on the two sides, we obtain

$$\mathcal{H}_0\Psi_0 = \varepsilon_0\Psi_0,$$
$$(\mathcal{H}_0 - \varepsilon_0)\Psi_1 + (\mathcal{H}_1 - \varepsilon_1)\Psi_0 = 0,$$
$$\vdots \tag{8.8}$$
$$(\mathcal{H}_0 - \varepsilon_0)\Psi_n + (\mathcal{H}_1 - \varepsilon_1)\Psi_{n-1} = \sum_{i=2}^{n}\varepsilon_i\Psi_{n-i}, \quad n \geq 2,$$

and then, after some rearranging,

$$\varepsilon_0 = \langle\Psi_0|\mathcal{H}_0|\Psi_0\rangle,$$
$$\varepsilon_1 = \langle\Psi_0|\mathcal{H}_1|\Psi_0\rangle,$$
$$\varepsilon_2 = 2\langle\Psi_0|\mathcal{H}_1|\Psi_1\rangle, \tag{8.9}$$
$$\varepsilon_3 = \langle\Psi_1|\mathcal{H}_1 - \varepsilon_1|\Psi_1\rangle,$$
$$\vdots$$

These results are special cases of a theorem [1] that knowledge of the wave function to Ψ_n determines the energy to ε_{2n+1}.

8.2.2 Applications to Helium-Like Ions

If we choose, in the notation of Equation 8.3,

$$\mathcal{H}_0 = \left(-\frac{1}{2}\nabla_1^2 - \frac{1}{s_1}\right) + \left(-\frac{1}{2}\nabla_2^2 - \frac{1}{s_2}\right), \tag{8.10}$$

$$\mathcal{H}_1 = \frac{1}{s_{12}}, \tag{8.11}$$

$$\lambda = \frac{1}{Z}, \tag{8.12}$$

then we can solve $\mathcal{H}_0 \Psi_0 = \varepsilon_0 \Psi_0$ exactly to give, for the *ground* state,

$$\Psi_0 = R_{1s}(r_1) R_{1s}(r_2) \tag{8.13}$$

$$\varepsilon_0 = -1, \tag{8.14}$$

where R_{1s} is the radial function of the *hydrogen* 1s orbital. Then

$$\varepsilon_1 = \frac{5}{8}. \tag{8.15}$$

If we solve the first-order equation for Ψ_1 and evaluate ε_2 and ε_3, we find that

$$\varepsilon_2 = -0.157666 \tag{8.16}$$

and

$$\varepsilon_3 = 0.008699. \tag{8.17}$$

Then the energy expression is

$$E = Z^2 \varepsilon = Z^2 \left[-1 + 0.625 \frac{1}{Z} - 0.157666 \frac{1}{Z^2} + 0.008699 \frac{1}{Z^3} + \cdots \right] \tag{8.18}$$

$$= -Z^2 + 0.625Z - 0.157666 + 0.008699 \frac{1}{Z} + \cdots. \tag{8.19}$$

See [2] for additional terms in the expansion.

A similar analysis of the Hartree–Fock equations was undertaken by Linderberg [3]. The zeroth-order function is identical to that for the exact Schrödinger equation. Hence ε_0 and ε_1 are the same.

Linderberg obtained

$$E^{\mathrm{HF}} = -Z^2 + 0.625Z - 0.111003 - 0.001055 \frac{1}{Z} + \cdots. \tag{8.20}$$

Hence the *correlation energy* is a slowly varying function of Z:

$$E^{\mathrm{corr}} = E^{\mathrm{exact}} - E^{\mathrm{HF}} = -0.046663 + 0.009754 \frac{1}{Z} + \cdots. \tag{8.21}$$

For example, when $Z = 2$, E^{corr} is approximately -0.0418 a.u.

When the zeroth-order function is not a pure hydrogenic function, such as in the case of the ^1S ground state of beryllium-like ions, where it takes the form of a configuration interaction function

$$\Psi_0 = a_1 \Phi_1 (1s^2 2s^2) + a_2 \Phi_2 (1s^2 2p^2) \tag{8.22}$$

with 1s, 2s, and 2p orbitals being described by hydrogen functions, the value of ε_0 in the exact and Hartree–Fock expansions are the same (the sum of the hydrogen energies of the orbitals), but the values of ε_1 are different in the two expansions. This is because in the Hartree–Fock approximation, the near-degeneracy effect of Equation 8.22 is omitted; the HF state is represented by the single configuration $\Phi_1(1s^2 2s^2)$. In such cases, the correlation energy expansion is not dominated by the Z-independent term, but rather by a *linear* dependence on Z.

8.3 Screening Parameters

If we use a trial function for the 1s orbital in neutral helium (or helium-like ions) in the form

$$R_{1s}(r) = k \exp(-\alpha r), \tag{8.23}$$

where k is a normalization constant, then the standard textbooks show that the variational principle gives the optimal α as $\alpha = Z - (5/16)$, and the energy is $-(Z - (5/16))^2$, in atomic units (Hartrees).

This is what we would obtain if we treated the helium-like ion as consisting of two independent electrons moving in a Coulomb field with potential $-(Z - (5/16))/r$. Effectively, the nucleus is partially screened by the electrons, so that the effective nuclear charge "seen" by one electron is reduced (by 5/16) due to the presence of the other electron.

This *screened hydrogenic* approximation is often used as a starting point for the Hartree–Fock iterations. If we consider the Z-dependence of the Hartree–Fock energy of the ground state of helium-like ions, we can write

$$E^{HF} = -Z^2 + \frac{5}{8}Z - 0.111003 - 0.001055\frac{1}{Z} + \cdots$$

$$= -\left(Z - \frac{5}{16}\right)^2 - 0.013347 - 0.001055\frac{1}{Z} + \cdots. \tag{8.24}$$

So the screened hydrogenic approximation accounts for the first two terms of the energy expansion and most of the third (constant) term. It is, therefore, useful to consider an expansion of various atomic properties in terms, not of the bare nuclear charge Z, but rather in terms of a screened nuclear charge $(Z - \sigma)$, where σ is known as the *screening parameter*. The values of σ will normally be different for different atomic states, but frequently transitions occur between valence electrons of atomic systems and these tend to have rather similar values of σ. So it is possible to talk meaningfully about expanding transition probabilities or oscillator strengths in powers of $(Z - \sigma)$.

8.4 Z-Dependence of Atomic Properties

8.4.1 Transition Energies in LS Coupling

We have already seen that we can write energies of atomic systems in the form

$$E = Z^2 \left[\varepsilon_0 + \frac{1}{Z}\varepsilon_1 + \frac{1}{Z^2}\varepsilon_2 + \frac{1}{Z^3}\varepsilon_3 + \cdots \right], \tag{8.25}$$

where ε_0 is the sum of the hydrogen energies of the orbitals in the main configuration.

Hence *transition energies* have a Z-dependence dominated by

Z^2, if the *n*-values of the orbitals are different (e.g., $1s^2 - 1s2p$)

Z, if the *n*-values of the orbitals are the same (e.g., $1s^2 2s^2 - 1s^2 2s2p$)

8.4.2 Expectation Values of Powers of *r*

If we make the substitution $s = Zr$, then the matrix elements of r^n with respect to different orbitals with radial functions R_1 and R_2 transform as

$$\frac{\int_0^\infty R_1(r)r^n R_2(r)r^2\,dr}{\int_0^\infty R_1(r)R_2(r)r^2\,dr} = \frac{(1/Z^{n+3})\int_0^\infty R_1(s)s^n R_2(s)s^2\,ds}{(1/Z^3)\int_0^\infty R_1(s)R_2(s)s^2\,ds} \tag{8.26}$$

The integrals, expressed in terms of s, are the same for all ions of an isoelectronic sequence. Putting $R_2 = R_1$, we see that $\langle r^n \rangle$ behaves as Z^{-n}.

8.4.3 Fine Structure Effects

The spin-dependent operators of the Breit–Pauli approximation produce a quite good representation of the fine structure splitting of terms in light to medium-sized atoms and ions. The dominant term is the *spin-orbit* interaction. The radial integral in this is of the form

$$\left\langle \frac{Z}{r^3} \right\rangle, \tag{8.27}$$

which therefore, from the previous section, behaves like Z^4.

The spin-other-orbit term, with operator

$$-\frac{\alpha^2}{2} \sum_{i \neq j} \left(\frac{\mathbf{r}_{ij}}{r_{ij}^3} \times \mathbf{p}_i \right) \cdot (\mathbf{s}_i + 2\mathbf{s}_j) \tag{8.28}$$

is dimensionally $[L]^{-3}$ and therefore scales as Z^3. For the most part, its effect is to reduce the size of the calculated fine structure splitting, but because of the stronger dependence on Z of the spin-orbit term, this reduction diminishes along an isoelectronic sequence (see, e.g. [4, Table III]).

The spin–spin term also contributes to the fine structure splitting. The operator is

$$\alpha^2 \sum_{i<j} \frac{1}{r_{ij}^3} \left(\mathbf{s}_i \cdot \mathbf{s}_j - 3 \frac{(\mathbf{s}_i \cdot \mathbf{r}_{ij})(\mathbf{s}_j \cdot \mathbf{r}_{ij})}{r_{ij}^3} \right). \tag{8.29}$$

Hence it also has linear dimensions of $[L]^{-3}$ and Z-dependence of Z^3. However, for most atomic systems, the contribution is two or three orders of magnitude smaller than the size of the contributions from the spin-orbit or spin-other-orbit operators, so that the trends are largely unchanged by its inclusion.

8.4.4 E1 Transition Rates and Oscillator Strengths

E1 transitions fall into two categories: (i) those which are allowed in LS coupling (from one LS term to another), and (ii) intercombination lines which become allowed only through the mixing of levels of the same J but different LS symmetries. The Z-dependence of transition rates or alternatively oscillator strengths is illustrated in the following four examples in Be-like ions. In all cases, the transition rate (in length form) can be expressed as

$$A = k\alpha^3 (\Delta E)^3 \left| \left\langle \Phi_1 \left| \sum_i \mathbf{r}_i \right| \Phi_2 \right\rangle \right|^2, \tag{8.30}$$

where Φ_i are the wave functions of the two states of the transition, α is the fine structure constant ($\approx 1/137$), ΔE is the transition energy, and k is independent of Z. The corresponding result for the oscillator strength is

$$f = k'(\Delta E) \left| \left\langle \Phi_1 \left| \sum_i \mathbf{r}_i \right| \Phi_2 \right\rangle \right|^2. \tag{8.31}$$

1. $2s^2\ ^1S - 2s2p\ ^1P^o$

   For this allowed transition, A reduces to the following expression where the terms explicitly dependent on Z are shown in detail.

   $$A = k''\alpha^3(\Delta E)^3 |\langle R_1|r|R_2\rangle|^2, \tag{8.32}$$

   where R_1 and R_2 are the radial functions of the orbitals involved explicitly in the transition. The preceding subsections lead to the conclusion that $\Delta E \propto Z$ and $\langle P_1|r|P_2 \rangle \propto Z^{-1}$. Combining these results, the transition rate $A \propto Z$. The oscillator strength contains ΔE rather than $(\Delta E)^3$, so that $f \propto Z^{-1}$.

2. $2s^2\ {}^1S - 2s3p\ {}^1P^o$

The only difference in this case is that ΔE is now proportional to Z^2 rather than to Z, so that $A \propto Z^4$ and f is approximately constant along an isoelectronic sequence.

3. $2s^2\ {}^1S_0 - 2s2p\ {}^3P_1^o$

These transitions arise because of the mixing, through the spin-dependent operators of the Breit–Pauli Hamiltonian shown in the previous subsection, of the ${}^3P_1^o$ and ${}^1P_1^o$ levels. The dipole operator is spin-independent and so the contributions to the transition rate come from the interaction between configurations with the same spin. Principally, in this case

$$A({}^1S_0 - {}^3P_1^o) = c^2 A({}^1S_0 - {}^1P_1^o), \tag{8.33}$$

where the ${}^3P_1^o$ wave function can be written approximately as

$$\Psi = b\Phi_1(2s2p\ {}^3P_1^o) + c\Phi_2(2s2p^1\ P_1^o) \tag{8.34}$$

and for fairly small values of Z, $b \simeq 1$.

The actual values of b and c form the eigenvector components of the Hamiltonian matrix

$$\begin{pmatrix} H_{11} & H_{12} \\ H_{21} & H_{22} \end{pmatrix}, \tag{8.35}$$

where $H_{ij} = \langle \Phi_i | H | \Phi_j \rangle$ and $H_{21} = H_{12}$.

Then approximately (i.e., when $|H_{12}| \ll |H_{11} - H_{22}|$)

$$\frac{c}{b} \simeq \frac{H_{12}}{H_{22} - H_{11}}. \tag{8.36}$$

The Z-dependence of these expressions is

$$\begin{aligned} H_{12} &= \langle \Phi_i | H_{so} | \Phi_j \rangle & \propto Z^4, \\ H_{22} - H_{11} &\simeq E(2s2p\ {}^1P_1^o) - E(2s2p\ {}^3P_1^o) & \propto Z. \end{aligned} \tag{8.37}$$

Hence $c \propto Z^3$ so that

$$A({}^1S_0 - {}^3P_1^o) \propto (Z^3)^2 \times Z = Z^7. \tag{8.38}$$

4. $2s^2\ {}^1S_0 - 2s3p\ {}^3P_1^o$

As with the allowed transitions, all that changes is that the transition energy ΔE is proportional to Z^2 rather than to Z. Hence for this transition, $A \propto Z^{10}$.

8.4.5 Forbidden Transitions

E1 transitions are between levels of opposite parity, and with J-values differing by at most 1. For other transitions, for example, between levels of the same parity or with J-values differing by more than 1, magnetic multipole operators Mλ or higher electric multipole operators Eλ take on a dominant role. The key factors which determine the overall size of the rates of such transitions are

$$A_{E\lambda} : \alpha^{2\lambda+1}(\Delta E)^{2\lambda+1}|\langle R_1|r^\lambda|R_2\rangle|^2 \tag{8.39}$$

and

$$A_{M\lambda} : \alpha^{2\lambda+1}(\Delta E)^{2\lambda+1}|\langle R_1|r^{\lambda-1}|R_2\rangle|^2. \tag{8.40}$$

Therefore, E2 and M1 transitions are between levels of the same parity, E3 and M2 transitions are between states of opposite parity, and so on. However, of transitions between the same two levels, those with the lowest power of λ tend to dominate over the other possibilities, because of the smallness of the overall factors $\alpha^{2\lambda+1}$. From the analysis of the previous subsections giving the Z-dependence of ΔE and $\langle r^n \rangle$ we find the following: for transitions between levels of different configurations,

$A_{E\lambda} \propto Z$ when R_1 and R_2 have the same n-value,

$A_{E\lambda} \propto Z^{2\lambda+2}$ when R_1 and R_2 have different n-values,

$A_{M\lambda} \propto Z^3$ when R_1 and R_2 have the same n-value,

$A_{M\lambda} \propto Z^{2\lambda+4}$ when R_1 and R_2 have different n-values,

while for transitions between the fine structure levels of the same configuration, with $\Delta E \propto Z^4$,

$$A_{E\lambda} \propto Z^{6\lambda+4} \quad \text{and} \quad A_{M\lambda} \propto Z^{6\lambda+6}.$$

In these expressions, if screening effects are taken into account, the powers of Z can be replaced by the same powers of $(Z - \sigma)$.

8.4.6 Examples of Trends

The data given by Hibbert et al. [4] for the Ne isoelectronic sequence allows us to test some of these derived trends.

Table 8.1 shows their calculated $2p^5 3s\ ^3P_2 - {}^3P_0$ fine structure separations. These two levels are chosen since they interact very little with other LS symmetries with the same J, whereas the $^3P_1$ level interacts strongly with $^1P_1$ and causes a substantial shift from the position of $^3P_1$ expected from the Landé interval rule. The trend follows very closely the predicted screened Z^4 trend:

$$\Delta E_{fs} = 0.37749(Z - 3.23554)^4, \tag{8.41}$$

where we have introduced the screening parameter of 3.23554.

TABLE 8.1

Fine Structure Separations of $2p^5 3s \, {}^3P_2 - {}^3P_0$ in Ne-Like Ions

Z	ΔE (cm^{-1})	Equation 8.41	Ratio
12	2212	2227	0.993
16	10,021	10,021	1.000
20	30,016	29,817	1.007
24	69,881	70,176	0.996
28	142,214	141,980	1.002
32	260,723	258,423	1.009
36	443,534	435,028	1.020

In Table 8.2, we show their f and A values for the allowed transition $2p^6 \, {}^1S_0 - 2p^5 3s \, {}^1P_1^o$. As expected from the above analysis, the oscillator strength is approximately constant, whereas the transition rate varies fairly steadily as $(Z - \sigma)^4$, with an approximate screening constant of $\sigma = 2.7$, particularly away from the neutral end of the sequence. Clearly, the fit is not so good as that seen in Table 8.1, but the overall trend is established.

8.4.7 Surprising Effects

It is clear from the above analysis that the Z-dependence of different types of transition varies considerably. This can have a significant effect on the calculated lifetimes of levels in ions. The classic example is the lifetime of the $1s2p \, {}^3P_2^o$ level in He-like ions. The main decay channel is an E1 transition to $1s2s \, {}^3S_1$, and for the lighter ions this is effectively the only contributor to the total transition rate whose reciprocal is the mean lifetime of the upper level. According to the above analysis, the transition rate grows linearly with Z. However, as Z increases, the M2 transition from $1s2p \, {}^3P_2^o$ down to the ground state $1s^2 \, {}^1S_0$ increases rapidly, because of its Z-dependence of Z^8. By $Z = 10$, the M2 rate is some 10% of that of the allowed E1 line, and this "forbidden" M2 transition becomes the dominant decay mode from $Z = 18$ onwards (see [5]).

TABLE 8.2

The $2p^6 \, {}^1S_0 - 2p^5 3s \, {}^1P_1^o$ Transition in Ne-Like Ions

Z	f	A (s^{-1})	$A/(Z - 2.7)^4$
12	0.236	2.93×10^{10}	3.92×10^6
16	0.205	2.68×10^{11}	8.41×10^6
20	0.146	7.86×10^{11}	8.78×10^6
24	0.116	1.73×10^{12}	8.41×10^6
28	0.125	4.24×10^{12}	10.35×10^6
32	0.128	8.52×10^{12}	11.56×10^6
36	0.128	1.51×10^{13}	12.28×10^6

A second example of higher multipole transitions providing an important contribution to lifetime determinations is found in the recent work of Lundin et al. [6] on the lifetimes of two metastable levels in Ar^+: $3s^2 3p^4(^1D)3d$ $^2G_{7/2,9/2}$. The natural decay route is through M1 and/or E2 transitions to the $3s^2 3p^4 3d\ ^2F,\ ^4F$ or $3s^2 3p^4 4s\ ^2P,\ ^2D$ levels. The use of these transitions alone resulted in calculated lifetimes of 4.50 s and 3.50 s for the $3s^2 3p^4(^1D)3d\ ^2G_{7/2,9/2}$ levels, respectively, in rather poor agreement with the experimental values of 3.0 ± 0.4 and 2.1 ± 0.1, respectively. However, when the E3 transition to the much lower ground $3s^2 3p^5\ ^2P^o_{1/2,3/2}$ levels, and also M2 to $3s^2 3p^5\ ^2P^o_{3/2}$, are included, these calculated lifetimes reduced to 2.70 s and 2.38 s, in good agreement with experiment. These higher multipole transitions contribute so significantly because their transition energies are much higher than those of the M1 and E2 transitions, even though the power of the fine structure constant involved is also higher.

8.4.8 Deviations from Z-Dependent Trends

The analysis in the preceding sections is based on the underlying assumption that the different LS terms of ions in an isoelectronic sequence remain well separated, compared, for example, with the breadth of the fine structure separations within any one term and that the interactions between different levels with differing LS symmetry but common J are relatively weak. The general trends which we have established will inevitably break down for high Z, not least because we have assumed an intermediate coupling regime for the angular momenta, whereas for higher Z, a jj coupling scheme will be necessary. An example of where the simple trends break down is shown in Table 8.3, again using the data of Hibbert et al. [4], this time for an intercombination line. Table 8.3 shows that, for the lower end of the sequence, the A values do indeed exhibit a $(Z - \sigma)^{10}$ trend, but this pattern is not followed further along the sequence. This trend was established (Section 8.4.4) on the assumption that the spin-orbit coupling of the $^3P_1$ with $^1P_1$ increased as Z^3. This is valid only when the conditions of Equation 8.36 are met. Clearly, the mixing cannot

TABLE 8.3

The $2p^6\ ^1S_0 - 2p^5 3s^3 P^o_1$ Transition in Ne-Like Ions

Z	A (s^{-1})	A/(Z − 2.7)^{10}	A/(Z − 2.7)^4
12	5.34×10^8	0.110	
16	1.99×10^{10}	0.115	
20	1.63×10^{11}	0.068	
24	5.71×10^{11}	0.029	2.774×10^6
28	1.10×10^{12}	0.010	2.685×10^6
32	1.98×10^{12}	0.004	2.687×10^6
36	3.40×10^{12}	0.002	2.765×10^6

TABLE 8.4

Energy Levels in Mg-Like Ions

Level	P IV	S V	Cl VI
$3p3d\,^3P_0$	281,425	346,816	411,117
$3p3d\,^3P_1$	281,331	346,663	410,835
$3p3d\,^3P_2$	281,166	346,456	410,409
$3s4p\,^3P_0$	254,888	347,878	451,721
$3s4p\,^3P_1$	254,937	347,929	451,872
$3s4p\,^3P_2$	255,048	348,124	452,190

grow indefinitely, and we find that by $Z \approx 24$, the mixing coefficient for the $^1P_1$ contribution to the $^3P_1$ level is fairly constant. Hence we would expect that, beyond this point, the rise of the A-value of this intercombination transition behaves as does that of the allowed transition, that is proportional to $(Z - 2.7)^4$. This is demonstrated in the final column of Table 8.3.

Another major deviation from the smooth trends we have established can occur when the energy order of configurations changes along a sequence. If we work in LS coupling, then for high Z, the energy order of configurations will follow the hydrogenic order. For example, those composed entirely of up to $n = 3$ orbitals will lie below those with even a single $n = 4$ orbital. For example, in Al-like ions, the states labeled $3s3p3d$ will for high Z have lower energies than, say, $3s^24s$. However, nearer the neutral end of the sequence this is not so, for example, for Ar V and even more so for Si II. This relative rearranging of the configurations can cause changes to the trends in oscillator strengths. Brage and Hibbert [7] have discussed the effect on the lifetimes of the $3s4p\,^3P_J^o$ lifetimes due to the interaction with the $3p3d\,^3P$ configuration for Mg-like ions. The energy levels (in cm^{-1}) are shown in Table 8.4 for three adjacent ions. In P IV, $3s4p\,^3P$ lies below $3p3d\,^3P$; in Cl VI, they are reversed in energy order; in S V, they are almost degenerate. The closeness of the $3s4p$ and $3p3d$ levels in S V causes a shift in the energy position of $3s4p\,^3P_1$ so that it comes very close to $3s4p\,^1P_1$, resulting in a strong mixing between the two levels, considerably enhancing the strength of the intercombination line from $3s4p\,^3P_1$ to the ground state, but there is no such enhancement for the ions P IV and Cl VI. Thus, the lifetimes of the $3s4p\,^3P_J$ levels show J-dependence, unlike the other two ions. This is displayed in Table 8.5.

TABLE 8.5

Lifetimes (ns) of $3s4p\,^3P$ Levels in Mg-Like Ions

Level	P IV	S V	Cl VI
$3s4p\,^3P_0$	1.12	0.627	0.286
$3s4p\,^3P_1$	1.12	0.487	0.284
$3s4p\,^3P_2$	1.12	0.602	0.280

8.5 Conclusion

The general Z-dependence which we have established for total energies, energy separations, and transition data allow us to explain the patterns which are apparent in isoelectronic trends. We have focused on these properties, but a similar analysis can be carried out for other atomic properties. We have also seen, through a small number of examples, that there are sometimes deviations from these general trends. The value of the trends is that they highlight where deviations occur, and this can be the first step in explaining the causes of such deviations.

References

1. A. Dalgarno and A.L. Stewart, *Proc. Roy. Soc. A* 238, 269, 1956.
2. J. Midtdal, *Phys. Rev.* 138, A1010, 1965.
3. J. Linderberg, *Phys. Rev.* 121, 816, 1961.
4. A. Hibbert, M. Le Dourneuf, and M. Mohan, *Atom. Data Nucl. Data Tables* 53, 23 1993.
5. G.W.F. Drake, *Astrophys. J.* 158, 1199, 1969.
6. P. Lundin, J. Gurell, L.-O. Norlin, P. Royen, S. Mannervik, P. Palmeri, P. Quinet, V. Fivet, and E. Biémont, *Phys. Rev. Lett.* 99, 213001, 2007.
7. T. Brage and A. Hibbert, *J. Phys. B: Atom. Mol. Opt. Phys.* 22, 713, 1989.

Part II

Investigations of Atomic Structure and Applications

9

Experimental Investigation of the Structure of Highly Ionized Atoms

Indrek Martinson

CONTENTS

9.1 Introduction

Contemporary atomic physics has been revitalized by several important experimental developments. The structures of neutral and singly ionized atoms can nowadays be very accurately determined by exciting the atoms using, for example, tunable lasers or synchrotron radiation. Another important line of research concerns the different atomic structure problems occurring in highly charged ions (HCIs), which has become one of the most dynamic areas of modern atomic physics. On the experimental side, powerful light sources such as laser-produced plasmas, magnetically confined fusion plasmas, and excited fast ions from particle accelerators have been further developed. At the same time, novel techniques, involving very highly stripped ions from advanced ion sources, have been successfully applied. A comparable development has taken place on the theoretical side. This includes a deeper understanding of the various physical effects that determine the structure of multiply charged ions, as well as the ever-increasing availability of great computational capabilities.

HCIs are of considerable basic atomic physics interest, because here a number of fundamental interactions, such as electron correlation and the effects of relativity, quantum electrodynamics (QED), and nuclear structure, may occur. Some of these effects can be nearly negligible in neutral atoms, but their influence increases strongly with the nuclear charge Z. Radiation from highly stripped ions is prominent in the solar corona and solar flares [1,2]. Furthermore, such ions may be abundant in plasma physics including fusion research, for example, by means of tokamaks. In such plasmas, the various ions can appear as unwanted impurities or deliberately introduced species for diagnostic purposes [3–5]. These two often quoted applications have largely stimulated the ongoing studies of the structure and interactions of highly stripped ions. Furthermore, the breakthroughs that have occurred in the research and development of x-ray lasers [6,7] are partly responsible for spectroscopic studies of some selected ionic systems, such as the Ne- or Ni-like ions.

Spectroscopic work of highly ionized atoms has been reviewed by Fawcett [8–10] and Edlén [11], whose last publication was a valuable summary which updated his famous handbook article [12] and a wealth of other instructive summaries. In addition to the excellent reviews by Sellin [13], Kononov [14], Drawin [15], Ivanov et al. [16], O'Sullivan [17], and Träbert [18], several other articles have also appeared [19–21]. Furthermore, there are also important books, which contain valuable information about the structure and spectra of HCIs [6,22–24].

9.2 Early Developments

It is well known that the spectra of multiply ionized atoms have been investigated for many years. More than 80 years ago, Bowen and Millikan, using a spark light source, observed two spectral lines, at 800 and 813 Å, which were related to Cl VII (six times ionized Cl) [25]. In the 1930s, decisive experimental developments were made at Uppsala, in the laboratory of Manne Siegbahn, where powerful spectrographs for the far-ultraviolet region were combined with an efficient light source (vacuum spark). By means of such equipment, Bengt Edlén, Folke Tyrén, and other young scientists were able to produce multiply charged ions and study their properties. For example, the spectra of the Na-like ions K IX – Cu XIX [26] and the Ne-like ions Cr XV – Co XVIII [27] were investigated. More than 20 times ionized atoms could be produced, for instance, Sb XXIII and Sn XXIV. The early studies were later reviewed by Edlén [28,29], who even mentioned that the work from 1930s was not pushed much further because "it seemed highly improbable that such HCIs would ever be found anywhere else on earth or in heaven" [29].

However, this assumption was slowly and systematically revised. An early indication came in 1939 when Bowen and Edlén [30] identified spectral lines in the spectrum of a star (Nova RR Pictoris) as the so-called forbidden transitions in Fe VII. At that time, this was the highest known degree of ionization observed in a cosmic light source. Some years later, Edlén [31] reported the startling discovery implying that atoms of 10–15 times ionized Ca, Fe, and Ni were present in the solar corona. From this observation, it could be inferred that the temperature of the corona was about 2×10^6 K, much higher than previously assumed.

9.3 Some Properties of HCIs

Highly ionized atoms have several properties that differ from those of neutral and lightly ionized atoms. Since the nuclear charge Z greatly exceeds that of the remaining electrons, the electrostatic attraction is predominantly central, which might mislead one to conclude that the structure will resemble that of neutral hydrogen. However, the high value of Z causes large relativistic shifts in the energies of all atomic orbitals that deeply penetrate the nuclear core. There also appear cases in which the magnetic effects become larger than the electrostatic ones, that is, "fine structure" splittings exceed electrostatic "gross structure" separations. The reason for this is as follows: Suppose that Z is the nuclear charge and s the screening parameter which is a measure of the screening of the nucleus by the passive electrons. It depends on the quantum number l but is approximately independent of Z. Then the energy difference between two levels scales approximately as $(Z - s)$ or $(Z - s)^2$ depending on whether the principal quantum number n of the two levels is the same or different, whereas the magnetic (spin-orbit) interaction energy is proportional to $(Z - s)^4$.

A similar change with increasing degree of ionization takes place for the rates of transitions between energy levels. In neutral and lightly ionized atoms, electric dipole ($E1$) or "allowed" transitions dominate. "Forbidden" transitions of other multipolarities, such as magnetic dipole ($M1$) or electric and magnetic quadrupole (*E2* and *M2*, respectively) are also possible but their transition probabilities are extremely low. However, the decay rates of forbidden transitions scale with higher powers of Z than those of *E1* processes, and situations can arise in which forbidden transitions dominate over allowed ones.

Moreover, the effects of QED, such as the Lamb shift (caused by the electron self-energy and vacuum polarization) which are small although of fundamental importance in neutral and lightly ionized atoms, also scale as a high power of Z – the first term is proportional to $(\alpha Z)^4$, where α is the fine-structure constant and they thus become quite substantial in highly ionized atoms.

The interplay of electron correlation, an effect of electrostatic interaction between the electrons, and relativistic as well as QED effects are thus particularly important in highly charged atoms. For example, the inner-shell electrons in such systems undergo relativistic correction which can greatly change the potential that the outer electrons move in, producing an indirect relativistic effect. Thus, by studying the inner electrons in a highly stripped atom, the specification of neutral atoms can be improved.

High-precision spectroscopic measurements are now available over a sufficiently large range of ionization stages that it is possible to treat the nuclear charge Z as a tuning parameter, which can be varied while all other properties of the system are held constant. This is the basis of the concept "isoelectronic sequence" in which data for ions with a given number of electrons are arranged in increasing order of Z.

For comprehensive discussions of theoretical methods, we refer to the excellent books by Sobelman [32], Cowan [33], Heckmann and Träbert [34], Froese Fischer et al. [35], Rudzikas [36], and Curtis [37].

9.4 Experimental Techniques

9.4.1 Light Sources

9.4.1.1 Sparks

In his review from 1963 about wavelength measurements in the vacuum ultraviolet (VUV), Edlén [38] mentioned that "for wavelengths shorter than 500 Å and for very high ionization stages a spark in a high vacuum remains the only practicable light source." Such a light source, from the 1920s and 1930s, was until the mid-1960s indeed the only practical facility for spectroscopic studies of HCIs. Here a capacitor of about 0.3–0.5 μF is connected to two electrodes in vacuum and a few mm apart. One of the electrodes is often a carbon rod, whereas the other also consists of the element to be studied. The capacitor is charged to high voltage, typically 70–80 kV, until breakdown occurs. The inductance in the circuit must be very low, if extreme ionization stages are investigated. Such were the light sources in the pioneering work [26–28].

This source was later complemented with the low-inductance vacuum spark [39]. Here the voltage was rather low, 10–20 kV, whereas the capacity was comparatively high 15–30 μF and the inductance typically 2–100 nH. With such a light source, Beier and Kunze [40] observed radiation from He- and Li-like Mo (Mo XLI and Mo XL). The vacuum spark is a relatively simple and inexpensive light source. Its main disadvantage is that the spectra typically obtained can sometimes be rather difficult to analyze, largely because several ionization degrees can be produced in the plasma during a discharge. The separation of these ionization degrees may present problems in the case of complicated spectra, for example, those of the iron-group elements. The varying reproducibility of the discharges could also cause questions. However, spark spectra were investigated in the 1980s by Aglitskii et al. [41] and Ishii and Ando [42] who examined He-like ions and Zr XIV, respectively. The technique can still be applied, especially when about 3–8 times ionized atoms are studied. See also Chapter 1 for more details on sparks.

An interesting method of "exploding wires" was used in the late 1970s by Burkhalter et al. [43,44]. Thin wires exploded by very sudden electric heating obtained with discharges of relativistic electrons. The plasmas produced exhibited transitions in He-like Ti and Fe, as well as Ni-like spectra of W XLII, Pt LI, and Au LII. At that time, these still belonged to the highest ionization stages ever observed in the laboratory.

9.4.1.2 Laser-Produced Plasmas

One of the most efficient and versatile ways of producing HCIs consists of using radiation from a powerful laser to obtain a hot plasma. Indeed, it has been stated that laser-produced plasmas are the most intense laboratory sources of extreme ultraviolet and x-ray emission. Such plasmas have been used in the spectroscopy of HCIs for more than 40 years. In one of the earliest experiments, Fawcett et al. [45] applied a ruby laser which gave pulses of 8 J energy with a half-width of about 15 ns. The laser output was focused to a spot on metallic targets (Fe and Ni) with a power density of about 10^{12} W cm^{-2}, which provided transitions in Fe XV, Fe XVI, Ni XVII, and Ni XVIII. Besides consisting of only a few ionization stages, the spectra showed no impurity lines (C, N, O, etc.) which may appear in spark spectra. Only a few charge states, which can be varied by changing the laser energy, simplify analyses of spectra. However, reference lines from well-known spectra may be quite valuable and even necessary for accurate wavelength determinations.

In the following years, pulsed lasers, for example, Nd(YAG) solid-state lasers which emit radiation at 1.06 μm wavelength have been generally used to produce the ions. Their power density can be quite high (10^{12} W/cm^2 or higher), and a hot and rapidly expanding plasma is formed. Comparatively modest laboratory lasers, with pulse energies of a few J and pulse lengths f 1–10 ns are already quite useful for producing ions with charges of +10 to +20. However, in the 1980s and 1990s, very highly ionized atoms were also produced and investigated. For example, Kononov et al. [46] could study the Na-like ions Cu XIX – Br XXV in this way. They even used wavelength standards (O III – O V) from a spark source for accurate measurements. With an energy of 160 J and a pulse length of 0.07 ns, Conturie et al. [47] could study the spectra Xe XLVI – Xe XLVIII. Furthermore, the work for Mg-like ions was extended to very high values of nuclear charge Z, that is, studies of spectra from Ge XXI to Cs XLIV by Ekberg et al. [48]. To produce such HCIs, the powerful (up to 4 TW) Omega laser facility in Rochester (USA) was utilized. Similar investigations have been carried out for many other sequences, for example, by extending the work to Na-like ions [49], as well as to Cu-like ions [50]. The plasma physics processes occurring in laser-produced plasmas have been discussed in several reviews, for example, Refs. [51,52].

9.4.1.3 Fusion Plasma Devices

Magnetically confined hydrogen (deuterium) plasmas, such as those in pinches, tokamaks, and stellarators, have for many years been used for the spectroscopy of HCIs. In an early such study, using the toroidal plasma device ZETA (UK), Fawcett et al. [53] observed transitions in highly ionized Ne, Kr, and Xe. Heavier elements can enter the hot plasma when it interacts with the surrounding walls of the vessel, discharge limiter, or divertor (the task of the latter is to remove the impurities). The tokamak is one of the most interesting light sources for the spectroscopy of highly ionized atoms [54,55]. The electron and ion temperatures in tokamaks can reach several keV (1 keV = 1.18 × 10^7 K), whereas the densities are orders of magnitude lower than in the light sources as discussed in Sections 1.2 through 1.5, 1.6.1.1, and 1.6.1.2. One of the most important properties of the tokamak light sources is that forbidden transitions (e.g., of the *M*1 or *E*2 type) can be observed. In denser light sources, collisional processes usually dominate over radiative decays. Much of our knowledge about forbidden lines originates from tokamak studies [56,57]. Interesting results for spectra of highly charged krypton, Kr XVII – Kr XXIX were obtained in the French TFR Tokamak in 1985 [58].

In the early 1980s, a new generation of tokamaks was introduced. These were TFTR (Tokamak Fusion Test Reactor) in Princeton (USA), JET in Culham (UK), and JT-60 in Naka (Japan). In these devices, the central electron and ion temperatures can exceed 10 and 30 keV, respectively. Together with an electron density of 10^{13} cm^{-3}, these tokamaks become comparable to solar flares and thus excellent tools for the spectroscopy of HCIs. However, in addition to these facilities (of which TFTR is no longer in operation), there exist several somewhat smaller tokamaks, for instance, D III-D in San Diego (USA), TEXT in Austin (USA), ASDEX in Garching (Germany), TEXTOR in Jülich (Germany), and Tore Supra in Cadarache (France).

A wealth of important results has been obtained over the years. Already in 1984, there was much enthusiasm about the contributions to atomic structure research that had been obtained with atomic spectroscopy of fusion plasmas, as eloquently expressed in the review by Fawcett [10] "It is difficult to overemphasize the value of tokamak as a source for atomic spectroscopy. In addition to facilitating direct observations, the tokamak has stimulated many vacuum-spark and laser-produced plasma studies of heavy-element spectra." One interesting example is the TEXT tokamak, where dedicated studies were carried out for a number of ions and isoelectronic sequences, including Mg-like [59], Al-like [60], and Si-like [61] ions. A review of the contributions to the structure of atomic ions which have been obtained by the spectroscopy of fusion plasmas has also appeared [62]. An interesting expert review of HCIs in tokamak plasmas is given by Hinnov [63], and there are also several additional reviews on these problems [3,5,54]. The parameters and various research programs at tokamaks have been described by Wesson [64]. A great deal of work is currently going on at tokamaks, in conjunction with other light sources, such as EBITs, and theoretical efforts to generate spectroscopic data of elements of ITER interest, such as tungsten (see Chapter 11).

9.4.1.4 Ion-Beam Techniques

Most of the spectroscopic techniques for HCIs including the methods so far described are often time-integrated. However, the ion-beam methods, which are also excellent for spectroscopic studies, add another important dimension, time resolution, to experimental investigations of HCIs.

In the early 1960s, Kay [65] and Bashkin [66] realized the atomic physics potential of fast ions from particle accelerators, for example, van de Graaff generators, which were primarily constructed for research in nuclear physics. The method of beam-foil spectroscopy (BFS) is based on directing fast monoenergetic ions from the accelerator through a very thin foil where the interaction with foil atoms leads to further ionization and excitation of the ions. The excited states decay on the downstream side of the foil, in high vacuum, and the emitted light can be recorded. The BFS technique is a universal method, and it has been applied to many different ions from H^- to U^{91+}. Thus, in 1994, Träbert [18] underlined that H-, He-, and Li-like spectra of U had already been studied in this way at Berkeley, Caen, and Darmstadt. A variety of ion accelerators have been utilized in BFS research. At low ion energies, typically less than 0.5 MeV, isotope separators are common, whereas the range 0.5–6 MeV can be covered with electrostatic accelerators (e.g., van de Graaff generators), the tandem versions of which may provide ion energies as high as 200–300 MeV. Even higher energies, up to several GeV, have been reached with a newer generation of heavy ion accelerators, such as the heavy ion linear accelerator UNILAC in Darmstadt and the BEVALAC accelerator facility in Berkeley. An interesting aspect of the BFS method is that multiply excited states in atoms and ions are profusely populated. Such states can be studied both by photon spectroscopy and electron spectroscopy, if they decay by autoionization. Another important class of levels that are prominent in beam-foil spectra are those with high n and l quantum numbers

(Rydberg orbits) in several times ionized atoms. Such levels have very large radii and are populated when the ions leave the foil. The properties of BFS are thoroughly outlined in Chapter 10, as well as in some earlier reviews [13,19–21]. Chapter 10 also discusses the measurements of lifetimes of excited states by BFS.

9.4.1.5 Advanced Ion Traps and Ion Sources

The introduction of electron beam ion trap (EBIT) facilities has strongly vitalized the spectroscopy of highly ionized atoms. Here a high-density beam of energetic electrons (1–200 keV) is directed along the axis of strong magnetic field. The space charge of electron beam traps ions, which will undergo stepwise ionization by collisions with the electrons. With the Livermore Super-EBIT, even U^{92+} has been produced. A review of this work was given by Beiersdorfer [67].

The physics and spectroscopy of EBITs is also discussed in detail in Chapter 2. In recent years, cooler and storage rings for beams of heavy ions have also been used for investigations of HCIs with great success (see, e.g., the reviews by Mokler and Stöhlker [68] and Larsson [69]). The properties, technical performance, and so on of these facilities will be discussed in detail in Chapter 10.

9.4.1.6 Astrophysical Light Sources

It has been known for more than 60 years that certain astrophysical objects emit radiation from HCIs. We have already mentioned the discovery of Fe VII lines in a star in 1939 [30] and Edlén's explanation of the solar corona lines [31]. However, observations of solar and stellar spectra in the VUV and soft x-ray regions (approximately in the wavelength interval 1–2000 Å) using rocket- or satellite-borne instrumentation were begun after World War II. These spectra showed many unidentified lines and provided a great impetus to the study of highly ionized atoms. Much of the work has been summarized by Fawcett [8–10], Doschek [70], Feldman [71], and Dupree [72].

More than 70 chemical elements have been identified in the solar spectrum (including the photosphere, chromosphere, and corona). The spectra of the solar photosphere mostly show absorption lines belonging to neutral and singly ionized species. Above the photosphere, the temperature first reaches a minimum and then increases to around 2×10^6 K in the solar corona. The electron density there varies between 10^{11} and 10^8 cm^{-3} and the light elements are completely ionized in the corona. In the case of iron, transitions in Fe XV – Fe XVII are pronounced there. Much higher ionization stages occur in solar flares where stored magnetic energy suddenly heats the plasma to the order of 2×10^7 K and there is an electron density of 10^{13} cm^{-3}. Elements up to Fe and Ni may become totally stripped in the solar flares [70,73]. Excellent solar spectra have also been reported from the Skylab mission [74].

Fairly recent space observatories such as the Hubble Space Telescope (HST) and the Solar and Heliospheric Observatory (SOHO)—launched in 1990 and 1995, respectively—have tremendously increased the quality of spectroscopic data of various astrophysical objects. In contrast to HST, which mainly records spectra from relatively low charged ions, the SOHO instruments cover wavelength regions where HCIs dominate. For instance, with the coronal diagnostic spectrometer (CDS), ions such as Fe^{15+}, Ni^{17+}, and Ca^{13+} have been recorded in the solar corona in the wavelength region 150–800 Å [75]. Observations with another instrument onboard SOHO, the SUMER (Solar UV Measurements of Emitted Radiation), have yielded excellent spectra of the solar corona and chromosphere, where about 40% of the observed lines were unidentified [76].

It can often be advantageous to combine a number of methods in order to get the best possible result. In an interesting case [77], beam-foil measurements, studies of spectra from tokamaks, and laser-produced plasmas resulted in the classification of over 40 spectral lines belonging to Fe X – Fe XIV spectra, of which 19 had also been observed in solar flare spectra but left unidentified [73].

9.4.2 Detection Systems

The light emitted by the various sources discussed above must first be dispersed with a suitable optical instrument. Highly ionized atoms may emit strong transitions in the x-ray region and a variety of x-ray instruments have therefore been used for such studies. High-resolution photon spectroscopy is mainly being carried out using monochromators or spectrographs equipped with concave gratings, since in the VUV the number of reflections has to be kept at a minimum. For spectral studies at higher wavelength, optical spectrometers or spectrographs are used. More detailed information about spectrometers is found in the reviews by Edlén [38] and Samson [78]. In early spectroscopic studies, photographic plates were commonly used. These were later followed and complemented by single-channel detectors, such as photomultipliers, channeltrons, and so on. More recently, photoelectric position-sensitive detectors such as multichannel plates (MCPs) and charge-coupled devices (CCDs) have widely replaced the older systems [79]. MCPs and CCDs are discussed in detail in the present volume (see Chapters 5 and 6).

9.5 Experimental Results for Energy Levels

In recent years, much work has obviously been performed to determine the structures of energy levels for highly ionized atoms. Only a fraction of these results can be discussed in this chapter. Following previous surveys by Edlén

[11,12,80–82], Fawcett [8–10], O'Sullivan [17], Martinson et al. [19–21] and others, we will now discuss the results with the emphasis and respect to iso-electronic sequences. Here data for ions with a given number of electrons are arranged in increasing order of Z. The various effects (e.g., interelectron repulsion, magnetic interactions, exchange interactions, and QED contributions) scale with different characteristic powers of $(Z - s)$.

9.5.1 Helium-Like Ions

Although they have been studied for many years, the spectra of helium-like ions continue to present interesting experimental and theoretical problems. Here the Schrödinger equation cannot be exactly solved. Various approximation methods [83–85] have resulted in very high theoretical accuracies which can only be equaled by the most careful experimental investigations, however. Many experimental studies have concentrated on the determination of QED effects, by observing transitions between low-lying levels, but also more extensive spectral analyses have appeared. As an example, Figure 9.1 shows the energy level diagram for He-like ions, which includes transitions from terms and levels with $n = 2$ to the ground term $1s^2\ ^1S_0$ and from the excited levels $1s2s\ ^1S_0$ and $1s2s\ ^3S_1$. In addition to the "allowed" $E1$ transitions, there is also a "spin-forbidden" $E1$ transition (from $1s2p\ ^3P_1$ to $1s^2\ ^1S_0$) which violates the $\Delta S = 0$ selection rule and some forbidden transitions (discussed later).

In 1981, Martin [86] pointed out that an extensive amount of experimental data was available for He I – C V, while the data were less complete for systems with higher Z. However, some years later, new results were reported for N VI [87] and O VII [88], based on analyses of laser plasma spectra and those from a tokamak at Oak Ridge, respectively.

Furthermore, experimental data for Na X – Ar XVII were also summarized and compared with theoretical predictions by Martin [86]. Feldman et al. [89]

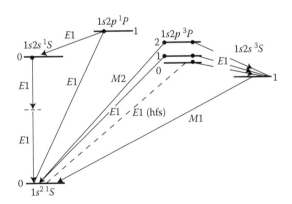

FIGURE 9.1
Allowed and forbidden transitions in He-like ions.

reported new experimental data for Si XIII and pointed out that these could even be relevant for the construction of x-ray lasers.

The wavelengths of the resonance transition $1s^2\ {}^1S_0 - 1s2p\ {}^1P_1$ have been very accurately determined by Beiersdorfer et al. [90] who studied He-like K, Sc, Ti, V, Cr, and Fe. Small differences between experiment and theory were noted which was concluded that additional theoretical analyses were needed. Much work has also been done to determine the wavelengths of the $1s2s\ {}^3S_1 - 1s2p\ {}^3P_{0,2}$ lines in He-like ions. Here the transition energies can be expressed as sums of several components, including QED contributions. However, here QED theory is more complicated than in the one-electron case, because electron–electron interactions must also be included. There are experimental studies of these triplet lines including the work by Beiersdorfer et al. [91] who used the EBIT facility in Livermore, USA. They recorded the $1s2s$ ${}^3S_1 - 1s2p\ {}^3P_2$ combination in He-like U^{90+} and found excellent agreement with theory. References to many earlier experiments, up to $Z = 54$ (Xe) can be found in the paper by Kukla et al. [92].

9.5.2 Ions with Three to Nine Electrons

The spectra belonging to the Li I – F I isoelectronic sequences have been quite extensively investigated over many years. The number of electrons is sufficiently low to permit quite accurate calculations of transition energies and also lifetimes. These systems with three to nine electrons have low configurations of the type $2s^m 2p^k$. Edlén has pointed out [80] that these configurations are especially important because transitions between various levels are strong in many light sources, including laboratory and astrophysical plasmas. A detailed knowledge of these levels is also necessary to firmly establish the higher levels of the individual term systems.

These structures are also of theoretical interest, providing good examples of configuration interaction and relativistic effects. Edlén [93,94] has made critical compilations of the experimental material, which originates from laboratory and astrophysical observations. Such data were compared with the results of extensive multiconfiguration Dirac–Fock (MCDF) relativistic calculations by Cheng et al. [95]. The difference Δ between the observed and theoretical energies σ was expressed by Edlén as

$$\Delta = \sigma_{\text{obs}} - \nu_{\text{th}} = A + B(Z - c)^{-1} + a(Z - s)^x.$$

Here A, B, c, a, and x were treated as adjustable parameters. The first two terms account for electron correlation, not included in the MCDF calculations, whereas the last term represents possible higher-order effects. In this way, the experimental data are smoothed and quite reliable interpolations and extrapolations to other values of the nuclear charge Z are possible. For all the transitions studied, Edlén derived Z-dependent functions to represent the relatively small differences between theoretical and experimental data.

From these functions, recommended and quite reliable values for transition wavelengths were obtained.

By combining laser-produced and tokamak spectra, Reader et al. [96] were able to study Li-like iron, Fe XXIV, obtaining accurate energy levels from the ground term $2s\,^2S$ up to $7d\,^2D$. In the case of Be-like ions, Denne and Hinnov [97] have determined the wavelengths of the $2s^2\,^1S_0 - 2s2p\,^1P_1$, resonance line and the $2s^2\,^1S_0 - 2s2p\,^3P_1$ intercombination line for several elements from Ti XIX to Kr XXXIII at the Princeton tokamak. The agreement with theory is very satisfactory. At the JET tokamak, transitions B-like Ni, Ge, Kr, and Mo as well as F-like Zr and Mo were studied [98]. In the case of B-like ions, the spectra of Ni XXIV, Ge XXVIII, Kr XXXII, and Mo XXXVIII were also studied, whereas of the F-like systems Zr XXXII and Mo XXXIV could be investigated.

While the $n = 2$ complexes are quite satisfactorily known, less information is available about the $n = 3$ and higher complexes in these sequences. In the case of Be-like ions, valuable theoretical calculations have appeared for the $n = 3$ complex, however [99]. Here, Edlén has made isoelectronic comparisons of the energies, from $Z = 4$ (Be I) till $Z = 23$ (Ti XIX), which agreed well with theory [100].

9.5.3 Heavier Systems with a few Valence Electrons

The structure of the ions belonging to Na I – Cl I isoelectronic sequences is more complicated than that of the Li I – F I sequences. There are additional configurations with $n = 3$ and the $3d$ electrons can complicate matters, but analyses of these structures are motivated by several factors. Besides the basic physics interest, such ions (especially Ti, Cr, Ni, and Co) appear in the spectra of solar corona and solar flares, and they are often present as plasma impurities in tokamaks and other fusion devices.

The Na I-like spectra have been investigated by Edlén for ionization stages as high as Mo XXII [101]. Here the structure is comparatively simple and the spectra can usually be interpreted without too much effort. Configuration interaction is small (because possible perturbing terms would involve inner-shell excitation. Very accurate expressions for level energies are now available. Some years later, Reader et al. [102] provided the $3s - 3p, 3p - 3d$, and $3d - 4f$ transitions for spectra up to Sn XL, using data from laser-produced plasmas and tokamaks.

For Mg-like ions, the $n = 3$ complex consists of the $3s^2$ ground configuration and the $3s3p$, $3p^2$, $3s3d$, $3p3d$, and $3d^2$ excited ones. In 1983, transitions from $3s3p$ and $3s3d$ levels have been observed up to $Z = 45$ (Rh) by Reader [103] who used a laser-produced plasma light source. In the following years, valuable material has been obtained for additional HCIs of this sequence. For instance, Ekberg et al. [104] provided energies for levels of the $3s3p$, $3p^2$, $3s3d$, $3p3d$, and $3d^2$ configurations up to Cs^{44+}. Work has also been extended to the $n = 4$ complex, the levels of which used to be fragmentarily known. Thus, using laser-produced plasmas, Kink et al. [105] were able to observe a large number of transitions between $n = 4$ and $n = 3$ levels in Mg-like Sc X – Fe XV.

As already mentioned, the structures of Mg-, Al-, and Si-like ions were studied by Sugar et al. [59–61] at the TEXT tokamak. However, these authors also investigated P-like [106], S-like [107], and Cl-like [108] ions between Cu ($Z = 29$) and Mo ($Z = 42$) in the same way. Complementary studies were carried out at JET by Jupén et al. for Na-, Mg-, and Al-like [109] and Si-like [110] Kr and Mo.

An electron structure that is rather simple for low values of Z in an isoelectronic sequence can become much more complicated as Z increases. The reason for this is that empty orbitals, for example, $3d$, may become partially filled for higher Z. An interesting example of this was demonstrated by Mansfield et al. [111], who investigated Mo XXIV, which belongs to the K sequence. The alkali-like structure $3s^2 3p^6 nl$ ($n = 4, 5, \ldots$) of K I and Ca II can change gradually by promotion of a $3p$ electron and later a $3s$ electron, resulting in configurations such as $3s^2 3p^5 3d^2$, $3s^2 3p^5 3d4s$, and $3s3p^6 3d^2$, all of which have many levels. Term analyses are then quite laborious. In Mo XXIV, several configurations which involve excitation of $3s$ or $3p$ electrons to the $3d$ shell were added. The K sequence has also been studied, for $Z = 25$–29, by Ramonas and Ryabtsev [112] and discussed by Edlén [10]. The spectrum is thus more difficult to analyze than the comparatively simple system of K I, as emphasized by Edlén.

9.5.4 Cu- and Zn-Like Ions

The structure of Cu- and Zn-like ions has been studied for many years, up to very highly charged ones. This activity was strongly encouraged by the observation of the $n = 4$, $\Delta n = 0$ resonance lines in Cu-like Mo XIV and Zn-like Mo XIII in the Princeton PLT tokamak [55]. The Zn-like ions, which are homologous to Mg-like ones, have two valence electrons outside the closed $3d^{10}$ shell. The two transitions, $4s^2\, {}^1S_0 - 4s4p\, {}^1P_1$ and $4s^2\, {}^1S_0 - 4s4p\, {}^3P_1$ (intercombination line) have been observed for a large number of elements up to U[61+] and Re[45+] [113,114], respectively. The experimental energies are in good agreement with relativistic calculations [115,116].

In analogy with the situation for Cu-like ions, the $4s^2\, {}^1S - 4s4p\, {}^1P$ resonance line in the Zn sequence has been followed through many ionization stages, up to Zn-like U [117], whereas detailed results, based on analyses of many transitions and covering wide spectral ranges have also appeared. For instance, Litzén and Reader [118] carried out a systematic spectral study of Rb VIII – Mo XIII. Higher ionization stages (Y X – Sn XXI) of Zn-like ions have also been investigated [119,120].

9.5.5 Heavier, Highly Ionized Atoms

The spectra of the iron-group elements (Sc – Ni) often show great complexity, because of the partially filled $3d$ shell. Here the lower ionization stages usually have the ground configurations $3s^2 3p^6 3d^k$ ($k = 1$–10) and higher configurations arise by promoting a $3p$ or $3d$ electron. In many cases, it can be fairly difficult to perform a complete analysis of anything but the lowest configurations, $3d^k$, $3d^{k-1} 4s$, and $3d^{k-1} 4p$. Here the three configurations, $3d^4$, $3d^{k-1}4s$, and $3d^{k-1}4p$ give rise to about 180 energy levels, practically all of which were established by classifying about 1000 spectral lines.

Very complicated spectra are observed in the case of highly ionized heavy elements. Two extensively studied systems are those with 46 (Pd sequence) and 47 (Ag sequence). For Pd-like ions, the ground state is $4d^{10}\,{}^1S_0$. Transitions from the $4d^9 5p$ and $4d^9 4f$ levels with $J = 1$ have been observed for a number of Pd-like ions from I VIII to Ho XXII by Sugar and Kaufman [121,122]. The Ag-like ions have one valence electron outside the closed $4d^{10}$ shell. For lower values of Z, the ground state is $4d^{10}5s$, the lowest excited one $4d^{10}5p$, while the $4f$ shell is empty. However, with Z higher than 61, the lowest configuration will instead be $4d^{10}4f$ and later also $4d^9 4f^2$ drives through the level system. It was shown by Sugar and Kaufman [123,124] that the structure of Ag-like W XXVIII is quite complicated because here the resonance multiplet ($4d^{10}4f - 4d^9 4f^2$) consists of a large number of transitions. Highly ionized tungsten was already observed in 1977 as a tokamak impurity at Oak Ridge [125] and Princeton [126].

Even heavier systems have been quite thoroughly analyzed, as can be seen from the reviews of Edlén [10], O'Sullivan [17], Fawcett [8–10], and others.

9.5.6 Intercombination and Forbidden Transitions

Intercombination or intersystem lines are electric dipole transitions that violate the $\Delta S = 0$ selection rule. A well-known example is the $1s^2\,{}^1S_0 - 1s2p\,{}^3P_1$ combination in He-like ions, made possible by the spin-orbit interaction which mixes the $1s2p\,{}^1P_1$ and ${}^3P_1$ levels (see Figure 9.1). The transition probabilities of intercombination lines are strongly Z-dependent, being proportional to $(Z - s)^7$ or $(Z - s)^{10}$ for $\Delta n = 0$ or $\Delta n = 1$ transitions, respectively. Here s is a screening constant (see above). (For allowed $\Delta S = 0$ transitions, the corresponding rates scale as $(Z - s)$ or $(Z - s)^4$.) Intercombination lines play an important role in low-density light sources and are useful for the diagnostics of laboratory and astrophysical plasmas, for example, in the determination of electron densities and temperatures. There are several reviews dealing with intercombination lines, for example, [127,128]. Of the various experimental methods available for studies of such transitions, BFS is nearly unique in providing wavelengths as well as lifetimes. This technique was applied to study highly charged iron and resulted in identifying intercombination transitions in Fe XIV and Fe XIII, with transitions between 440 and 510 Å [129]. Several of the lines studied have also been observed in the spectra of solar flares [73]. An interesting case was found by Jupén and Curtis [130] for the $3s^2 3p$ ${}^2P - 3s3p^2\,{}^4P$ intersystem multiplet in the spectra P III – Mo XXX of Al-like ions. This multiplet consists of five fine-structure components which have been observed for most ions between P III and Mo XXX, using a variety of techniques, including tokamak plasma spectroscopy.

The "traditionally" forbidden decay processes, such as $M1$, $M2$, $E2$, and two-photon decay $2E1$, which are often negligible in neutral atoms, can become quite important in highly ionized atoms. This is because their decay rates scale as $(Z - s)^6 - (Z - s)^{10}$, and thus much faster than those of the allowed transitions. Here s is the screening parameter. As already mentioned, the forbidden lines in astrophysics (mentioned in Section 9.1) are due to magnetic dipole ($M1$) transitions between fine-structure levels of a term [11,12].

In the solar corona, more than 50 forbidden lines have been classified. Some of these are $M1$ transitions within the ground configuration $2s^2 2p^k$ of Ca XII – Ca XV and $3s^2 3p^k$ of Fe X – Fe XIV and Ni XII – Ni XVI with k ranging from 1 to 5 [11,12]. Later also $M1$ transitions between excited metastable configurations, such as $3p^4 3d$ (Fe X, Ni XII) and $3p^5 3d$ (Fe IX, Ni XI) have been assigned to corona lines [131].

The solar corona and solar flares are similar to tokamak plasmas in the sense that they have high temperatures and low electron densities. It is therefore natural that "corona lines" are also observed in fusion plasmas. Thus, in 1978, Suckewer and Hinnov [56] identified a forbidden $M1$ transition belonging to Fe XX in the PLT tokamak at Princeton. From the Doppler width of this line, a record ion temperature (at that time) of about 50×10^6 K could be determined. Somewhat later, they reported $M1$ transitions in Ti, Cr, Fe, and Ni in the same tokamak [132]. A survey of such lines in tokamak plasmas was thereafter given by Edlén [133]. An $E2$ transition in Ni-like Mo XV has been observed in a French tokamak by Klapisch et al. [134]. A particularly interesting case of forbidden lines was reported by Träbert et al. [135,136]. Using the Livermore EBIT, they were able to observe $M1$, $E2$, and $M3$ of long-lived levels in the Ni-like ions Xe^{26+}, Cs^{27+}, and Ba^{28+}. The advent of EBITs has led to a wealth of spectroscopic information on the more exotic transitions such as $M2$, $E2$, $M3$, and so on. Thus, the work done by Träbert et al. [135,136] has also provided the lifetimes for the $M3$ decay in Ni-like Cs, Ba, and Xe, and hence being the first work to measure $M3$ lifetimes.

Several other studies of forbidden lines have, in recent years, been carried out with EBIT facilities. An interesting example concerns Ti-like HCIs (which have 22 electrons). Here the lowest term is $3d^4\ {}^5D$ with ${}^5D_0$ being the ground state. The $M1$ transition ${}^5D_3 \to {}^5D_2$ has a surprisingly small energy variation with the nuclear charge Z. Thus between Xe^{32+} and U^{70+}, the wavelength of this line stays in the interval 3200–4000 Å [137,138]. One would normally expect the fine-structure separations to increase strongly with Z. This anomalous behavior is expected to be important in the diagnostics of high-temperature plasmas.

Finally, it is quite obvious that experimental wavelength of forbidden lines provide important information about fine-structure separations. These quantities can be difficult to calculate accurately.

9.5.7 Multiply Excited States

States that involve the excitation of more than one electron may be populated in several light sources. In their spark spectra of carbon, Edlén and Tyrén [139] observed transitions of the type $1s^2nl - 1s2pnl$ in C IV as satellites to the C V resonance line, $1s^2\ ^1S - 1s2p\ ^1P$. It was later shown that such satellites are of fundamental importance for the diagnostics of astrophysical [70] and laboratory plasmas (e.g., tokamaks) [140] where they are mainly populated by dielectronic recombination but sometimes also by direct inner-shell excitation.

The first beam-foil studies of doubly excited states [141,142] showed that levels such as $1s2snl$ and $2s2pnl\ ^4L$ in Li-like ions could be studied with this technique. Indeed many such experiments have carried out over the years, including a study of Li-like N V where more than 20 terms, from ^4S to ^4G were established [143]. For instance, the transition $1s2s2p\ ^4P - 1s2p^2\ ^4P$ has been very carefully studied in several Li-like ions. Similar work has been done to investigate the multiplet $1s2s2p^2\ ^5P - 1s2p^3\ ^5S$ in Be-like ions. These multiply excited levels often show strong effects of electron correlation which clearly complicate theoretical analyses. For example, odd ^4P terms in Li-like ions may arise from several series ($1s2snp$, $1s2pns$, and $1s2pnd\ ^4P$). In addition to radiative decays, the ^4L states in Li-like ions may autoionize via the spin-orbit interaction mechanism which mixes quartet and doublet levels. The latter usually decay to the doublet continuum by electron emission, which can be investigated by electron spectroscopy [144].

9.5.8 Hydrogen-Like States

In many light sources, it is difficult to excite states with high n, l quantum numbers in multiply ionized atoms. The beam-foil method seems here to be an exception and transitions between such states appear quite strong in beam-foil spectra. The term values of such states can be expressed as

$$T = T_{\mathrm{H}} + \Delta_{\mathrm{P}}.$$

Here T_{H} is the hydrogenic value (corrected for relativistic effects) and Δ_{P} the polarization energy, usually expressed as

$$\Delta_{\mathrm{P}} = \alpha_{\mathrm{d}}R\langle r^{-4}\rangle + \alpha_{\mathrm{q}}R\langle r^{-6}\rangle.$$

Here α_{d} and α_{q} are the dipole and quadrupole polarizabilities of the core, R the Rydberg constant, and $\langle r^{-4}\rangle$ and $\langle r^{-6}\rangle$ the expectation values of the radial wave functions. There is a great interest in obtaining values for α_{d} and α_{q} in HCIs. Accurate wavelength measurements using beam-foil spectra have yielded much data on these quantities. Other experimental methods for investigating high n, l transitions are based on electron capture low-energy collisions between highly-charged ions and atoms. Most often the ions for such experiments are provided by electron cyclotron resonance (ECR) ion sources. A description of such ECR ions sources are given in Chapter 1.

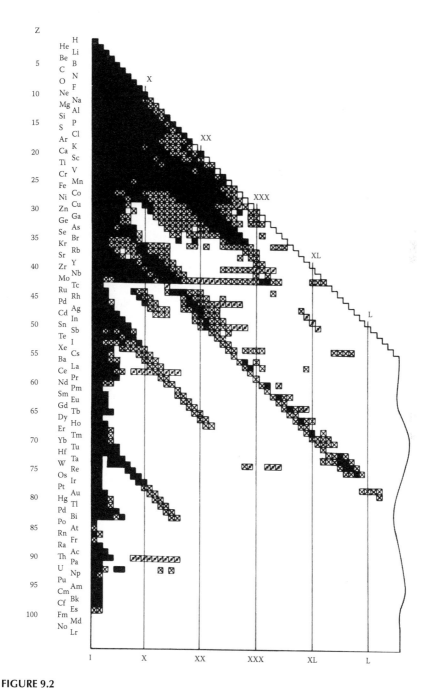

FIGURE 9.2

The present situation, at least as reported in 1999 concerning the spectra of atoms and ions. (Adapted from I. Martinson and I. Kink, In J. Gillaspy (Ed.), *Trapping Highly Charged Ions: Fundamentals and Applications.* Huntington: Nova Science Publishers, Inc., 1999, p. 365.)

9.6 Conclusion and Prospects

The spectroscopy of highly stripped ions is undoubtedly a very dynamic and important part of development. In this chapter, I have discussed the atomic spectra and structures of HCIs. As noted, these studies have been going on for many years which has resulted in an impressive number of publications and numerous monographs. Furthermore, immense progress has been possible in the late 1980s and 1990s, mainly because of the introduction of new experimental facilities such as ion sources, traps, and cooler rings. All this has resulted in an impressive number of publications and monographs. Only a fraction of this material has been mentioned here. The present situation concerning the spectra of atoms and ions is shown in Figure 9.2 (taken from [21]). This figure was developed at NIST, the US National Institute of Standards and Technology. At NIST, and its predecessor the National Bureau of Standards, NBS, there has been an Atomic Spectroscopy group for nearly 90 years, which still exists [138]. The scientists there are involved in production, evaluation, and compilation of atomic spectroscopic data. Now there exists the NIST Atomic Spectra Database (ASD; http://physics.nist.gov/asd).

For details, the valuable review by Wiese can be consulted [145].

There are also many conferences on the topic "Physics of Highly Charged Ions," of which the latest have been held in Vienna (Austria, 1994), Omiya (Japan, 1996), Bensheim (Germany, 1998), Berkeley (the United States, 2000), Caen (France, 2002),Vilnius (Lithuania, 2004), and Belfast (the United Kingdom, 2006).

Acknowledgments

I am grateful to Professors Laorenzo J. Curtis and Roger Hutton for valuable advice and criticism.

References

1. U. Feldman, *Phys. Scripta* 24, 681, 1981.
2. U. Feldman, G. A. Doschek, and J. F. Seely, *J. Opt. Soc. Am. B* 5, 2237, 1988.
3. R. C. Isler, *Nucl. Fusion* 24, 1599, 1984.
4. B. Denne and E. Hinnov, *Phys. Scripta* 35, 811, 1987.
5. N. J. Peacock, *Astrophys. Space Sci.* 237, 341, 1996.
6. H. F. Beyer, H.-J. Kluge, and V. P. Shevelko, *X-Ray Radiation of Highly Charged Ions*. Berlin: Springer, 1997.
7. R. C. Elton, *X-Ray Lasers*. Boston: Academic Press, 1990.
8. B. C. Fawcett, *Adv. At. Mol.* 10, 223, 1974.

9. B. C. Fawcett, *Phys. Scripta* 24, 663, 1981.
10. B. C. Fawcett, *J. Opt. Soc. Am. B* 1, 195, 1984.
11. B. Edlén, In H. J. Beyer and H. Kleinpoppen (Eds.), *Progress in Atomic Spectroscopy, Part D.*, New York: Plenum Press, 1987, p. 271.
12. B. Edlén, In S. Flügge (Ed.), *Handbuch der Physik*, Vol. 27. Berlin: Springer, 1964, p. 80.
13. I. A. Sellin, *Adv. At. Mol.* 12, 215, 1976.
14. E. Ya. Kononov, *Phys. Scripta* 17, 425, 1978.
15. H. W. Drawin, *Ann. Phys. Fr.* 7, 417, 1982.
16. L. N. Ivanov, E. P. Ivanova, and E. V. Aglitsky, *Phys. Rep.* 164, 315, 1988.
17. G. O'Sullivan, *Comments Atom. Mol. Phys.* 24, 213, 1992.
18. E. Träbert, *Nucl. Instrum. Meth. Phys. Res. B* 98, 10, 1995.
19. I. Martinson, *Rep. Prog. Phys.* 52, 157, 1989.
20. I. Martinson and L. J. Curtis, *Contemp. Phys.* 30, 173, 1989.
21. I. Martinson and I. Kink, In J. Gillaspy (Ed.), *Trapping Highly Charged Ions: Fundamentals and Applications*. Huntington: Nova Science Publishers, Inc., 1999, p. 365.
22. I. Sobelman, *X-Ray Plasma Spectroscopy and the Properties of Multiply-Charged Ions*. Commack, New York: Nova Science Publishers, 1988.
23. R. K. Janev, L. P. Presnyakov, and V. P. Shevelko, *Physics of Highly Charged Ions*. Berlin: Springer, 1985.
24. H. F. Beyer and V. P. Shevelko, *Atomic Physics with Heavy Ions*. Berlin: Springer, 1999.
25. I. S. Bowen and R. A. Millikan, *Phys. Rev.* 25, 295, 1925.
26. B. Edlén, *Z. Physik* 100, 621, 1936.
27. F. Tyrén, *Z. Physik* 111, 314, 1938.
28. B. Edlén, *Physica* 13, 545, 1947.
29. B. Edlén, *Phys. Scripta T* 3, 5, 1983.
30. I. S. Bowen and B. Edlén, *Nature* 143, 374, 1939.
31. B. Edlén, *Z. Astrophysik* 22, 30, 1942.
32. I. I. Sobelman, *Introduction to the Theory of Atomic Spectra*. Oxford: Pergamon Press, 1972.
33. R. D. Cowan, *The Theory of Atomic Structure and Spectra*. University of California Press, Berkeley, CA, 1981.
34. P. H. Heckmann and E. Träbert, *Introduction to the Spectroscopy of Atoms*. Amsterdam: North-Holland, 1989.
35. C. Froese Fischer, T. Brage, and P. Jönsson, *Computational Atomic Structure—An MCHF Approach*. Bristol: IOP, 1997.
36. Z. Rudzikas, *Theoretical Atomic Spectroscopy*. Cambridge University Press, Cambridge, 1999.
37. L. J. Curtis, *Atomic Structure and Lifetimes*. Cambridge University Press, Cambridge, 2003.
38. B. Edlén, *Rep. Prog. Phys.* 26, 181, 1963.
39. U. Feldman, M. Swartz, and L. Cohen, *Rev. Sci. Instrum.* 38, 1372, 1967.
40. R. Beier and H.-J. Kunze, *Z. Physik A* 285, 347, 1978.
41. E. V. Aglitskii, P. S. Antisferov, and A. M. Panin, *Opt. Spectrosc.* 58, 604, 1985.
42. K. Ishii and K. Ando, *J. Opt. Soc. Am. B* 3, 1193, 1986.
43. P. G. Burkhalter, C. M. Dozier, and D. J. Nagel, *Phys. Rev. A* 15, 700, 1977.

44. P. G. Burkhalter, R. Schneider, C. M. Dozier, and R. D. Cowan, *Phys. Rev. A* 18, 718, 1978.
45. B. C. Fawcett, A. H. Gabriel, F. E. Irons, N. J. Peacock, and P. A. H. Saunders, *Proc. Phys. Soc.* 88, 1051, 1966.
46. E. Ya. Kononov, A. N. Ryabtsev, and S. S. Churilov, *Phys. Scripta* 19, 328, 1979.
47. Y. Conturie, B. Yaakobi, U. Feldman, G. A. Doschek, and R. D. Cowan, *J. Opt. Soc. Am.* 71, 1309, 1981.
48. J. O. Ekberg, U. Feldman, J. F. Seely, C. M. Brown, B. J. MacGowan, D. R. Kania, and C. J. Keane, *Phys Scripta* 43, 19, 1991.
49. C. F. Seely, C. M. Brown, U. Feldman, J. O. Ekberg, C. J. Keane, B. J. MacGowan, D. R. Kania, and W. E. Behring, *At. Data Nucl. Data Tables* 47, 1, 1991 and references therein.
50. J. F. Seely, J. O. Ekberg, C. M. Brown, U. Feldman, W. E. Behring, J. Reader, and M. C. Richardson, *Phys. Rev. Lett.* 57, 2924, 1986.
51. M. H. Key and R. J. Cutcheon, *Adv. Atom. Mol. Phys.* 16, 202, 1980.
52. P. K. Carroll and E. T. Kennedy, *Contemp. Phys.* 22, 61, 1981.
53. B. C. Fawcett, B. B. Jones, and R. Wilson, *Proc. Phys. Soc.* 78, 1223, 1961.
54. E. Hinnov, *Phys. Rev. A* 14, 1533, 1976.
55. H. W. Drawin, *Phys. Rep.* 37, 125, 1978.
56. S. Suckewer and E. Hinnov, *Phys. Rev. A* 20, 578, 1979.
57. B. Denne, *Phys. Scripta T* 26, 42, 1989.
58. J. F. Wyart and TFR Group, *Phys. Scripta* 31, 539, 1985.
59. J. Sugar, V. Kaufman, P. Indelicato, and W. L. Rowan, *J. Opt. Soc. Am. B* 6, 1437, 1989.
60. J. Sugar, V. Kaufman, and W. L. Rowan, *J. Opt. Soc. Am. B* 5, 2183, 1988.
61. J. Sugar, V. Kaufman, and W. L. Rowan, *J. Opt. Soc. Am. B* 7, 152, 1990.
62. I. Martinson and C. Jupén, *Phys. Scripta T* 68, C123, 2003.
63. E. Hinnov, In R. Marrus (Ed.), *Atomic Physics of Highly Ionized Atoms*. Plenum Press, Plenum, New York, 1983, pp. 48–74.
64. J. Wesson, *Tokamaks*. Oxford: Clarendon Press, 1997.
65. L. Kay, *Phys. Lett.* 5, 36, 1963.
66. S. Bashkin, *Nucl. Instrum. Methods* 28, 88, 1964.
67. P. Beiersdorfer, *Phys. Scripta T* 120, 40, 2005.
68. Mokler and Th. Stöhlker, *Adv. Atom. Mol. Opt. Phys.* 37, 297, 1996.
69. M. Larsson, *Rep. Prog. Phys.* 58, 1267, 1995.
70. G. A. Doschek, In A. Temkin (Ed.) *Autoionization*. Plenum Press, Plenum, New York, 1985, pp. 171–256.
71. U. Feldman, *Phys. Scripta* 24, 681, 1981.
72. A. K. Dupree, *Adv. Atom. Mol. Phys.* 14, 393, 1978.
73. K. P. Dere, *Astrophys. J.* 221, 062, 1978.
74. K. Wilhelm et al., *Solar Phys.* 170, 75, 1997.
75. T. Kato, *Phys. Scripta T* 73, 98, 1997.
76. U. Feldman, W. Behring, W. Curdt, U. Schühle, K. Wilhelm, and T. M. Moran, *Astrophys. J. Suppl. Ser.* 113, 195, 1997.
77. C. Jupén, R. C. Isler, and E. Träbert, *Mon. Not. R. Astron. Soc.* 264, 627, 1993.
78. J. A. R. Samson, In W. Mehlhorn (Ed.), *Handbuch der Physik*, Vol. 31. Berlin, Springer, 1982, pp. 123–213.
79. J. Reader and N. Acquista, *J. Opt. Soc. Am.* 69, 1285, 1979.
80. B. Edlén, *Phys. Scripta* 7, 93, 1973.

81. B. Edlén, *Atoms, Molecules and Lasers.* Vienna, IAEA, 1974, pp. 93–131.
82. B. Edlén, In I. A. Sellin and D. J. Pegg (Eds.), *Beam-Foil Spectroscopy*, Vol. I. New York: Plenum Press, 1976.
83. G. W. F. Drake, *Phys. Rev.* A 19, 1387, 1979.
84. G. W. F. Drake, *Can. J. Phys.* 66, 586, 1988.
85. U. I. Safronova, *Phys. Scripta* 23, 253, 1981.
86. W. C. Martin, *Phys. Scripta* 24, 725, 1981.
87. A. M. Malvezzi, *Phys. Scripta* 27, 413, 1983.
88. R. C. Isler, C. Jupén, and I. Martinson, *Phys. Scripta* 47, 32, 1993.
89. U. Feldman, J. O. Ekberg, C. M. Brown, J. F. Selley, and M. C. Richardson, *J. Opt. Soc. Am.* B 4, 103, 1987.
90. P. Beiersdorfer, M. Bitter, S. von Goeler, and K. W. Hill, *Phys. Rev.* A 40, 150, 1989.
91. P. Beiersdorfer, S. R. Elliott, A. Osterheld, Th. Stöhlker, J. Autrey, G. V. Brown, A. J. Smith, and K. Widmann, *Phys. Rev.* A 53, 4000, 1996.
92. K. W. Kukla, A. E. Livingston, J. Suleiman, H. G. Berry, R. W. Dunford, D. S. Gemmell, E. P. Kanter, S. Cheng, and L. J. Curtis, *Phys. Rev.* A 51, 1905, 1995.
93. B. Edlén, *Phys. Scripta* 28, 51, 1983.
94. B. Edlén, *Phys. Scripta* 31, 345, 1985 and references therein.
95. K. T. Cheng, Y.-K. Kim, and J. P. Desclaux, *At. Data Nucl. Data Tables* 24, 111, 1979.
96. J. Reader, J. Sugar, N. Acquista, and R. Bahr, *J. Opt. Soc. Am.* B 11, 1930, 1994.
97. B. Denne and E. Hinnov, *Phys. Scripta* 35, 811, 1987.
98. R. Myrnäs, C. Jupén, G. Miecznik, I. Martinson, and B. Denne-Hinnov, *Phys. Scripta* 49, 429, 1994.
99. M. S. Safronova, W. S. Johnson, and U. I. Safronova, *J. Phys. B: At. Mol. Opt. Phys.* 30, 2375, 1997.
100. I. Martinson, *Nucl. Instrum. Methods Phys. Res.* B 43, 323, 1989. It includes Edlén's unpublished data for Be-like ions.
101. B. Edlén, *Phys. Scripta* 17, 565, 1978.
102. J. Reader, V. Kaufman, J. Sugar, J. O. Ekberg, U. Feldman, C. M. Brown, J. F. Seely, and W. L. Rowan, *J. Opt. Soc. Am.* B 4, 1821, 1987.
103. J. Reader, *J. Opt. Soc. Am.* 73, 796, 1983.
104. J. O. Ekberg, U. Feldman, J. F. Seely, C. M. Brown, B. J. MacGowan, D. R. Kania, and C. J. Keane, *Phys. Scripta* 43, 19, 1991.
105. I. Kink, M. Tunklev, and U. Litzén, *J. Opt. Soc. Am.* B 14, 722, 1997.
106. J. Sugar, V. Kaufman, and W. L. Rowan, *J. Opt. Soc. Am.* B 8, 22, 1991.
107. J. Sugar, V. Kaufman, and W. L. Rowan, *J. Opt. Soc. Am.* B 7, 1169, 1990.
108. J. Sugar, V. Kaufman, and W. L. Rowan, *J. Opt. Soc. Am.* B 6, 1444, 1989.
109. C. Jupén, B. Denne, and I. Martinson, *Phys. Scripta* 41, 669, 1990.
110. C. Jupén, I. Martinson, and B. Denne-Hinnov, *Phys. Scripta* 44, 562, 1991.
111. M. W. D. Mansfield, N. J. Peacock, C. C. Smith, M. G. Hobby, and R. D. Cowan, *J. Phys. B* 11, 1521, 1978.
112. A. A. Ramonas and A. N. Ryabtsev, *Opt. Spectrosc.* 48, 348, 1980.
113. J. F. Seely, J. O. Ekberg, C. M. Brown, U. Feldman, W. E. Behring, J. Reader, and M. C. Richardson, *Phys. Rev. Lett.* 57, 2924, 1986.
114. J. Sugar, V. Kaufman, D. H. Baik, Y. K. Kim, and W. L. Rowan, *J. Opt. Soc. Am.* B 8, 1795, 1991.
115. F. Seely and A. Bar-Shalom, *At. Data Nucl. Data Tables* 55, 143, 1994.
116. H.-S. Chou, H.-C. Chi, and K.-N. Huang, *Phys. Rev.* A 49, 2934, 1994.

117. C. M. Brown, J. F. Seely, D. R. Kania, B. A. Hammel, C. A. Back, R. W. Lee, A. Bra-Shalom, and W. E. Behrig, *At. Data Nucl. Tables* 58, 203, 1994.
118. U. Litzén and J. Reader, *Phys. Scripta* 36, 895, 1987.
119. F. Wyart, P. Mandelbaum, M. Klapisch, J. L. Schwob, and N. Schweitzer, *Phys. Scripta* 36, 224, 1987.
120. S. S. Churilov, A. N. Ryabtsev, and J.-F. Wyart, *Phys. Scripta* 38, 326, 1988.
121. J. Sugar, *J. Opt. Soc. Am.* 67, 1518, 1977.
122. J. Sugar and V. Kaufman, *Phys. Scripta* 26, 419, 1982.
123. J. Sugar and V. Kaufman, *Phys. Rev.* 21, 2096, 1980.
124. J. Sugar and V. Kaufman, *Phys. Scripta* 24, 726, 1981.
125. R. C. Isler, R. V. Neidigh, and R. D. Cowan, *Phys. Lett.* 63A, 295, 1977.
126. E. Hinnov and M. Mattioli, *Phys. Lett.* 66A, 109, 1978.
127. D. G. Ellis, I. Martinson, and E. Träbert, *Comments At. Mol. Phys.* 22, 241, 1989.
128. E. Träbert, *Phys. Scripta* 48, 699, 1993.
129. E. Träbert, R. Hutton, and I. Martinson, *Mon. Not. R. Astr. Soc.* 227, 27, 1987.
130. C. Jupén and L. J. Curtis, *Phys. Scripta* 53, 312, 1996.
131. B. Edlén and R. Smitt, *Solar Phys.* 57, 329, 1978.
132. E. Hinnov and S. Suckewer, *Phys. Lett.* 79A, 298, 1980.
133. B. Edlén, *Phys. Scripta T* 8, 5, 1984.
134. M. Klapisch, J. L. Schwob, M. Finkenthal, B. S. Graenkel, S. Egert, A. Bar-Shalom, C. Breton, C. De Micaelis, and M. Mattioli, *Phys. Rev. Lett.* 41, 403, 1978.
135. E. Träbert, P. Beiersdorfer, G. V. Brown, K. Boyce, R. L. Kelley, G. A. Gilbourne, F. S. Porter, and A. Szymkowiak, *Phys. Rev. A* 73, 2006.
136. E. Träbert, P. Beiersdorfer, G. V. Brown, S. Terracol, and U. I. Safronova, *Nucl. Instrum. Meth. Phys. Res. B* 235, 23, 2005.
137. F. G. Serpa, E. S. Meyer, C. A. Morgan, J. D. Gillaspy, J. Sugar, J. R. Roberts, C. M. Brown, and U. Feldman, *Phys. Rev. A* 53, 2220, 1996.
138. C. A. Morgan, F. G. Serpa, E. Takács, E. S. Meyer, J. D. Gillaspy, J. Sugar, J. R. Roberts, C. M. Brown, and U. Feldman, *Phys. Rev. Lett.* 74, 1716, 1995.
139. B. Edlén and F. Tyrén, *Nature* 140, 940, 1939.
140. C. De Michelis and M. Mattioli, *Rep. Prog. Phys.* 47, 1233, 1984.
141. W. S. Bickel, I. Bergström, R. Buchta, L. Lundin, and I. Martinson, *Phys. Rev.* 178, 118, 1969.
142. J. P. Buchet, A. Denis, J. Desesquelles, and M. Dufay, *Phys. Lett.* 28A, 529, 1969.
143. H. P. Garnir, Y. Baudinet-Robinet, P.-D. Dumont, E. Träbert, and P. H. Heckmann, *Nucl. Instrum. Meth. B* 31, 161, 1988.
144. N. Stolterfoht, *Phys. Rep.* 146, 315, 1987.
145. W. L. Wiese, *Phys. Scripta T* 105, 85, 2003.

10

Atomic Lifetime Measurements of Highly Charged Ions

Elmar Träbert

CONTENTS

10.1 Introduction

Atomic line spectra reveal atomic structure and the fact that energy is quantized. Only transitions between the "fixed" excitation levels of an atom are possible. However, not all combinations of levels occur in actual spectra and not all spectral lines are of similar intensity. These observations point to selection rules (invoking parity, angular momentum, spin, etc.) and to the concept of transition probability, or the "A factor" A_{ki} for a transition from level k to level i. The mean lifetime τ that appears in the exponential decay law of an excited level is the reciprocal of the sum of all transition probabilities from a given level:

$$\tau_k = \frac{1}{\Sigma A_{ki}}.$$

Atomic lifetime measurements on electric dipole (E1) transitions yield information on atomic wavefunctions that supplements the insights gained from atomic energy levels alone. The E1 transition rate depends on the transition energy and an extra power r, because the electric dipole operator er explicitly depends on r, the vector from the nucleus to the position of the electron. Transitions between fine structure levels of a given term ("forbidden transitions," magnetic dipole (M1), and electric quadrupole (E2) transitions) are supposedly insensitive to this, as they connect levels with similar radial wave functions. However, complex wave functions as well as relativistic effects in highly charged ions modify this simple picture.

Interpreting τ as the time constant of a damped oscillator, there are two ways to measure this parameter. Firstly, by the (Lorentzian) line width and secondly, by measuring the line intensity as a function of time and fitting an exponential curve to the data. (Lorentzian line profile and exponential decay curve are Fourier transforms of each other.) Classical spectroscopy is hardly sufficient to observe the natural line width of atomic levels. Narrow-band laser spectroscopy, however, has achieved this goal. Also, with (nanosecond-) pulsed laser excitation, it is nowadays quite feasible to selectively excite and then to measure the typical nanosecond lifetimes of most low-lying levels of neutral atoms and of some levels of singly charged ions. This option pertains to levels that decay by electric dipole (E1) radiation.

E1 transition rates of transitions which involve a change in principal quantum number n ($\Delta n \neq 0$) scale as Z^4. For multiply charged ions, the level lifetimes therefore are much shorter than those for neutral ions. They are out of the reach of lasers both because of the laser photon energy being insufficient to reach the excited levels and excite them selectively and because the decay time usually is too short for classical electronic timing measurements. An alternative is provided by fast ion beams that experience excitation when being passed through a thin foil [1]. The ions in the beam lose a fraction of their energy, but apart from that, the ion beam leaves the foil largely unharmed and continues its trajectory. Distance of the ions from the rear side of the foil translates into time after the end of excitation. Therefore, one can record the spatial decrease in the light intensity emitted by the ion beam as a function of distance from the foil and convert that to a time measurement in the picosecond to many-nanosecond range.

Atoms with levels that are particularly short-lived may have a natural line width that is greater than the Doppler and instrumental line widths that are typical for beam-foil spectroscopy (BFS). If autoionization is a direct competitor to the radiative decay one observes, then the intensity of the radiative branch and thus the signal rate suffers, of course. Fortunately, there are cases in which the lower level of a transition autoionizes and thus broadens the final level, whereas the radiative signal benefits from an unbranched radiative decay. Given the typical autoionization rates of the order of 10^{14} s^{-1}, the typical lifetimes studied in this way are in the range of a few femtoseconds (e.g., see [2,3]).

Electron beam ion traps (EBITs) intuitively suggest that lifetimes to be studied there would be long (else there would be no need for extended storage). We will discuss how such measurements of long lifetimes in the microsecond to many-millisecond range can be done with EBITs and electronic timing. We will also explain how short lifetimes (again in the femtosecond range) can be addressed by a line width measurement, and how all these techniques relate to other types of lifetime measurement, for example, by beam-foil spectroscopy. Reviews of measurements of long atomic lifetimes in various types of ion traps have been presented elsewhere [4–6].

With so much work already done, and theory providing atomic structure properties (including level lifetimes) cheaply (on the basis of atomic structure algorithms that need only seconds of run time on modern personal computers), is it really worth doing such experiments? Ever so often it has, indeed, been suggested to leave all further atomic lifetime work to theory. For the multitude (tens of thousands) of high-lying levels and the vast number (hundreds of thousands) of transitions between them that are required in viable collisional-radiative models which are used to provide line intensity references for specific plasma observations, theory cannot be replaced by experiment, because of the sheer quantity of the necessary data and the practical inaccessability of most decays. However, besides this vast amount of levels and transitions that are not needed to be known individually with high accuracy, there usually are some key transitions that need to be known well and that can be measured and should be measured as a benchmark for the theoretical model. Sometimes it is found that experiment has underestimated systematic errors, and sometimes theoretical results are clearly inappropriate. In the interplay of experiment and theory, both can and do evolve. Mutual challenges are very helpful in this context.

Occasionally, it has been suggested that theory is now so good that surely no further effort (money) should be wasted on atomic lifetime experiments. I have heard such comments from an eminent theoretician when I began to measure a particularly long-lived class of levels. Decades earlier, the same person had been named as living proof for the superiority of theory (and she has done excellent work). In my case, I was happy to report that I found— by way of experiment—some calculations by this eminent expert to be very good, indeed, and others far from satisfactory. The latter happened to be for transitions that this very person had stated as needing no experimental work, because the calculations were so good already. Much of what used to be atomic structure theory has migrated to theoretical chemistry, where thousands of molecules are handled in amazing detail by enormous computations. However, the techniques are different. A theoretical chemistry colleague heard of our (beam foil) intercombination transition lifetime measurements and offered to solve the problem by using their superb algorithms. The results took much longer than anticipated and never came close to our experimental data. An additional note of caution will be reflected in the "Examples" section: until very recently, the fine structure intervals in multiply charged multielectron ions were not reliably calculated by anyone. Seeing this, a customary resort was to replace the calculated energy splitting by the experimentally much better known one, before calculating the transition rate. Only in the last few years have *ab initio* calculations been successful in deriving such splittings and then transition probabilities for electric-dipole forbidden transitions without having to resort to "semiempirical" corrections. Such complex computations may take weeks on a present-day personal computer, including checks for convergence and reliability. And a last note: Calculations that come after the experimental fact should not be considered as having proven any predictive power. Key atomic lifetime experiments will continue to be of high value.

10.2 Experimental Techniques

The concepts of A values (Einstein coefficients), oscillator strengths f, and line strengths S are about a century old, predating the actual capability to measure atomic-level lifetimes. The first techniques aiming for this goal employed the absorption of light to derive oscillator strengths and exploited the insights that absorption (f_{ik}) and emission (f_{ki}) oscillator strengths between lower level i and upper level k are equal (but the statistical weights $g = 2J + 1$ have to be taken into account). Atomic absorption spectroscopy is still a viable analytic technique, and occasionally absorption may still be used to determine very small oscillator strengths. In astrophysics, absorption spectra dominate, but they are rarely, if ever, used to determine lifetimes. Nowadays, emission techniques dominate, especially so since photoelectric detection has enabled linear measurements of signal strengths over wide dynamic ranges. Progress in vacuum technology has made it possible to produce and, if necessary, store ever more highly charged ions. Progress in technical tools has made it possible to produce and ever more selectively excite such ions. A variety of techniques to measure atomic lifetimes has been developed over the years, but they all belong to just a few categories that exploit two aspects of radiative decay.

    A. The natural line width is related to the upper-level lifetime.

    B. If excitation suddenly stops, an excited level will decay exponentially, with the level lifetime τ appearing as the time constant of the exponential $dN/dt = AN_0\, e^{-(t-t_0)/\tau}$, where N_0 is the upper-level population at time $t = t_0$ and dN/dt is the decay rate.

Aspects (A) and (B) are mathematically equivalent, implying that a Lorentzian line profile (with a certain width) is the Fourier transform of an exponential function. In practice, line widths can only be measured if they are sufficiently large (very short-lived levels, say, in the femtosecond lifetime range). Exponential decays have been measured from the range of a few femtoseconds to the range of minutes, but this wide range cannot be measured by one and the same technique. The techniques actually used are mainly based on electronic timing (from the nanosecond range upward) or on spatial measurements on fast ion beams, translating lateral displacement to time via the speed of the ions.

Historically, some of the techniques have produced excellent results with ions in low charge states or with neutral atoms. However, referencing all the good work done would overburden the this chapter, and I will restrict my presentation to multiply and highly charged ions. I will thus leave out, for example, the excitation of atomic beams by a beam of electrons or laser light, pulsed excitation of gases by energetic electron beams, or a detailed description of various tricks employed in the laser excitation of beams of singly charged ions. In multiply charged ions, the first excitation step (from the ground state or a low-lying metastable level) is usually too large to be bridged by present-day lasers, but synchrotron radiation may step into the gap at some stage and, therefore, the principles will be discussed in Section 10.2.1. The very coarse ordering principle of the techniques presented in Section 10.2.1 is by the atomic-level lifetime range covered, using either ion beams or ion trapping, or both. Of the line width measurements there are very few, and I will discuss them in the context of the overarching beam-foil and EBIT work.

10.2.1 Fast Ion Beams: Beam-Foil, Beam-Laser, Beam-Gas-Laser, Beam-Foil-Laser, Line Width Measurement of Autoionization

Ion accelerators started out as devices for nuclear physics, sending energetic particles at others that often were exposed to the ion beam in the form of thin foils or as a coating on a thin foil of, for example, carbon. The target foil needs to be thin so that the energy loss of the fast ions inside the foil is not so large as to wash out the energy dependence of a nuclear excitation resonance or whatever is being measured. Reaction products are to be measured for their energy, which is another reason to keep target foils thin. This, however, implies that the original ion beam is minimally affected and largely travels on unharmed, that is, at about the same energy (speed) and in the same direction as before. In several laboratories, people noticed that there was light emission from the ion beam, but only two researchers [7–12] realized the atomic physics options that were opened by this beam-foil technique (sketched in Figure 10.1): excitation of whatever single elemental species, effected at an adjustable energy, and the time distribution of atomic decays being drawn out in space. (In the 1920s, Wilhelm Wien had had some key ideas along these lines, but his experimental plans were far ahead of the necessary vacuum and other technologies.)

In terms of atomic lifetime measurements, there would be a discussion of what time interval describes the excitation process, whether it is the transit time through the foil or any time scale associated with interactions near the rear surface of the foil. The only cleanly defined time interval seemed to be the transit time spent inside the foil. At a mass density of $10\,\mu g/cm^2$, a carbon foil

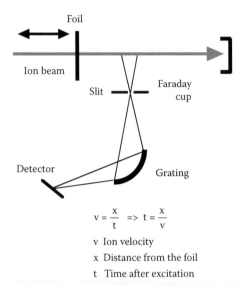

FIGURE 10.1
Schematics of beam-foil spectroscopy. (Reprinted with permission from E. Träbert, In S. M. Shafroth and J. C. Austin (Eds.), *Accelerator-based Atomic Physics—Techniques and Applications.* p. 567. Washington, DC. Copyright (1997), American Institute of Physics.)

is about 30 nm thick, and a typical ion beam of energy 0.5 MeV/amu (atomic mass unit), that is 500 keV protons or 14 MeV Si ions, needs about 3 fs to traverse this foil. At this time, no ions with valence electron lifetimes of this order of magnitude were of interest or produceable. However, the interaction of the ions with the foil material was a matter of discussion, whether the excitation process of swift ions was dominated by collisions with the atomic cores of the foil material atoms or with the bound or quasi-free electrons. This question was studied by observing x-rays from inside the foil and outside the foil, that is, emitted by ions with (short-lived) inner-shell vacancies. The vacancy lifetimes were estimated to be compatible to the transit time, and it was found that with thicker foils the x-ray yield increased. This effect was eventually exploited to model ionization and electron capture and to derive K vacancy level lifetimes in the femtosecond range [13]. This was a demonstration of principle, but had no chance to yield a precise measurement. A major obstacle lies in the fact that it was (and is) difficult to measure accurately the thickness (and density) of such foils (which are about 1/800th the thickness of ordinary writing paper).

Beam-foil spectroscopy has emerged as a generic name for many techniques that make use of fast ion beams for atomic physics (and "fast-beam spectroscopy" might be a less restrictive title). All of these use the interaction of the fast ion beam with a medium (solid or gas) to establish a charge state distribution and provide excitation. However, a laser might be used instead of the material target foil or in addition to it. I will start from the basic arrangement and then explain some of the variants.

In the basic scheme, an ion beam of whatever energy (traveling in a high vacuum environment) passes through a thin foil. Depending on the beam energy, the ions may capture or lose electrons, and a new charge state distribution emerges that depends mostly on the ion energy; its width may also depend on the foil material (for simplicity and handling, the standard material is carbon, but beryllium, aluminum, gold, and whatever else have also been employed). The ions experience a statistically distributed small energy loss and also some angular scattering. In the early years of beam-foil spectroscopy, when the ion beam energies were rather moderate, much effort was spent on characterizing nuclear versus electronic energy loss (from collisions with the screened nuclear Coulomb field of the target foil ions or with the quasi-free electrons), but at higher energies, the first part becomes negligible, as does large-angle scattering, the process by which ions deviate from the beam trajectory and get lost. Because of the mass ratio, the collisions with the light electrons are strongly peaked forward (small-angle scattering). The energy loss in matter (which is important also for various nuclear physics experiments) has been extensively studied, parameterized, and tabulated [14]. However, as stated before, the thickness of a foil cannot routinely be determined with good accuracy and, therefore, an energy-loss correction of the ion speed by a few percent may well itself be uncertain within a sizable fraction of the correction (say 10–20%). Moreover, under ion irradiation, crystallographic reordering can take place and change foil properties such as the areal density.

It is best to measure the ion energy after the interaction with the target. For light ion beams with energies up to a few hundred keV, a magnet or an electrical sector field can be employed to deflect the ion beam and to control the beam energy via feedback to the accelerator. For high-energy ion beams, this effort is impractical. In some accelerator laboratories that operate pulsed ion beams, time-of-flight techniques have been implemented. To make such measurements precise, a long flight path is required, which usually precludes implementation after the ions have passed through an exciter foil at the experiment. Since the beam velocity is a crucial part of the lifetime measurement using beam-foil excitation (see later this section), the uncertainty of the ion velocity makes for a serious limitation of the lifetime measurement precision. It is no surprise then that the decisive factor in the most accurate beam-foil lifetime measurement reported so far [15], on neutral helium atoms, has employed an in-beam technique to establish the time scale, in this case by concurrently observing a well-calculated quantum beat pattern in the decay of another level of the same atomic species, which serves as a built-in atomic clock. Such a quantum beat frequency, however, increases with the fine structure splitting, which increases with the fourth power of the nuclear charge Z. In highly charged ions, the spatial frequency of the quantum beats is therefore too high to be practically employed for reference there.

Following the above example of 0.5 MeV/amu, a typical ion beam speed is about 1 cm/ns (light travels 30 cm in a nanosecond), or 10 μm per picosecond. Owing to the constant velocity of the ions after leaving the exciter foil, distance from the foil is strictly proportional to time after excitation. A detector can simply travel along the excited ion beam and register the light intensity (suitably filtered or observed through a spectrometer; see Figure 10.1). The spatial curve relates to a development in time. A spatial displacement of the field of view by 1 μm (easily achieved mechanically) corresponds to a time interval of 100 fs, and no high-speed clock is required to measure atomic lifetimes of only a few picoseconds [16].

The minimum mechanical displacement interval, however, does not define the minimum lifetime that can be measured. This has several reasons. The ion beam usually has a diameter of several millimeters, and the exciter foil is not necessarily flat over the full beam cross-section. A detector viewing the ion beam sideways normally captures a divergent cone of light, not a parallel pencil of light, and thus very fine spatial structures of the decay curve are smeared out in the observation. In order to collect enough light so that the signal can exceed the detector noise, the field of view at the location of the ion beam should not be too narrow. This intrinsic integration over a section of the ion beam favors the observation of longer lifetimes over the very short ones, that is, of decays that die out before the ions have even passed through the full field of view. (For a similar reason, a wide field of view would integrate over many quantum beat oscillations, making the contrast suffer.) However, the width of the field of view influences the relative intensity collected of a decay component, but has a limited influence on the time resolution. For this, the detail of the function that represents the field of view matters, that is, the wings of the usually trapezoidal "window function" are essential for probing any spatial/temporal detail [17–20]. Taking this feature into account, lifetimes have routinely been measured down to a few picoseconds.

However, time resolution is a problem that limits the range of objects to study. The transition rate of electric dipole transitions between shells scales with Z^4. In highly charged ions, most of the levels that have such decay channels are too short-lived for practical measurements. These very transitions are basically the same as in hydrogen-like ions and thus they can be calculated with confidence. There is no expectation that any foreseeable measurement could reach the accuracy necessary to test the quality of the calculations. The interest in employing a window function in most cases is somewhat different from time resolution: The very fact of a structured detection zone distorts the purely exponential decay curve (usually a superposition of several exponentials, due to cascade repopulation). Any evaluation without a window function has to cut off the data channels of highest signal or suffer systematic error. With a complete curve analysis (taking the part that is affected by the window function into account), this waste is avoided, and the actual foil position (time zero) can be recovered as a bonus. This knowledge is helpful when evaluating the relative intensities of the various decay components and provides physically meaningful constraints in the modeling of complex decays.

A much more important problem than time resolution is posed by complexity. The very advantage of beam-foil spectroscopy, the fact that practically any atomic level can be excited in collisions under high-density conditions, has the attached downside of all of these levels de-exciting and thus repopulating lower levels the decay of which may be of immediate interest in a given measurement. Hence all beam-foil lifetime measurements deal with multiexponential decays. A number of techniques have been derived which aim at selective excitation of the level of interest (seeking a single-exponential decay curve, which is much easier to evaluate), as will be discussed later in this section. In straightforward BFS, there will be complex decay curves, but some of the experimental situations yield access to decay curves of manageable complexity, and these cases are of great practical importance.

Least-squares fitting of exponential curves to decay data is nonlinear (i.e., there is no mathematical inversion process that leads to a unique solution); a systematic variation of the parameters is undertaken to minimize the (quadratic) deviation of a (synthetic) fit curve from the data. In parameter space, minima of a hypersurface are being sought, for various models (one exponential plus background, two exponentials plus background, etc.), with success (small χ^2 value) being measured in relation to the statistical scatter of the data. It can be shown that certain combinations of amplitudes and lifetimes of only two decay components result in data curves that cannot be analyzed reliably. Even the combination of a single exponential with a flat detector background poses serious problems, if the decay curve does not comprise a sufficiently long section of background after the decay has died out. Decay curves well represented by three exponentials can be analyzed, if the time constants of the exponentials differ from each other by at least a factor of 3. Analyses with more than three exponentials usually remain ambiguous, unless one or more lifetimes and/or amplitudes can be constrained on the basis of other measurements or theoretical insights. It helps if the short-lived components have the relatively larger amplitudes—this situation corresponds to about equal initial-level populations.

The cleanest technique of analyzing decay curves for the lifetime of a specific level requires not only the measurement of the decay of that level, but also of all cascades into that level, so that they can be subtracted out from the decay curve of interest. This ANDC technique (arbitrarily normalized direct cascades) [21–23] does not need to measure the actual cascade transitions, but can work with other decay branches of the same feeding levels, because all decay curves of a given level have the same pattern (plus individual background levels of the individual detectors). In cases with two or three dominant cascades (such as the $nsnp\ {}^1P_1^\circ$ levels in Be- or Mg-like ions), the ANDC technique has been highly successful. In the face of cascade level lifetimes very close to that of the primary decay, the technique has recovered the correct primary lifetime with good accuracy. In contrast, a naive multiexponential fit of these curves returns a result that is systematically too long by 30–50%. The high fraction of BFS results with exactly this error situation illustrates the need for ANDC analysis whenever possible, as well as a historic lack of attention to atomic structure detailed in a number of laboratories which has become apparent in systematic error studies [24,25].

Unfortunately, ANDC is not always feasible, because several or all of the major cascades may occur in a different spectral range (e.g., x-ray vs. EUV) for which no spectrometer is available locally. Also, the cascade level lifetimes may be so short that only the cascade tail (a number of cascade steps away from the wanted direct cascade) can be measured with sufficient time resolution, reducing the veracity of the cascade input to ANDC. In such cases, cascade modeling based on semiempirical assumptions about level populations can provide an approximation that, at least, is better than a fit of too few exponentials to the original data. Cascade modeling would benefit from a small set of population parameters that describe the n- and l-dependence of the level population after ion–foil interaction. However, the search for generally valid simple population laws has not met with success.

Since the inter-shell transitions (see above) are so hydrogen-like, fast, and better calculated than measured, lifetime measurements using BFS have largely dealt with level lifetimes that, for one reason or another, are much longer. Examples are intercombination transitions in He-like ions (at rather low Z up to about $Z = 16$), the magnetic dipole decay of the $1s2s\,^3S_1$ level (from $Z = 16$ to 54), and the $2E1$, $M1$, and $M2$ decays of other $n = 2$ levels in one- and two-electron ions, up to $Z = 92$ [26]. Next, there are $\Delta n = 0$ resonance transitions in Li- (up to $Z = 92$ [27]), Na-, and Cu-like ions, resonance, and intercombination transitions in Be- [28], Ne-, Mg-, Al-, Si- [29], Zn-, Ga-, and Ge-like ions, and so on. A number of these sequences have been systematized by Curtis [30] who also has found ways to combine resonance and intercombination transition rates in a joint representation. Such work is very helpful in detecting inconsistencies and systematic errors and ultimately establishes the basis for consistent pictures of our knowledge of atomic structure and dynamics.

The measurement of these relatively long-lived level lifetimes is actually helped by cascades. The dominant cascade pattern in most cases may be seen as one of a single electron outside a core, that is, a hydrogenic model. The branching ratios of most decays favor a change of the orbital angular momentum quantum number l by -1, and the energy scaling favors a maximum change of principal quantum number n. This results in an evolution of the level population toward the yrast line of levels of maximum l for a given n. Once there, the further decays need many steps of $\Delta n = 1$ to reach the low-lying levels. In highly charged ions, the low-lying excited levels are very short-lived, while high-lying yrast levels can be very long-lived. Thus, a pattern emerges: the low-lying excited levels quickly repopulate the possibly long-lived levels one is interested in and practically empty the associated reservoir; this population boost enhances the signal that one observes from the decay of level of interest. However, there is always a tail of many slow yrast contributions, and it is worth including this tail in an analysis [31], although the individual cascade amplitudes are very small. The superposition of those many slow exponential contributions with their underlying steady progression of lifetimes can be described by a power law, and a time dependence such as $t^{-1.5}$ has been repeatedly observed [32].

The cascade problem in lifetime measurements could be avoided, if selective excitation of only the level of interest was possible. In atoms, of course, single or multiple laser excitation is a standard technique. There are also schemes for combining fast atom beams and lasers, producing the fast atoms from a beam of singly charged ions that capture electrons from a dilute gas target; such an isotopically pure beam of same-velocity atoms has certain advantages over experiments in which a laser is pointed at a gas cell. For rare gas atoms, the electron capture offers the excitation of metastable levels which are rather lying too high for most lasers. Visible laser light would then excite the atoms from there to resonance levels, and the subsequent decay to the ground state can be monitored almost free of background contributions in the vacuum ultraviolet. Such schemes have also been applied to multiply charged ions, for example, starting from the metastable $2s$ level of one-electron ions and seeking to induce by resonance with laser light the transition to one of the $2p$ levels, in the quest for accurate Lamb shift determinations. Short-wavelength light at high power levels is available at synchrotron light sources. In fact, in one experiment, synchrotron light was employed to ionize and excite Ar, and a lifetime of a level in Ar^+ was obtained with high accuracy [33]. Possibly this approach will, some day, reach multiply ionized species.

Only one experiment, however, appears to have combined foil-excitation with subsequent laser excitation of an ion in order to measure atomic-level lifetimes [34]. This experiment on two levels of singly charged N^+ demonstrates important points. The beam–foil interaction results in a shift and broadening of the velocity distribution in the beam, which makes it difficult to exploit high-resolution laser techniques. The level is being excited by the ion–foil interaction even without the help of the laser. The experiment is therefore less clean than one might imagine under "selective excitation": decay curves obtained without the laser are subtracted from decay curves obtained with the laser on resonance (each of which have statistical scatter). The result was a single exponential decay, as was hoped for, but it was difficult to achieve, facing such problems as the sensitivity of the frequency match for resonant excitation of fast ions (Doppler effect) whose velocity depends on (changing) properties of the exciter foil. In multiply charged ions, most level splittings exceed, by far, the photon energies of practical lasers and then the selective excitation by laser resonance is just out of reach. In highly charged ions, however, forbidden transitions in the ground configuration, or even the hyperfine splitting of the lowest levels of very highly charged few-electron ions, can be large enough so that laser techniques are of interest again, and will be discussed in the section on heavy-ion storage rings.

There is one very different lifetime measurement technique that involves fast ion beams, although the primary role is that of the ion–foil interaction which amply populates also multiply excited states. When observing the light emission of an ion beam at the rear surface of the exciter foil, especially in the EUV and x-ray ranges, there is almost a continuum of radiation from very short-lived core-excited ions. (The not so short-lived ones among them, e.g., the Li-like ions in the $1s2s2p\ {}^{4}P^{o}_{5/2}$ state, can be studied by standard foil-displacement measurement techniques.) Moving away from the foil, the spectrum rapidly gets cleaner, with fewer and fewer lines surviving. However, there exist curious cases of ions which decay to an autoionizing state which hence is extremely short-lived and thus broadened, and the lower-level broadening can be seen in the line width of the transition leading to the respective states. Fast-beam observations with a fast spectrometer (large solid angle of acceptance) usually suffer notable Doppler broadening, which for some spectrometer designs can be countered by refocusing [35–39] which, unfortunately, is likely to distort the line profile somewhat. Observations of the line broadening due to the shortening of a level lifetime by autoionization [2,3] had results in the ball park (10^{-14} s) that theory predicted; however, the remaining disagreement was never fully resolved as to which part might be blamed on experimental problems and which on the shortcomings of theory.

10.2.2 Slow Ion Beam: Recoil Ions

BFS encounters a physical limit when striving for the study of very long-lived levels, in the sheer size of the decay path associated with that. When Marrus and Schmieder [40–42] attempted to measure the about 200-ns lifetime of the $1s2s\ {}^{3}S_{1}$ level in the He-like ion Ar^{16+}, their 8 MeV/amu ion beam from SuperHILAC traveled about 8 m per atomic lifetime. It is difficult to control the geometry of an ion beam over such distances, with energy and angle straggling in the exciter foil adding to the problem. Moreover, a given decay happens only once per ion and thus the signal per unit of ion beam path is very low for long-lived levels. Under these circumstances, it is quite understandable that Marrus and Schmieder did not recognize a major systematic error, the presence of core-excited ions with a spectator electron [43], which caused their measurement to yield a lifetime 15% short of expectation. (Work by the same group years later found a result that agreed with calculations— theory also needed some sorting out.) Intrigued by the problem of such long level lifetimes, my own group made two attempts to employ slower beams. One attempt called for beam-foil production and excitation of the right charge state (requiring an ion beam energy of about 2–3 MeV/amu) and then substantially decelerating the ion beam (in this case, Cl^{15+}, with a calculated lifetime of about 400 ns) in a post-accelerator switched to deceleration mode. Such an energy variation by more than an order of magnitude had been demonstrated at the Heidelberg MPI-K Institute, but our atomic physics experiment missed the time window before the TSR storage ring was being built (which later on

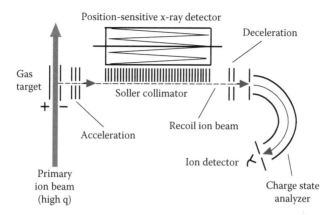

FIGURE 10.2
Slow ion beam ion–photon coincidence schematics. (Adapted from E. Träbert, in S. M. Shafroth and J. C. Austin (Eds.), *Accelerator-based Atomic Physics—Techniques and Applications.* Washington, DC: American Institute of Physics, 1997, p. 567.)

permitted to measure lifetimes many orders of magnitude longer). Our other approach [44–46] employed a beam of highly charged ions at GSI Darmstadt. A primary beam of highly charged uranium ions at an energy of 1.4 MeV/amu was passed through an Ar gas target where it produced a wide range of mostly low-energy (few-eV) recoil ions. With a weak electric field, ions were extracted sideways and then accelerated to about 1 keV per charge (Figure 10.2). The only long-lived level capable of emitting x-rays was the $1s2s\ {}^3S_1$ level in the He-like ion Ar^{16+}, and its x-ray decay was monitored by a 10-cm long position-sensitive detector along the recoil ion beam path—a detector idea proposed decades earlier by Mowat. The recoil ion beam was then charge-separated in a magnetic sector field, and the x-rays that were detected in (delayed) coincidence with the correct charge state ions were counted, the events being stored as a function of position (corresponding to flight time) in the x-ray detector. The pilot experiment reached an accuracy of about 5%; its results were compatible with later fast-beam work at Berkeley. Incidentally, photon–ion coincidences have been applied very recently with fast ion beams, too: the lifetime of the hyperfine-quenched $1s2p\ {}^3P_0$ level in the He-like ion of Au ($Z = 79$) was measured at GSI Darmstadt, separating the ion charge states after the photon observation, and cleaning the x-ray spectrum by filtering out only those x-ray events that coincided with the right ion charge state ions [47].

Nowadays, such an ion–photon coincidence measurement scheme for long-lived atomic lifetimes (in the microsecond range) would profitably be set up using the relatively slow ion beam available from an electron cyclotron resonance ion source or an electron beam ion source. Such an arrangement would yield experimental access to atomic lifetimes in between the beam-foil (straight fast ion beam) experiments discussed above, which can reach up to about 100 ns, and the ion trap experiments (discussed in Sections 10.2.4 and

10.2.5) which work best from many microseconds to many seconds, although measurements down to 700 ns [48] have been reported. Depending on the spectral range of the emitted radiation, the position-sensitive x-ray detector in the above scheme would have to be replaced by some detector for the EUV or UV range. Position-sensitive EUV detectors of the size of the x-ray detector may be rather expensive; the equipment cost could be lowered at the expense of measurement time by placing a regular detector (or a small, fast spectrometer) a traveling mount that can be displaced alongside the recoil ion trajectory. Such a scheme may remind the reader of the basic layout of BFS, but at velocities (specific energies) that are about two (four) orders of magnitude lower. The decisive point is the decoupling of the production of highly charged ions from the ion beam energy, whereas in BFS the two are strongly linked.

10.2.3 Stored Ion Beam: Heavy-Ion Storage Ring

At level lifetimes of a few hundred microseconds, fast ions travel about a kilometer during one such lifetime. A thread of yarn or a wire of great length are easily stored if rolled on a spool. The same can be done with fast ion beams. Bending the ion beam trajectory around to form an approximate circle, the same elements of the beam guidance system are used over and over again. From one turn to the next, the trajectory is slightly shifted, so that by this stacking procedure a beam current can be accumulated that (at TSR Heidelberg) is higher than the original one by up to a factor of 30. Then the injection into the storage ring vessel is stopped, and the beam can be left cruising, or be further accelerated, or be cooled by (partial) superposition of a cold (low-energy spread) electron beam. Depending on the ion charge state and energy, and, most importantly, on the extremely good vacuum (10^{-14} bar), the beam can last for seconds, minutes, or hours. The circumference of the TSR heavy-ion storage ring TSR at Heidelberg is 55 m, and the ions typically need a few microseconds per turn. Other such storage rings (ASTRID at Aarhus, CRYRING at Stockholm) are rather similar in size, ESR at GSI Darmstadt and the new Lanzhou facility in China are twice as large.

Various schemes of lifetime measurements have been employed at heavy-ion storage rings. Conceptually, the simplest is to use the ions as they are provided by the injecting accelerator (Figure 10.3). That machine uses a gas or foil stripper to produce ions in high charge states, and there are many levels of interest in such ions that have lifetimes way beyond the time the ion beam needs to travel to the storage ring (some 5 μs) and to settle down in a stable beam configuration after injection and stacking (less than 1 ms). All one has to do is to synchronize the data recording cycle to the ion injection and to store the signal of a stationary detector as a function of time. The ions will pass by the detector every few μs, and the signal rate will go down over time with the number of ions surviving in excited states. There are no moving parts and no geometry changes (as there are in BFS) during the measurement. Detectors for photons can be mounted outside the storage ring vessel, observing the

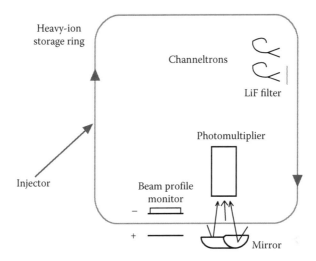

FIGURE 10.3
Layout of atomic lifetime experiments using passive observations at a heavy-ion storage ring.

ion beam through a window, or inside (for EUV radiation). In this way, a great number of intercombination transitions in light ions and $E1$ forbidden transitions in the ground configurations of Fe-group ions [49–53] have been studied at Heidelberg, with lifetimes from 0.5 ms [54] to some 50 s [55].

Most of these lifetimes are shorter than the process of electron cooling would require (several seconds at least), which is why electron cooling has not been used in most measurements. However, the electron beam of the electron cooler (or a second such device acting as an electron target) can be tuned away from the ion velocity to an energy that brings the electrons in dielectronic resonance (DR) with the ions. Then inner excitation can take place, and the ion can bind an electron and forms a multiply excited system that either reverts to its previous state (giving up the former beam electron) or stabilize radiatively. Then the ion has changed charge state and will be caught by a specific detector after the next bending magnet. The cross-sections for the DR process are so small that the DR signal acts as a level-specific, almost noninvasive, probe into the surviving population of excited ions circulating in the ring.

Another probe is provided by laser light that can be used to pump population from a long-lived level to a short-lived one that either decays back or to another level; in any case, the fluorescence is proportional to the population of the long-lived level. Laser excitation can be quite narrow-band, especially for co- and counterpropagating beams that combine to effect two-photon processes. The Doppler shift between these laser beams as seen by the fast moving ions can be very helpful in shifting the laser stray light out of the bandpass of the detector, or in Doppler shifting visible laser light into the near-UV for excitation. At the ESR storage ring at GSI Darmstadt, in the observation of the ground-state hyperfine structure of H-like $^{209}\mathrm{Bi}^{82+}$ ions, counterpropagating visible laser light of 489 nm was seen in the ion rest frame (at more than half the speed of light) as having a wavelength of about 243 nm [56–58]. Pulsed laser excitation resulted in delayed fluorescence with a time constant of nearly 0.3 ms.

Another way of employing a laser has been found at CRYRING (Stockholm), where, however, mostly singly charged ions are being studied. The laser pulse is so bright that it pumps all the population out of the long-lived level. A decay curve is built up over many experimental cycles; in each cycle, the laser-induced fluorescence signal at a given time after injection is obtained as a single data entry that tells us how much excited level population was left. In order to study repopulation processes (much more likely at the relatively low ion beam energies used at Stocholm than at the energies used elsewhere), the population may be quenched early, and then probed again later, checking whether there is a significant amount of ions in a given metastable level again [59].

In all these decay curve experiments, the observed decay rate represents the sum of the radiative decay rate and the ion loss rate from the stored ion beam, that is, the decrease of the ion beam current over time. The beam current can be monitored in various ways; a practical one is the use of the beam profile monitor that observes the spatial distribution of particles of the residual gas that have been ionized in collisions with the fast ion beam. A planar electric field draws these low-energy recoil ions to a microchannel plate detector with a position-sensitive read-out. The overall signal rate depends on the ion beam current and will dwindle with the beam. The slope of the current signal reveals the ion loss rate, since only one charge state species (one isotope, one beam velocity) is stored in the ring. After this technique had produced highly accurate lifetime data on a number of relatively low charge state ions, the quest continued with higher charge state ions. The low-charge state ions had usually be produced by gas stripping in the injector accelerator serving the Heidelberg storage ring. Higher charge states required the use of a foil stripper. The latest experiments suggest that then the bane of BFS, cascades from enormously long-lived higher-lying levels [60], may affect lifetime measurements at the storage ring, too.

10.2.4 Classical Ion Traps: Kingdon, Penning, Paul

The heavy-ion storage ring is, in fact, a large ion trap, with ions of energies of a few dozen keV to many MeV. There also are ion traps for less energetic ions. These are, for example, the cylindrical-symmetric electrostatic trap (Kingdon), the Penning trap in which a strong magnetic field constrains the ion motion across the field, and a static voltage on some drift tubes limits the ion motion along the magnetic field lines. In the radiofrequency (Paul) trap, a ring and two cap electrodes at a static voltage and a superimposed radiofrequency field form a quadrupole field that can trap ions. There are optimum geometries of the trap electrodes (hyperboloidal surfaces), and many simpler designs (all electrodes being part of a common cylinder, or a flat arrangement) that offer benefits of access or manufacture. There also are various other designs, for example, the electrostatic mirror [61,62]. All that is needed in the present context are the common principles, not the details. Since most traps operate with conservative fields, the ions of interest have either to be produced inside the trap to stay there, or the traps have to be opened to admit particles, and then be closed in order to constrain them. A problem then is to determine how many ions are being stored, or how large the loss rate is. The Zajfman trap [61, 62] relies on charge exchange (CX) processes that will ultimately lead to low charge states or even neutral particles that leak out through the electrostatic mirror toward a detector on the symmetry axis. Other traps are being opened after a while, and a detector outside captures a fraction of the ejectiles from which the surviving number of ions is being estimated—with very limited information on the charge state distribution in the sample. Laser probing of the population of metastable levels is an option, but only in ions that are not highly charged. There are limits because of the energy level structure, but also because it is difficult to cool highly charged ions sufficiently to take advantage of narrow-band lasers. Since the ions are confined spatially, but their energy is not very low (and thus the velocity distribution is not very narrow), ordinary lasers will not be intense enough to cause much fluorescence.

From a combination of many such factors, lifetime measurements on highly charged ions in classical ion traps are few, and in hindsight many of them appear to suffer from systematic errors larger than recognized. Most of these errors are being taken into account in a new generation of electrostatic ion storage rings (ELISA at Aarhus, DESIREE at Stockholm, and CSR at Heidelberg), some of which are to operate at extremely low temperature in order to obtain a near-perfect vacuum and to suppress the influence of black body radiation. The first such devices are dedicated to low charge state ion work (including simple molecules or even large biomolecules), but in the Heidelberg CSR project, operation with highly charged ions is also foreseen, and it will include lifetime measurements, for example, with the aim of finding out whether the absence of magnetic fields is important, or whether the extremely low pressure permits the measurement of very long atomic lifetimes (into the many minute range?). In a few years, we will know more.

10.2.5 Electron Beam Ion Trap

There is one variant of the classical Penning trap that is capable of operating with highly charged ions, and, in fact, it can do so with all charge states of all elements, as it breeds them internally. This is the EBIT [63–65] that combines the Penning trap principle with a strong, extremely well-collimated electron beam along the magnetic field (of typically 3–5 T field strength B), thus also defining an axis of symmetry. The electron beam is compressed by the field, to a diameter of about 60 μm. The Penning parts of the trap are completed by drift tubes on different potentials that keep ions in the trap volume axially confined. The electron beam serves several purposes: the electrons collisionally ionize atoms from the ambient gas or from a gas flow injected ballistically on purpose [e.g., a neutral gas stream of a density corresponding to a pressure of as low as 10^{-10} mbar may be crossing the electron beam trajectory under ultra-high vacuum (UHV) conditions ($<10^{-11}$ mbar)], or low-charge ions injected along the magnetic field lines from an external MeVVA ion source. Instead of a gas injector, neutral atoms may also be provided by laser-produced ablation of an external target. These freshly produced ions are then confined by the trap fields and can be hit by fast electrons over and over again. If the electron–ion collisions are sufficiently frequent (that is why the electron beam needs to be so tightly focused) and energetic, stepwise ionization to ever higher charge states can proceed. This process is moderated by the interplay of ionization, recombination, and charge-changing collisions with the residual gas (this determines the need for UHV). The charge state limit is given by the electron energy and the increasingly high ionization potentials of highly charged ions. The second job of the electron beam is a compensation of the space charge of the cloud of positive ions that is being built up in the trap. Even with a strong magnetic field for radial confinement, the ions would repel each other and move away from the location of the electron beam, if the attractive potential of the beam electrons and the space charge compensation were absent.

The electron–ion collisions ionize the target ions until the beam energy can no longer overcome the ionization potential. Therefore, the highest charge state in the cloud of trapped ions can be predetermined. Since the electron beam can be varied in current and energy, this offers ways to make lifetime measurements.

10.2.5.1 Line Width Measurements

As discussed above, the spectral resolution of some BFS observations has been high enough to measure the spectral line broadening that resulted from autoionization, a consequence of the uncertainty principle by which the extremely short level lifetime corresponds to an increased level width. The same lifetime range of a few femtoseconds applies also to electric dipole transitions of few-electron ions in the mid-Z range of the periodic table, which have transition wavelengths in the x-ray range. Measuring the line width in such ions takes several steps, beyond the provision of ions of the desired charge state. A spectrometer of sufficient resolving power is only one of the prerequistes. Ions produced by frequent collisions with multi-keV electrons have kinetic energies in the keV range, and this does not change by the ions being trapped. Trapping is improved by evaporative cooling, which is best achieved by mixing in a light ion species. The Doppler broadening of spectral lines emitted by keV-ions largely camouflages the underlying natural line width (Lorentz curve). Only when the trap was made very shallow, evaporative cooling of the stored ion cloud achieved to bring down the ion motion sufficiently to permit lines from long-lived levels to shrink substantially in width below that of short-lived levels. The pilot experiment on this topic [66] worked with Cs ($Z = 55$) and intercompared the widths of lines from $3s$ and $3d$ levels of the Ne-like ion Cs^{45+}. Similar to this Ne-like ion species, Graf et al. [67] have evaluated EBIT measurements of the He-like ion Fe^{24+}. Here the resonance line "w" was investigated in comparison to the intercombination line "y" and the forbidden line "z." Both of these studies yielded lifetime results within a factor of 2 of well-established predictions. It is not yet clear what limits the accuracy; one of the problems is the general difficulty of the analysis of a Voigt line profile (convolution of Lorentzian and Gaussian profiles). In order to determine the Lorentzian part well, the low-intensity line wings are important that, however, overlap with those of the neighboring lines.

10.2.5.2 Magnetic Trapping

When the electron beam contributes to the trapping, an EBIT is said to operate in electronic trapping mode. However, when the beam is switched off, a Penning trap remains, and EBIT is said to operate in magnetic trapping mode [68], with ion storage being effected for many seconds (as has been ascertained by ion cyclotron resonance frequency observations at LLNL) [68,69]. The vast majority of lifetime measurements at EBITs has been obtaining decay curves in the magnetic trapping mode. The basic arrangement is simple. A detector system views the center of the trap through slots in the drift tubes. For detectors that are sensitive to stray magnetic fields (most EBITs operate with superconducting magnets), such as photomultiplier tubes, an optical system images the trap onto the detector which sits at some distance [70] (Figure 10.4). The electron beam is switched on for a few hundred milliseconds or so, until the proper charge state ions have been bred and the signal level has risen to show this. Then the electron beam is switched off for the duration of several times the atomic lifetime of interest, so that the delayed emission signal

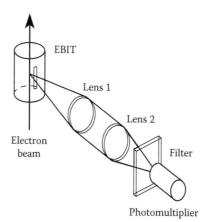

FIGURE 10.4
Optical observation at an EBIT. (From E. Träbert and P. Beiersdorfer, *Rev. Sci. Instrum.* 74, 2127, 2003. With permission.)

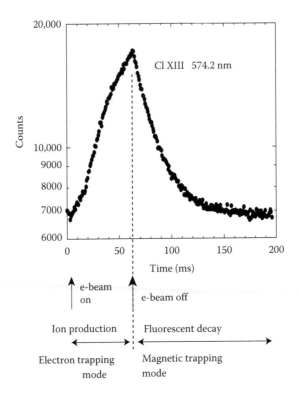

FIGURE 10.5
Timing pattern of a measurement with the "stovepipe" spectrometer. (From E. Träbert and P. Beiersdorfer, *Rev. Sci. Instrum.* 74, 2127, 2003. With permission.)

can reach the background level (Figure 10.5). Then the ions are purged from the trap, by lowering the potential of one of the drift tubes, and the cycle is repeated, thousands of times.

This basic scheme has experienced many variations; for some measurements of the metastable level in He-like ions, the electron beam has not been switched off completely, but only lowered in energy below the excitation threshold [71]. Some Heidelberg measurements strongly reduced the electron beam current, but did not fully cut it. The purge does not have to occur in each cycle, but not much is won by saving ions from a purge: the ion cloud expands once the electron beam is gone, and it does not shrink back to its previous size when the beam is restored.

10.2.5.3 Ion Loss Measurements/CX

The optical signal decay rates observed on the ion cloud in an ion trap represent the sum of the true atomic decay rate and of all other loss processes, notably the loss of ions from the stored sample. Hence it is of paramount importance to determine the ion storage time constant (the inverse of the loss rate). Only when the loss rates are known can the radiative transition rate be determined from the apparent decay rate. In contrast to heavy-ion storage rings, where only ions of a single charge state are being stored, so that a measure of the change of ion beam current is at the same time a measure of the change of ion number, most other ion traps may hold several ion species (elements and charge states) simultaneously. The number of ions of a given charge state may increase by CX collisions of a higher charge state ion with the residual gas. This effect has been seen drastically in lifetime measurements on He-like ions of Ne [72,73]. Those observations strongly suggest to make the charge state of interest the highest of the charge state distribution or, in other words, to keep the electron beam energy low enough so that no higher charge state is produced.

CX also reduces the number of ions of the proper charge state, and it is not easy to determine the loss rate. A general signature of CX is delayed photon emission by the ion that gained an electron, most often capturing into a high n, high l state. This excited ion will decay eventually, often in many steps. It is difficult to selectively observe the photons from typical steps along the decay chains, because the number density of CX-excited ions is small, and narrow-band optical detection of spontaneous emission is fairly inefficient. For a variety of ions, the optical determination of the CX loss rate is simply not practical. Only if the decay chain reaches the x-ray range, energy-dispersive detectors of large detection solid angle become applicable. Typical solid-state detectors have a resolution that does not distinguish individual charge states, and the x-ray energy detection threshold of about 1 keV is also limiting the information that can be gained. Assuming that the CX processes largely reflect the density of the residual gas, one might resort to storing more highly charged, few-electron ions under the same vacuum conditions, and use their x-ray emission to determine the CX loss rate. This transfer of information gained on, say, ions with a K-shell vacancy to others with an open L-shell is based on the assumption that the CX cross-sections follow simple scaling rules, which they are very likely to follow for ions with a constant number of electrons. The interpretation of L-shell processes in terms of K-shell processes, however, is burdenend with uncertainties. Modern microcalorimeters, in contrast to the traditional x-ray diodes, have a lower detection threshold (several hundred eV) and a much higher resolution (say, 5 eV). With such a device, the CX signal of an individual charge state ion species can be followed, reducing the associated systematic error.

As an alternative control measurement, the residual gas density in the trap can be varied. This should change the CX rate accordingly, which provides an *in situ* probe. Neglect of this test has been suggested [74] as a possible reason to explain the unsatisfactory results of some earlier experiments.

10.2.6 Examples

The lifetimes of more than 80 levels in multiply charged ions have, by now, been measured with trapping techniques, after hundreds were measured by

BFS. In the process, the calculations of radial wave functions and of multiplet mixing, nondiagonal matrix elements, relativistic effects on wave functions and transition operators, multipole expansions of the radiation field, configuration interaction, hyperfine interaction, multipole mixing, approximations of the Breit operator, autoionization decay, spin-orbit, spin–spin, spin-other-orbit interactions, and whatever else were tested for their influence on atomic level lifetimes. More often than not, several entries of this list act in combination, and their individual influences cannot always be determined by experiment. The best chance to distinguish the contributions usually is the study of isoelectronic sequences, because of the different Z-dependences of the individual effects. The following examples are roughly ordered by increasing the number of electrons, be they the total number or the number in the valence shell.

Electric dipole transitions in He-like ions have been studied up to U ($Z = 92$), including lifetime measurements that were almost exclusively performed by beam-foil techniques. In He-like U, the x-ray decay of the $1s2s\,^3S_1$ level is very fast; however, there is a slower in-shell $E1$ cascade from one of the $1s2p$ levels. Transition rates depend on the transition energy, and in this case the transition energy contains a major fraction from QED effects, the Lamb shift. Thus a lifetime measurement has revealed the then most accurate information on the $n = 2$ Lamb shift in a He-like very highly charged ion [26].

The aforementioned $1s2s\,^3S_1$ level decays by M1 radiation, a process that can be understood only as a relativistic effect on the transition operator. In fact, before the decay was actually recognized in astrophysical spectra [75], it had—on theoretical grounds—been declared not to exist. Such an M1 transition rate scales with Z^{10}, and it is then no surprise that the level lifetime is too short to be measured in U^{90+}, although the calculated level lifetime in neutral He is of the order of 6000 s. Nevertheless, this decay has been studied over a lifetime range of 15 orders of magnitude, up to Xe^{52+} where the lifetime is only a few picoseconds [76] (Figure 10.6). To cover all that range, the various techniques described above were applied in turn. The measurement on He [77] exploited the emission from a well-diagnosed gas discharge and thus is not really a lifetime measurement technique. The experiments on Li^+ employed a classical ion trap [78] and DR resonances in a heavy-ion storage ring [55]. Measurements on Be through N were done at the Heidelberg heavy-ion storage ring, recording the ion beam composition via the residual gas ionization signal [79] or monitoring the DR signal [80]. Experiments at the Livermore EBIT obtained lifetime data on He-like ions from N through S (see [48,71,72,81]), which overlaps with the range in which beam-foil experiments have been tried. At the lower end of this range ($Z = 16$–18), the lifetimes range from 200 to 700 ns, and it is here that a slow beam experiment (on Ar) was appropriate at the time [45], because it suffered from other systematic errors than the fast beam measurements, helping to clarify the situation. This clarification process included the discovery of sign errors in theory, and eventually theory progressed so far that the best calculations differ by less than

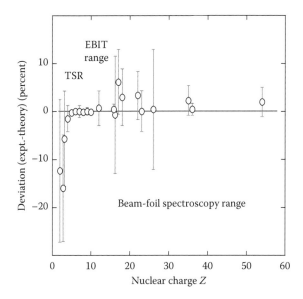

FIGURE 10.6
Isoelectronic trend of the transition rate data for the $1s^2$ $^1S_0 - 1s2s$ $^3S_1$ magnetic dipole ($M1$) transition in He-like ions. Data points for $Z = 4$–7 are fom the Heidelberg heavy-ion storage ring TSR, data for $Z = 7$–10 have been obtained at the Livermore EBIT-II. Data on heavier ions are from the use of (mostly fast) ion beams. All experimental data have been scaled by normalization to the results of the fully relativistic calculation by Johnson et al. [84], and only the deviations from this prediction are displayed. (From W. R. Johnson, D. R. Plante, and J. Sapirstein, In B. Bederson and H. Walther (Eds.), *Advances of Atomic, Molecular and Optical Physics*, Vol. 35, San Diego, CA: Academic Press, 1995, p. 255. With permission.)

0.1% in their prediction of this $M1$ transition rate [82–84]. Presently, experiment cannot afford the extreme accuracy needed to discriminate between the leading calculations for the He isoelectronic sequence. However, for now, we may consider theory to be successful in describing the two-electron ions, and then we have a ruler against which to test experimental techniques. If a measurement—in the case of this particular isoelectronic sequence—does not agree with calculation, there is likely a problem with the experiment.

The $ns - np$ resonance line series of the alkalis evolve into a branch ($np_{1/2}$) that, up to high Z, remains almost nonrelativistic, whereas the Z^4 dependence of the fine-structure splittings moves the $np_{3/2}$ level away and thus the components of the fine-structure doublet apart. At high Z, the $ns - np$ wavelengths differ by more than a factor of 2, and the lifetimes by a high multiple. The high-Z beam-foil lifetime results with their typical 5–10% uncertainties are individually less meaningful than the extremely accurate lifetime data obtained on neutral Li and Na [85–87]. Nevertheless, such a level of accuracy is sufficient to test many of the isoelectronic trends. For example, it was found that relativistic corrections to the matrix elements [84] were necessary to match theory and experiment in something as seemingly simple as the Li isoelectronic sequence (see [28]).

Also in the Li isoelectronic sequence, the $1s2p^2$ ^4P_J ($J = 1/2, 3/2, 5/2$) levels feature in-shell $E1$ decays to the $1s2s2p$ $^4P^\circ_J$ ($J = 1/2, 3/2, 5/2$) levels the rate of which scales linearly with Z. There also are higher multipole order x-ray decays to the $1s^2 2p$ doublet levels (the rates increase with high powers of Z), and autoionization processes that depend on spin-orbit, spin-other orbit, and spin–spin interactions (and Z-dependences such as Z^3 or Z^4). Over the years, measurements on ions from neutral Li (where almost only the in-shell radiative decay matters) to Mg, Al, and Si [89,90] have been made, and the isoelectronic trends of the level lifetimes helped to unravel the individual contributions [91]. It was found, for example, that the in-shell $E1$ transition rate was underestimated by some 20% in some early calculations; nowadays, some calculations can do much better.

Be-like ions (four electrons in total, two of them in the valence shell; see also the rather similar Mg-like ions, etc.) feature a wide range of transition types even among the lowest lying levels: typical $E1$, $E1$ with change of spin, $M2$, $M1$ between fine-structure levels, hyperfine-induced decays in some isotopes. The transition rate of the $E1$ resonance line $2s^2$ $^1S_0 - 2s2p$ $^1P^\circ_1$ has been claimed to be calculable with an accuracy (10^{-4}) [92] that, by far, exceeds experimental capabilities. Experiment, in fact, is hampered by the cascade from the $2p^2$ $^1S_0$ level which differs in lifetime from the resonance level by about 30%. (In Be, this displaced level lies above the ionization limit, and the cascade is absent.) Neglect of this growing-in cascade in beam-foil lifetime measurements has brought about a fair number of systematic overestimates of the $2s2p$ $^1P^\circ_1$ level lifetime by 30–50%. Critical reviews and the technique of isoelectronic smoothing of the line strength data [24,25] have helped to identify the more thoughtfully executed evaluations. Although the systematic shift error could be avoided, an evaluation of all sensibly evaluated lifetime data (Be ($Z = 4$) through Kr ($Z = 36$)) revealed that the authors opted for cautious error estimates in the 5–10% range, while the scatter of the data corresponds to no more than a 3% error [28].

For the $2s^2$ $^1S_0 - 2s2p$ $^3P^\circ_1$ intercombination transition in the same atomic system, the lifetime data used to reach from Ne ($Z = 10$) with a 50% error estimate to Xe ($Z = 54$) with a 10% measurement error estimate. Many calculations were available, especially for the C^{2+} ion of astrophysical interest, but the results scattered by about 20% around the mean. The advent of heavy-ion storage ring lifetime measurements [93] and of measurement uncertainties below 1% changed the situation thoroughly—now theory has problems keeping up [28]. For the specific case of C^{2+}, computationally extensive relativistic configuration interaction calculations with about 200,000 wave functions [94] appear to be the most accurate so far, with a quoted uncertainty of 0.5%. The calculated result and the measurement disagree by slightly more than the combined errors, an incentive to continue the research at this high level of accuracy.

B-like ions are the simplest ions with fine-structure intervals in the ground term. Between such fine-structure levels, $M1$ and (much weaker) $E2$ transitions take place. Since the same term is involved in both levels, the transition rate does not depend on radial wave functions, but only on Racah algebra factors (atomic geometry, in the form of the so-called line strength S) and the transition energy (ΔE^3 for $M1$ and ΔE^5 for $E2$ transitions). This simple picture applies to the single configuration, nonrelativistic limit; otherwise, the calculation can be much more difficult. Interestingly, theory has massive problems predicting fine-structure intervals in many-electron systems. Most calculations are found to be insufficient in the end, and the calculated energy differences are replaced by the much more accurately known experimental data. Only very recently have *ab initio* calculations yielded fine-structure intervals close to experiment [95].

These excited levels in the ground configurations of many ions are very interesting for plasma diagnostics. The appearance of the lines indicates the presence of charge state and thus the electron energy (temperature) of the plasma. The transitions are of relatively low energy and consequently the spectral lines in the visible, near, or extreme ultraviolet can be well resolved and identified with individual charge states. Moreover, the radiative transition rates compete with collisional excitation and de-excitation rates, at least in low-density plasmas such as tokamaks or stellar coronae. Collisional-radiative modeling can be applied to identify line ratios that depend on the particle density of the plasma, because of radiative and collisional processes affecting the lower-level population. The level lifetimes of these levels are among the very few testable parameters of such modeling exercises. It has been great progress when a number of these lifetimes were measured, using EBITs or heavy-ion storage rings, with uncertainties of only one or a few percent. Such experiments have also included cross-checks of the two techniques applied to the same atomic system [74]. At the same time, it has been revealing to see the scatter of theoretical predictions of the same entities in multielectron systems, ranging from a few percent in some cases to a factor of 3 or 5 in others.

For most applications in astrophysical or terrestrial plasma diagnostics, an uncertainty of radiative decay rates of 1% seems sufficient, because most other measurement parameters will be known with poorer precision. However, intellectually, it is intriguing to test how precise such measurements can be. Various experiments on the (electric dipole) resonance lines of alkali atoms have reached an uncertainty of about 0.1%. The aforementioned measurement of the spin-changing $E1$ (intercombination) transition in C^{2+} [93] carried an error bar of 0.14%. The same small uncertainty can test something unprecedented in the $M1$ transition rate between fine-structure levels in many-electron ions. Beyond the simple picture mentioned above, that the transition rate depends only on the line strength and the transition energy (and the transition energy calculation may require a QED correction), QED also requires a correction of the transition operator. This correction relates to the electron anomalous magnetic moment, or the (g-2) value. The correction increases the $M1$ transition rate by about 0.45%, and the Heidelberg EBIT results for Ar^{13+} [96,97] and Fe^{13+} [98] (while being compatible with less precise earlier measurements at the Livermore EBIT [99]) agree less well with the corrected than with the uncorrected predictions. Judged by the published uncertainties of experiment and calculation, the disagreement corresponds to several standard deviations. There is no tested idea yet of what might explain the discrepancy, whether theory missed something in the interplay of quantum mechanics and QED, or whether the experiments suffered from unrecognized systematic errors. Independent measurements are called for.

Among the $M1$ transitions in more complex systems is one in the $3d^4$ configuration of Ti-like ions. Whereas regular fine-structure intervals increase with the fourth power of the nuclear (or ion) charge, this particular interval varies very little. For more than 30 elements, the transition gives rise to a line in the near-ultraviolet, which has been considered as a most promising plasma diagnostic tool [100]. Calculations have met limited success when trying to predict the transition wavelength accurately (see [101]); however, various calculations agree in showing a discontinuity of the level energy trend near $Z = 55$. It is possible that lifetime measurements can help to elucidate this problem. So far only a single lifetime measurement [102] has been done on an ion in this isoelectronic sequence, and the accuracy of a sensitive test would need to be higher.

Higher-multipole order transition rates usually scale more steeply with the nuclear charge than the low orders. In various ions of Kr ($Z = 36$), $M2$ transitions have been found to contribute as much to the decay rates of certain levels as all others [103], whereas at low Z, the $M2$ contributions may be negligible. Some atomic clock schemes depend on high multipole order decays ($E3$) having a very low rate in atoms or singly charged ions. In highly charged ions, a magnetic octupole ($M3$) has been identified in the spectra of Ni-like ions of Th and U [104], and its transition rate has more recently been measured in Ni-like Xe^{26+} [105]. In odd isotopes, the hyperfine interaction can mix sublevels of the $3d^94s\ ^3D_3$ and $^3D_2$ levels, thus mixing $M3$ and $E2$ decays. The effect has been measured at the Livermore EBIT, the experimental level lifetimes corroborating theory [106].

The recognition of hyperfine mixing as a contributor to certain decay channels happened in the 1930s and was put on a quantitative basis by Mrozowski [107]. The first atomic lifetime measurements showing the effect in He-like ions were reported from Berkeley [108], followed by systematic studies at Lund [109,110]. The hyperfine-quenched lifetime in Ag^{45+} ions served to determine the spectroscopically unresolved fine-structure splitting of $1s2p$ triplet levels near a level crossing along the isoelectronic sequence [111]. In the He isoelectronic sequence, the most striking hyperfine lifetime effect is that of the $1s2p\ ^3P_0^o$ level mixing with the $1s2p\ ^3P_1^o$ level which in turn is (relativistically) mixed with the $1s2p\ ^1P_1^o$ level, and the latter's very short level lifetime (say, in the subpicosecond range for not so heavy ions) results in nanosecond lifetimes for the hyperfine-quenched $^3P_0^o$ levels—just right for BFS. The same physics applies to Be-like ions, but the reference level $2s2p$ $^1P_1^o$ may have a lifetime of hundreds of picoseconds, giving the $2s2p$ $^3P_0^o$ level a lifetime in the millisecond range—clearly any measurement of such lifetimes needs ion trapping. In N^{3+}, the expected lifetime of the $2s2p$ $^3P_0^o$ level is of the order of 20 min, and modeling of planetary nebula spectra agrees with this estimate [112]. The first storage ring measurement of such hyperfine quenching concerned Be-like ions of $^{47}Ti^{18+}$ and employed DR as a probe of the level population [113]. The lifetime result was much shorter (about 60%) than predicted. In a photon detection experiment at the same heavy-ion storage ring, but studying the rather similar Mg-like ions of $^{63,65}Cu^{17+}$, a similar deviation from old calculations was noted [114]. However, newer calculations yield results some 20% shorter than the old ones (yes, theory does evolve), and the search for systematic errors in specific atomic systems and their excitation process is not finished yet (yes, experiments may still be incomplete).

Last, but not least, relativistic effects affect the wave functions. What in a low-Z atomic system may be rather similar fine-structure components of an *LS*-coupling term can acquire quite different characters at high Z, due to *j*-dependent relativistic distortions of the wave functions. The individual fine-structure levels may then differ drastically from each other in lifetime, and the branch fractions of their decays can be very different. Measurements of these branch fractions as well as of individual level lifetimes can give detailed insights into relativistic effects [115].

10.3 Conclusion and Prospects

The published lifetime measurements of multiply charged ions reach from femtoseconds to seconds, the wavelength ranges from the visible (several eV photon energy) to the x-ray range. The short lifetime range is close to the physical limit. For the measurement of long lifetimes, singly charged ion lifetimes have already exceeded 1 min. In most cases, there is reasonable agreement between measured data and computed lifetimes. However, there often are several sets of data and predictions, not all of which are compatible. Sometimes experiment erred and was made aware of that only by calculations, and in a number of cases theory obtained acceptable results only after good measurements had shown the way, that is, with little predictive power. Fortunately, incited by mutual challenge, atomic lifetime experiment and theory are evolving.

Selective excitation of specific levels in highly charged ions is virtually impossible to achieve by exposing neutral atoms to a radiation field. The situation is much more promising, if ions of a specific charge state can be prepared by different means, and photons are required only for the excitation of a given ion species. In this vein, selective laser excitation of ions stored in an EBIT has been tried at Oxford, at Livermore, and at Heidelberg, with rather limited success. However, at the FLASH facility at Hamburg, a Heidelberg-built EBIT exposed Li-like Fe ions to the ultraviolet light from a synchrotron light source, achieving selective excitation [116]. Such a scheme would also offer benefits for certain atomic lifetime measurements. On such resonance transitions as excited at Hamburg, the EBIT produces the ions that are to be excited. The measurement of their lifetimes, which are in the range of about a nanosecond, however, does not require any cycling of the trap voltages, only fast (subnanosecond) timing of the detected photons in relation to the very short excitation pulses from the light source. Considering the low photon yield of such experiments, they are best suited for short-lived levels, in measurements that exploit the high repetition frequency of synchrotron light sources. The measurement of long-level lifetimes would suffer from a poor duty cycle imposed by the target atom lifetime. However, continuing development of ever brighter EUV and soft-x-ray light sources may change this perspective and open up yet another avenue leading to more insights about atomic structure and dynamics.

Acknowledgments

The author appreciates thankfully the hospitality and support experienced at the Livermore EBIT Laboratory, as well as travel support by Deutsche Forschungsgemeinschaft (DFG). Peter Beiersdorfer kindly provided advice on the manuscript. This work was performed in part under the auspices of the U.S. Department of Energy by Lawrence Livermore National Laboratory under Contract DE-AC52-07NA27344.

References

1. E. Träbert, In S. M. Shafroth and J. C. Austin (Eds.), *Accelerator-based Atomic Physics—Techniques and Applications*. Washington, DC: American Institute of Physics, 1997, p. 567.
2. H. Cederquist, M. Kisielinski, and S. Mannervik, *J. Phys. B: At. Mol. Phys.* 16, L479, 1983.
3. H. Cederquist, M. Kisielinski, and S. Mannervik, *J. Phys. B: At. Mol. Phys.* 17, 1969, 1984.
4. E. Träbert, *Phys. Scripta* 61, 257, 2000.
5. E. Träbert, *Phys. Scripta T* 100, 88, 2002.

6. E. Träbert, *Can. J. Phys.* 80, 1481, 2002.
7. L. Kay, *Phys. Lett.* 5, 36, 1963.
8. S. Bashkin and A. B. Meinel, *Astrophys. J.* 139, 413, 1964.
9. S. Bashkin, *Nucl. Instrum. Meth.* 28, 88, 1964.
10. S. Bashkin, *Science* 148, 1047, 1965.
11. L. Kay, *Nucl. Instrum. Methods B* 9, 544, 1985.
12. S. Bashkin, *Nucl. Instrum. Methods B* 9, 546, 1985.
13. H-D. Betz, F. Bell, H. Panke, G. Kalkoffen, M. Welz, and D. Evers, *Phys. Rev. Lett.* 33, 807, 1974.
14. L. C. Northcliffe and R. F. Schilling, *Nucl. Data A* 7, 233, 1970.
15. G. Astner, L. J. Curtis, L. Liljeby, S. Mannervik, and I. Martinson, *Z. Phys. A* 279, 1, 1976.
16. L. Barrette and R. Drouin, *Phys. Scripta* 10, 213, 1974.
17. E. Träbert, H. Winter, P. H. Heckmann, and H. v. Buttlar, *Nucl. Instrum. Methods* 135, 353, 1976.
18. Y. Baudinet-Robinet, H. P. Garnir, P. D. Dumont, and A. E. Livingston, *Phys. Scripta* 14, 224, 1976.
19. W. Ansbacher, E. H. Pinnington, and J. A. Kernahan, *Can. J. Phys.* 66, 402, 1988.
20. L. Engström and P. Bengtsson, *Phys. Scripta* 43, 480, 1991.
21. L. J. Curtis, *Am. J. Phys.* 36, 1123, 1968.
22. L. J. Curtis, H. H. Berry, and J. Bromander, *Phys. Lett. A* 34, 169, 1971.
23. L. Engström, *Nucl. Instrum. Methods* 202, 369, 1982.
24. N. Reistad and I. Martinson, *Phys. Rev. A* 34, 2632, 1986.
25. E. Träbert, *Z. Physik D* 9, 143, 1988.
26. C. Munger and H. Gould, *Phys. Rev. Lett.* 57, 2927, 1986.
27. J. Schweppe, A. Belkacem, L. Blumenfeld, N. Claytor, B. Feinberg, H. Gould, V. E. Kostroun et al., *Phys. Rev. Lett.* 66, 1434, 1991.
28. E. Träbert and L. J. Curtis, *Phys. Scripta* 74, C46, 2006.
29. E. Träbert, P. H. Heckmann, R. Hutton, and I. Martinson, *J. Opt. Soc. Am. B* 5, 2173, 1988.
30. L. J. Curtis, *Phys. Rev. A* 40, 6958, 1989.
31. E. Träbert, J. Doerfert, J. Granzow, R. Büttner, J. Brauckhoff, M. Nicolai, K.-H. Schartner, F. Folkmann, and P. H. Mokler, *Phys. Lett. A* 202, 91, 1995.
32. H. D. Betz, J. Rothermel, D. Röschenthaler, F. Bell, R. Schuch, and G. Nolte, *Phys. Lett. A* 91, 12, 1982.
33. S. Lauer, H. Liebel, F. Vollweiler, H. Schmoranzer, B. M. Lagutin, Ph. V. Demekhin, I. D. Petrov, and V. L. Sukhorukov, *J. Phys. B: At. Mol. Opt. Phys.* 32, 2015, 1999.
34. Y. Baudinet-Robinet, H. P. Garnir, P. D. Dumont, and A. El Himdy, *Phys. Scripta* 39, 221, 1989.
35. J. O. Stoner Jr and J. A. Leavitt, *Opt. Acta* 20, 435, 1973.
36. J. A. Leavitt, J. W. Robson, and J. O. Stoner Jr, *Nucl. Instrum. Methods* 110, 423, 1973.
37. K. E. Bergkvist, *J. Opt. Soc. Am.* 66, 837, 1976.
38. N. A. Jelley, J. D. Silver, and I. A. Armour, *J. Phys. B: At. Mol. Phys.* 10, 2339, 1977.
39. L. Kay and D. A. Sadler, *Phys. Scripta* 51, 459, 1995.
40. R. Marrus and R. W. Schmieder, *Phys. Rev. Lett.* 25, 1689, 1970.
41. R. Marrus and R. W. Schmieder, *Phys. Rev. A* 5, 1160, 1972.
42. H. Gould, R. Marrus, and R. W. Schmieder, *Phys. Rev. Lett.* 31, 504, 1973.
43. D. L. Lin and L. Armstrong Jr, *Phys. Rev. A* 16, 791, 1977.

44. G. Hubricht, E. Träbert, and H. M. Hellmann, *Z. Physik D* 4, 209, 1986.
45. G. Hubricht and E. Träbert, *Z. Physik D* 7, 243, 1987.
46. G. Hubricht and E. Träbert, *Phys. Scripta* 39, 581, 1989.
47. S. Toleikis, B. Manil, E. Berdermann, H. F. Beyer, F. Bosch, M. Czanta, R. W. Dunford et al., *Phys. Rev. A* 69, 022507, 2004.
48. J. R. Crespo López-Urrutia, P. Beiersdorfer, and K. Widmann, *Phys. Rev. A* 74, 012507, 2006.
49. E. Träbert, G. Gwinner, A. Wolf, E. J. Knystautas, H-P. Garnir, and X. J. Tordoir, *Phys. B: At. Mol. Opt. Phys.* 35, 671, 2002.
50. E. Träbert, A. G. Calamai, G. Gwinner, E. J. Knystautas, E. H. Pinnington, and A. Wolf, *J. Phys. B: At. Mol. Opt. Phys.* 36, 1129, 2003.
51. E. Träbert, G. Saathoff, and A. Wolf, *J. Phys. B: At. Mol. Opt. Phys.* 37, 945, 2004.
52. E. Träbert, G. Saathoff, and A. Wolf, *Eur. Phys. J. D* 30, 297, 2004.
53. E. Träbert, S. Reinhardt, J. Hoffmann, and A. Wolf, *J. Phys. B: At. Mol. Opt. Phys.* 39, 945, 2006.
54. E. Träbert, G. Gwinner, A. Wolf, X. Tordoir, and A. G. Calamai, *Phys. Lett. A* 264, 311, 1999.
55. A. A. Saghiri, J. Linkemann, M. Schmitt, D. Schwalm, A. Wolf, T. Bartsch, A. Hoffknecht et al., *Phys. Rev. A* 69, R3350, 1999.
56. I. Klaft, S. Borneis, T. Engel, B. Fricke, R. Grieser, G. Huber, T. Kühl et al., *Phys. Rev. Lett.* 73, 2425, 1994.
57. P. Seelig, S. Borneis, A. Dax, T. Engel, S. Faber, M. Gerlach, C. Holbrow et al., *Phys. Rev. Lett.* 81, 4824, 1998.
58. H. Winter, S. Borneis, A. Dax, S. Faber, T. Kühl, D. Marx, F. Schmitt et al., *GSI Scientific Report*, 1998, p. 87.
59. S. Mannervik, *Phys. Scripta T* 105, 67, 2003.
60. E. Träbert, *Phys. Scripta* 23, 253, 1981.
61. D. Zajfman, O. Heber, L. Vejby-Christensen, I. Ben-Itzhak, M. Rappaport, R. Fishman, and M. Dahan, *Phys. Rev. A* 55, R1577, 1997.
62. M. Dahan, R. Fishman, O. Heber, M. Rappaport, N. Altstein, D. Zajfman, and W. J. van der Zande, *Rev. Sci. Instrum.* 69, 76, 1998.
63. M. A. Levine, R. E. Marrs, J. R. Henderson, D. A. Knapp, and M. B. Schneider, *Phys. Scripta T* 22, 157, 1988.
64. M. A. Levine, R. E. Marrs et al., *Nucl. Instrum. Meth. B* 43, 431, 1989.
65. R. E. Marrs, S. R. Elliott, and D. A. Knapp, *Phys. Rev. Lett.* 72, 4082, 1994.
66. P. Beiersdorfer, A. L. Osterheld, V. Decaux, and K. Widmann, *Phys. Rev. Lett.* 77, 5353, 1996.
67. A. Graf, P. Beiersdorfer, C. L. Harris, D. Q. Hwang, and P. A. Neill, In C. A. Back (Ed.), *CP645, Spectral Line Shapes, Vol. 12, Proceedings of the 16th ICSLS*, Washington, DC: American Institute of Physics, 2002.
68. P. Beiersdorfer, L. Schweikhard, J. Crespo López-Urrutia, and K. Widmann, *Rev. Sci. Instrum.* 67, 3818, 1996.
69. P. Beiersdorfer, B. Beck, St. Becker, and L. Schweikhard, *Int. J. Mass Spectrom. Ion Proc.* 157/158, 149, 1996.
70. E. Träbert and P. Beiersdorfer, *Rev. Sci. Instrum.* 74, 2127, 2003.
71. B. J. Wargelin, P. Beiersdorfer, and S. M. Kahn, *Phys. Rev. Lett.* 71, 2196, 1993.
72. E. Träbert, P. Beiersdorfer, G. V. Brown, A. J. Smith, M. F. Gu, and D. W. Savin, *Phys. Rev. A* 60, 2034, 1999.

73. E. Träbert, P. Beiersdorfer, and S. B. Utter, *Phys. Scripta T* 80, 450, 1999.
74. E. Träbert, P. Beiersdorfer, G. Gwinner, E. H. Pinnington, and A. Wolf, *Phys. Rev. A* 66, 052507, 2002.
75. A. H. Gabriel and C. Jordan, *Mon. Not. R. Astron. Soc.* 145, 241, 1969.
76. R. Marrus, P. Charles, P. Indelicato, L. de Billy, C. Tazi, J.-P. Briand, A. Simionovice, D. D. Dietrich, F. Bosch, and D. Liesen, *Phys. Rev. A* 39, 3725, 1989.
77. H. W. Moos and J. R. Woodworth, *Phys. Rev. Lett.* 30, 775, 1973.
78. R. D. Knight and M. H. Prior, *Phys. Rev. A* 21, 179, 1980.
79. E. Träbert, G. Gwinner, E. J. Knystautas, and A. Wolf, *Can. J. Phys.* 81, 941, 2003.
80. H. T. Schmidt, P. Forck, M. Grieser, D. Habs, J. Kenntner, G. Miersch, R. Repnow et al., *Phys. Rev. Lett.* 72, 1616, 1994.
81. J. R. Crespo López-Urrutia, P. Beiersdorfer, D. W. Savin, and K. Widmann, *Phys. Rev. A* 57, 238, 1998.
82. G. W. F. Drake, *Phys. Rev. A* 3, 908, 1971.
83. C.D. Lin, Ph.D. Thesis, Columbia University, 1975.
84. W. R. Johnson, D. R. Plante, and J. Sapirstein, In B. Bederson and H. Walther (Eds.), *Advances of Atomic, Molecular and Optical Physics*, Vol. 35, San Diego, CA: Academic Press, 1995, p. 255.
85. W. I. McAlexander, E. R. I. Abraham, N. W. M. Ritchie, C. J. Williams, H. T. C. Stoof, and R. G. Hulet, *Phys. Rev. A* 51, R871, 1995.
86. W. I. McAlexander, E. R. I. Abraham, and R. G. Hulet, *Phys. Rev. A* 54, R5, 1996.
87. U. Volz and H. Schmoranzer, *Phys. Scripta T* 65, 48, 1996.
88. L. J. Curtis, *Phys. Rev. A* 51, 251, 1995.
89. A. E. Livingston, J. E. Hardis, L. J. Curtis, R. L. Brooks, and H. G. Berry, *Phys. Rev. A* 30, 2089, 1984.
90. H. Hellmann and E. Träbert, *Nucl. Instrum. Methods B* 9, 611, 1985.
91. E. Träbert, *Z. Physik A* 321, 51, 1985.
92. P. Jönsson, C. F. Fischer, and E. Träbert, *J. Phys. B: At. Mol. Opt. Phys.* 31, 3497, 1998.
93. J. Doerfert, E. Träbert, A. Wolf, D. Schwalm, and O. Uwira, *Phys. Rev. Lett.* 78, 4355, 1997.
94. M. H. Chen, K. T. Cheng, and W. R. Johnson, *Phys. Rev. A* 64, 042507, 2001.
95. M. J. Vilkas and Y. Ishikawa, *Phys. Rev. A* 68, 012503, 2003.
96. A. Lapierre, U. D. Jentschura, J. R. C. López-Urrutia, J. Braun, G. Brenner, H. Bruhns, D. Fischer et al., *Phys. Rev. Lett.* 95, 183001, 2005.
97. A. Lapierre, J. R. C. López-Urrutia, J. Braun, G. Brenner, H. Bruhns, D. Fischer, A. J. González-Martínez et al., *Phys. Rev. A* 73, 052507, 2006.
98. G. Brenner, J. R. C. López-Urrutia, Z. Harman, P. H. Mokler, and J. Ullrich, *Phys. Rev. A* 75, 032504, 2007.
99. P. Beiersdorfer, E. Träbert, and E. H. Pinnington, *Astrophys. J.* 587, 836, 2003.
100. U. Feldman, P. Indelicato, and J. Sugar, *J. Opt. Soc. Am. B* 8, 3, 1991.
101. E. Biémont, E. Träbert and C. Zeippen, *J. Phys. B: At. Mol. Opt. Phys.* 34, 1941, 2001.
102. F. G. Serpa, C. A. Morgan, E. S. Meyer, J. D. Gillaspy, E. Träbert, D. A. Church, and E. Takács, *Phys. Rev. A* 55, 4196, 1997.
103. E. Träbert, P. Beiersdorfer, G. V. Brown, H. Chen, D. B. Thorn, and E. Biémont, *Phys. Rev. A* 64, 042511, 2001.
104. P. Beiersdorfer, A. L. Osterheld, J. Scofield, B. Wargelin, and R. E. Marrs, *Phys. Rev. Lett.* 67, 2272, 1991.

105. E. Träbert, P. Beiersdorfer, and G. V. Brown, *Phys. Rev. Lett.* 98, 263001, 2007.
106. K. Yao, M. Andersson, T. Brage, R. Hutton, P. Jönsson, and Y. Zou, *Phys. Rev. Lett.* 97, 183001, 2006 [erratum 2007 Phys. Rev. Lett. 98 269903 [Ni-like ions].
107. S. Mrozowski, Z. *Physik* 108, 204, 1938.
108. H. Gould, R. Marrus, and P. J. Mohr, *Phys. Rev. Lett.* 33, 676, 1974.
109. L. Engström, C. Jupén, B. Denne, S. Huldt, W. T. Meng, P. Kaijser, U. Litzén, and I. Martinson, *J. Phys. B: At. Mol. Phys.* 13, L143, 1980.
110. B. Denne, S. Huldt, J. Pihl, and R. Hallin, *Phys. Scripta* 22, 45, 1980.
111. B. B. Birkett, I.-P. Briand, P. Charles, et al. *Phys. Rev. A* 47, R2454, 1993.
112. T. Brage, P. G. Judge, and C. R. Proffitt, *Phys. Rev. Lett.* 89, 281101, 2002.
113. S. Schippers, E. W. Schmidt, D. Bernhardt, D. Yu, A. Müller, M. Lestinsky, D. A. Orlov, M. Grieser, R. Repnow, and A. Wolf, *Phys. Rev. Lett.* 98, 033001, 2007.
114. E. Träbert, M. Grieser, J. Hoffmann, C. Krantz, S. Reinhardt, R. Repnow, A. Wolf, and P. Indelicato, *New J. Phys.* 13, 023017, 2011.
115. L. J. Curtis, S. R. Federman, K. Torok, M. Brown, S. Cheng, R. E. Irving, and R. M. Schectman, *Phys. Scripta* 75, C1, 2007.
116. S. Epp, J. R. C. López-Urrutia, G. Brenner, V. Mäckel, P. H. Mokler, R. Treusch, M. Kuhlmann et al., *Phys. Rev. Lett.* 98, 183001, 2007.

11

Importance of Tungsten Spectroscopy to the Success of ITER

R. Neu, T. Pütterich, C. Biedermann, R. Dux, and R. Radtke

CONTENTS

11.1 Introduction

The International Thermonuclear Experimental Reactor (ITER), which began construction in 2007, will demonstrate the physical and technological feasibility of fusion for power production [1]. It will employ tungsten at areas with high particle and power load (divertor entrance and baffles) owing to its high-energy threshold for sputtering and its low sputtering yield compared to the low-Z materials such as C and Be, which will be used in parallel. Although the requirements for the plasma-facing components (PFCs) in ITER will already be higher than in the present-day devices, the step to a quasisteady-state DEMO reactor will still considerably increase the particle fluencies to the PFCs. Therefore, it is foreseen that tungsten will be used as an armor material for all PFCs [2] in future fusion reactors.

Tungsten was already used as a plasma-facing material (PFM) in early devices, as PLT [3] and ORMAK [4], and it soon became clear that central W radiation was occasionally as large as the ohmic input power and consequently led to hollow temperature profiles in the plasma discharges [5]. In these cases, the W concentrations were estimated to be in the range 10^{-3} [6], consistent with estimations using present-day atomic data. Such high values would prevent ignition and burning in a fusion reactor as can be deduced to form zero-dimensional power balance calculations, taking into account also the inevitable losses through a finite confinement time of the He fusion ash [7] (see Figure 11.1). Unlike these early devices, which used a material limiter to keep the plasma away from the vacuum vessel walls, present-day tokamaks use a so-called divertor configuration, set up by external magnetic fields allowing us to separate the hot bulk plasma from the regions of plasma wall interaction (see, e.g., [8]). By this means, the plasma temperature in front of PFCs is reduced considerably leading in turn to a strong reduction of the erosion yield for high-Z ions. This opened up again the possibility to use W as a PFM. In order to prepare the plasma physical database for operation with W PFCs, the mid-sized tokamak ASDEX Upgrade [9] has started a dedicated W programme and is equipped with 100% W PFCs since the experimental campaign in 2007 [10]. Further investigations are conducted in the TEXTOR tokamak [11] and are foreseen in the framework of the ITER-like Wall Project at JET [12], which will use W PFCs in the divertor and Be as PFM in the main chamber. A review on experiments with high-Z PFCs can be found in [13].

For these obvious reasons, tungsten is moving back into the focus of spectroscopy for fusion plasmas, and a large data set of spectroscopic information on the radiation from tungsten ions is required for diagnostics. This information includes accurate atomic physics data on atomic transition wavelengths, line intensities, and rates for ionization, excitation and recombination in a range of electron temperatures from a few eV up to about 20 keV.

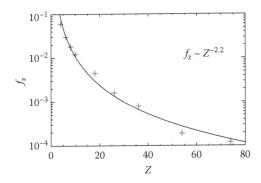

FIGURE 11.1
Allowed impurity concentrations in ignited plasmas from zero-dimensional power balance calculations, accounting for an effective He confinement time of $\rho = \tau_{He}/\tau_E = 5$ (τ_{He} gives the effective He confinement time, τ_E is the energy confinement time).

At fusion devices, the spectroscopic information gathered by measurements of radiation emitted from the plasma has to tackle a variety of challenges which need sophisticated interpretation of the particular situations of the measurement setup and the conditions at these devices. Passive spectroscopy integrates the radiation along a line of sight (LOS) through the plasma and the emission may originate from regions with different excitation conditions given by the radial profile of electron temperature and density. Thus, many ionization stages may coexist within the observation region, and a wide range of levels can be excited simultaneously leading to a multitude of emission features. The atomic structure of the high-Z element tungsten with plenty of electrons in up to six different shells complicates the identification of spectra. Additionally, plasma transport processes influence the occurrence of a large number of ionization stages emitting at the same time making the interpretation and analysis of spectra very difficult [14–17]. Moreover, although the edge and divertor temperatures in present-day fusion devices are similar to that of ITER, the central temperatures are usually below 10 keV. Therefore, W ionization stages from neutral up to around W^{56+} exist in present-day tokamaks, whereas up to around W^{68+} will be reached in a reactor [18].

In order to close this obvious gap in the accessible temperature range and to facilitate the investigations, electron beam ion traps (EBITs) are indispensable. EBITs can not only produce a well-selected ensemble of ions in specific charge stages, but also excite and confine them for extended periods of detailed spectroscopic investigation. Several devices (LLNL EBIT at Lawrence Livermore National-Laboratory, NIST EBIT at National Institute of Standards and Technology, Berlin EBIT at Max-Planck-Institut für Plasmaphysik, and Humboldt University, Berlin, Tokyo EBIT at University of Electrocommunication Tokyo) have already started with this task (see, e.g., [19–21]) and contributed significantly to the understanding of the W spectra observed in fusion devices [22,23].

However, not only the highly charged ions have to be addressed, but also the neutral W atom as well as low-lying ionization stages are important in order to quantify the W influx from PFCs. As already stated above, present-day devices can access the ITER-relevant plasma parameter range, but the large number of the transitions (see, e.g., [24]) leads, in general, to rather weak single spectral lines and the necessary atomic data (ionization and excitation rates) are scarce [25].

This chapter is structured as follows. In the next section, the principle of influx measurements in fusion devices is briefly described and the status of current research is presented. Section 11.3 introduces the principle of the determination of impurity concentrations with some special emphasis to tungsten and reviews the related investigations on W spectroscopy of highly charged W ions in EBIT as well as in fusion devices. Finally, Section 11.4 concludes the paper and highlights further necessary investigations.

11.2 Spectroscopy for W-Influx Measurements

11.2.1 Principle of Influx Measurements

Spectroscopy is the only method to measure particle fluxes arising from PFCs in real time. Typically, the regions of interest are monitored by spectroscopic systems viewing perpendicular on the surfaces, similar as shown in Figure 11.2. In order to use spectral lines of transitions in atoms or low charge stage ions as a quantitative measure for the particle influx, the adequate rate coefficients have to be known. If recombination can be neglected, then the influx can be determined rather directly from the measured photon flux as described by Behringer et al. [26]. Integrating along an LOS, the particle influx Γ_q^{in} from a material surface is given by

$$\Gamma_q^{in} = \int_0^R dl\, n_e S_q n_q, \tag{11.1}$$

where S_q denotes the ionization rate coefficient for ionization from charge stage q to $q+1$ and n_q denotes the density of the qth ion stage.

In the coronal equilibrium, which is mostly valid for low-density fusion plasmas, the collisional electron excitation is balanced by a spontaneous emission

$$\sum_{f<i} A_{if} n_q^i = n_e X^i n_q, \tag{11.2}$$

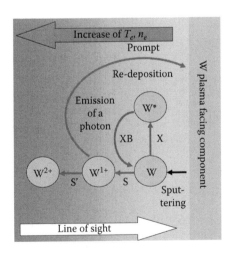

FIGURE 11.2
Schematics of the S/XB method. W^{q+} denotes the tungsten ionization stage, S the ionization rate, X the excitation rate, and B the branching ratio.

with the Einstein coefficients A_{if} and the rate coefficient X^i for collisional electron excitation to the level i. In principle, excitations from metastable states within one charge state have to be treated separately, but this is omitted here for simplicity. Thereby, one gets for the line emissivity $\varepsilon_{ph}^{i \rightarrow j}$ (photons per volume unit)

$$\varepsilon_{ph}^{i \rightarrow j} = Bn_e X^i n_q, \tag{11.3}$$

with the branching ratio B defined as (the indices are omitted for simplicity)

$$B = \frac{A_{i,j}}{\sum_f A_{i,f}}. \tag{11.4}$$

Integrating $\varepsilon_{ph}^{i \rightarrow j}$ over the LOS yields the photon flux Γ_{ph}^i for the observed transition:

$$\Gamma_{ph}^i = B \int_0^R dl\, n_e X^i n_q. \tag{11.5}$$

By dividing Equation 11.1 through Equation 11.5, one ends up with

$$\frac{\Gamma_q^{in}}{\Gamma_{ph}^i} = \frac{\int_0^R dl\, n_e S_q n_q}{B \int_0^R dl\, n_e X^i n_q}. \tag{11.6}$$

Since S_q and X^i behave in a very similar way along the LOS (similar temperature dependence) and the ion (or atom) with charge q exists only in a narrow temperature range, one can make the following approximation:

$$\Gamma_q^{in} = \frac{S_q}{X^i B} \Gamma_{ph}^i. \tag{11.7}$$

Therefore, the so-called inverse photon efficiency S/XB directly connects the particle flux to the measured photon flux. S/XB is a function of the plasma temperature and under typical conditions (influx of particles against a temperature gradient) it has to be evaluated at $T_e \approx 1/3E_{ion}^q$, with E_{ion}^q as the ionization potential of the ion with charge q. A recent review on the spectroscopic diagnostic of tokamak edge plasmas using this procedure is given in [27]. The measurement has to be performed in the direction of the impurity influx and for a laterally nonisotropic particle source the whole emission cloud has to be observed. This is illustrated in Figure 11.2 for the case of W particles which are sputtered from a PFC. As is indicated in Figure 11.2, the spectroscopic measurements using a spectral line of neutral W (WI) can only yield the gross W-influx, because a considerable amount of the initial W-influx may be redeposited promptly after ionization due to the large gyroradius of the W ion compared to its ionization length [28]. This caveat can be circumvented to some extent by using spectral lines from higher ionization stages, but to date no suitable lines were found (see Section 11.2.2).

11.2.2 W-Influx Measurements in Tokamaks

A thorough diagnostic of the W-influx into the plasma is of utmost importance in order to interpret the W behavior in the plasma. However, theoretical data for the inverse photon efficiency S/XB are scarce or even often missing in the case of tungsten. First measurements at the PSI-1 plasma generator in Berlin for a WI transition $(5d^5 6s^7 S_3 – 5d^5 6p^7 P_4)$ at 400.875 nm [29] yielded an experimental data set for S/XB, which was verified at ASDEX Upgrade under fusion plasma conditions [30]. Recently, a comprehensive effort has been started at the tokamak TEXTOR [25] to compare experimental spectroscopic results with theoretical ones, in order to provide reliable data for the determination of tungsten fluxes and to extend the S/XB method to other transitions. This is especially important because the wavelength of the above-mentioned line does not allow long transfer lines via fibers and additionally the transmission of fiber guides in this wavelength range rapidly degrades under neutron and gamma-ray irradiation. On the other hand, transitions with even shorter wavelengths might be of advantage to reduce blending by thermal radiation from the surfaces. Table 11.1 presents the lines, which were identified to be best suited for influx measurements because of their atomic structure and brightness. The calculations in [25] were performed using the ATOM code [31] for the ionization rates and the semiempirical van Regemorter formula [32] for excitation rates. Additionally, cascades into the upper level of a specific transition were estimated by using intensities from [24]. The calculations of the S/XB ratio for the WI spectral line at 400.9 nm show good agreement (better than a factor of 2) with the experimental data in the temperature range from 2 to 25 eV (see [25] and references therein), when using the Boltzmann formula with an effective temperature of ≈ 0.3 eV for the relative level population of the $^5 D_J$ and the $^7 S_3$ ground and metastable levels in neutral W. According to these data and analyses, $S/XB \approx 20$ should be used for edge temperatures

TABLE 11.1

Selected Lines of Neutral W for Influx Measurements in Fusion Devices and Calculated S/XB Values for $T_e = 20$ eV

	E (cm^{-1})		Transition		
λ (nm)	Lower	Upper	Lower	Upper	S/XB
255.135	0.00	39183.20	a_$^5 D_0$	x_$J = 1$	22
268.142	2951.29	40233.97	b_$^7 S_3$	x_$J = 4$	(54−)
400.875	2951.29	27889.68	b_$^7 S_3$	d_$^7 P_4$	51
429.461	2951.29	26229.77	b_$^7 S_3$	d_$^7 P_2$	93
505.328	1670.29	21453.90	a_$^7 D_1$	c_$^7 D_1$	(321+)

Source: Adapted from I. Beigmann et al., *Plasma Phys. Control. Fusion* 49, 1833–1847, 2007.

Note: a, $5d^4 (^5 D) 6s^2$; b, $5d^5 (^6 S) 6s$; c, $5d^4 (^5 D) 6s 6p$; d, $5d^5 (^6 S) 6p$; x, unidentified. The calculated S/XB values in brackets are inconsistent by more than a factor of 3 with the measurements (+: too high, −: too low; for details, see text).

between 10 and 20 eV with a very moderate increase toward higher temperature. A consistency check of the atomic physics calculations was performed, by a simultaneous measurement of the different spectral lines. It turned out that for the independently measured plasma temperature of 20 eV, no simultaneous agreement with all measured data could be achieved. Although the reason for this failure is not yet resolved, difficulties in the absolute calibration in the UV spectral range and population processes by cascading and collisional transfer were suggested by Beigmann et al. [25].

ASDEX Upgrade uses the WI transition at 400.9 nm for influx measurements since the first installation of a tungsten divertor in 1996 [29,33]. It could be shown already at that time that the major contribution to the W source is W sputtering by multiple charged light intrinsic impurity ions as C^{3+} and O^{4+}. Comparing net (erosion probes) and gross (spectroscopy) erosion measurements for different plasma edge parameters yields values in agreement with the model of prompt redeposition of W ions. By proceeding from single W sources to a device with complete W PFCs, a comprehensive set of LOSs is needed to measure the total W-influx and to allow to quantify its impact on the plasma performance. Figure 11.3 shows the temporal and spatial evolution of the W-influx (color-coded) from a low-field side guard limiter in ASDEX Upgrade, which is sketched in the insertion on the right-hand side.

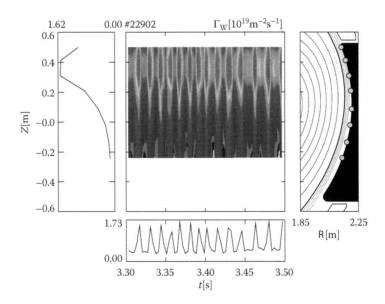

FIGURE 11.3
Fast W-influx measurements at the limiter of ASDEX Upgrade during discharge #22902. The 2D color-coded figure in the center highlights the temporal evolution of the W-influx density along the contour of a low-field side guard limiter, sketched in the right insert. The projections on the left-hand side and on the bottom show the integration of the signal over time and space, respectively.

The measurement is performed on eight LOSs and yields an averaged erosion profile depicted on the left-hand side. The strongly peaked profile highlights the importance of the plasma position in relation to PFCs. The insertion at the bottom gives the projection of all spatial channels on the time axis. The strong modulation of the W-influx is due to frequent instabilities at the plasma edge which eject particles and energy leading to strongly increased W-influxes. Details on these investigations can be found in [34] and references therein.

11.3 Spectroscopy for W-Concentration Measurements

11.3.1 Principle of Concentration Measurements

Neglecting plasma transport, the ionization equilibrium in the core of a fusion plasma can usually be described similarly to the coronal equilibrium, but with a small density-dependent correction to ionization and recombination rates. The total ionization rate is dominated by direct electron ionization and contributions from autoionization of inner-shell excited states. The total recombination rate is mainly given by radiative and dielectronic recombination. In the latter case, the energy of the electron is transferred to the recombining ion leading to double excitation, which is stabilized by a radiative transition. Usually, the stabilization process is more likely than the autoionization process for highly charged ions, since the relaxation rates increase with Z^4.

Including transport effects, the continuity equations in coronal equilibrium can be written as a set of Z equations of the form

$$\frac{\partial}{\partial t}n_q + \nabla\vec{\Gamma}_q = n_e(n_{q-1}S_{q-1} + n_{q+1}\alpha_{q+1} - n_qS_q - n_q\alpha_q) \tag{11.8}$$

with the particle flux Γ_q. Usually, it is described by using a diffusive and convective term

$$\vec{\Gamma}_q = D_q\nabla n_q + v_q n_q. \tag{11.9}$$

For charge stage-resolved transport calculations, as they are necessary for the interpretation of spectroscopic measurements, Z-coupled equations of the form of Equation 11.8 have to be solved. This is done numerically using a transport code as, for example, STRAHL [35,36] or MIST [37]. The particle transport in fusion plasma is still a matter of current research. It depends strongly on the background plasma parameters and usually contains terms caused by Coulomb collisions (neoclassical transport) and fluctuations (anomalous transport). The neoclassical transport parameters (D_q^{neo}, v_q^{neo}) show a pronounced q dependence and therefore are of special interest for the description of the W transport. A recent comprehensive overview of experimental results for impurity transport is given in [38]. For the investigations presented here, it is important to note that the impurity transport at the center of a fusion plasma is usually very small. Therefore, the influence on the deduced quantities, as, for example, the ionization equilibrium, is quite low. However, even small drifts (v_q) can change the total impurity density profile, which has to be taken into account when interpreting spectroscopic measurements.

As in the case of influx measurements, also the measurements of the impurity density in the plasma center are usually based on a direct view (LOS), similar to that described in Section 11.2.1. Generally, fusion plasmas show a monotonically increasing electron temperature from the plasma edge toward its center, leading to a shell-like structure of the ionization stages with an increasing charge number from the edge to the center. Therefore, the observed emission may originate from regions with different excitation conditions given by the radial profiles of electron temperature and density and leading to a multitude of emission features. In principle, these can be reconstructed from independent measurements of the background plasmas, when the adequate atomic physics data as well as the transport parameters are at hand. Then the spectral radiance for a specific transition i from an ion with charge q can be obtained similar to Equation 11.5 as

$$I_q^i = \frac{h\nu}{4\pi} \int_\ell n_q n_e X_{\text{eff}}^i B \, dl, \qquad (11.10)$$

where ℓ denotes the length of the LOS and $h\nu$ the energy of the emitted photon. X_{eff}^i is the so-called photon emissivity coefficient of the transition i. It accounts not only for electron impact excitation, but also other feeding mechanisms resulting from calculations within a collisional radiative model. In principle, it depends on the electron temperature and density, but for the conditions in a fusion plasma the density dependence can usually be neglected, especially since most of the important transitions are resonance lines. As stated above, n_q can be calculated from the coronal equilibrium, that is, $n_q = c_W f_q n_e$, with f_q being the fractional abundance of the ionization stage W^{q+} and $c_W = n_W / n_e$ the total W concentration. By using this relation, the (local) W concentration can be deduced from the LOS integrated measurement of I_q^i in the following way:

$$c_W = \frac{4\pi I_q^i}{h\nu \int_\ell f_q n_e^2 X_{\text{eff}}^i B \, dl}. \qquad (11.11)$$

Note that due to plasma transport, c_W need not to be a constant over the whole LOS and the value extracted using the method described is valid only in the temperature region where the specific transition has a significant emissivity. Due to the fact that the abundance of the W ionization stages depend strongly on T_e, which itself is a function of plasma radius (see Figure 11.4), localized measurements using passive spectroscopy are possible. At the same time, this means that for diagnosing the W-concentration profile, the spectral emissions of many different ionization stages need to be taken into account.

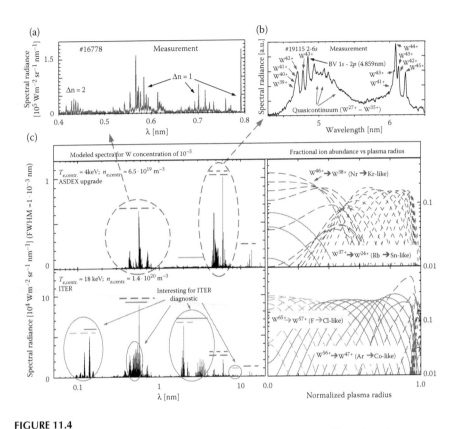

FIGURE 11.4
(a) Strong spectral feature emitted by Br-like W^{39+} to about Mn-like W^{49+} in the soft x-ray range (ASDEX Upgrade); (b) Emissions of Ag-like W^{27+}–Y-like W^{35+} blended by emissions from Br-like W^{39+}–Cu-like W^{45+} (ASDEX Upgrade); (c) On the left-hand side, modeled spectra for a typical ASDEX Upgrade discharge ($T_{e,\text{central}} \approx 4.0\,\text{keV}$) and a ITER discharge ($T_{e,\text{central}} \approx 18\,\text{keV}$) are depicted (details in text). On the right-hand side, the underlying ionization balance versus normalized plasma radius is presented taking estimations for impurity transport into account. The spectral radiance is calculated in both cases for a W concentration of $c_W = 10^{-5}$.

11.3.2 Highly Charged W Ions in Tokamaks

Spectral lines from highly charged tungsten ions have been found in the vacuum ultraviolet (VUV) at least since the use of a W limiter in PLT and ORMAK [5,6,14]. However, concentrations of W in fusion plasmas were mostly calculated from the radiation loss parameter using an average ion model [39,40]. Therefore, quantitative spectroscopy has to be developed to extract W densities from spectroscopic measurements. Ideally, lines from all existing ionization stages should be evaluated in order to compensate for the missing radial resolution of the LOS measurements (see Section 11.3.1). According to the temperature, which could be achieved in the tokamaks of the 1970s, the first spectral features identified in fusion plasmas originated from charge stages from Ag-like W^{27+} to Y-like W^{35+} and are known as the W quasicontinuum (QC) around 5 nm [6,14]. A more recent study of the tungsten impurity spectrum at the ASDEX Upgrade tokamak [17] reported that for electron temperatures above 2 keV, many isolated spectral lines appear in the spectrum which superimpose the QC emission band. These lines represent transitions in the higher charged tungsten ions (up to Cu-like W^{45+}; see Figure 11.4b). The behavior of the W QC could be investigated in more detail using discharges which exhibit the well-characterized phenomenon of impurity accumulation [41]. There, the tungsten impurity density at the center of the plasma (i.e., inside the inner 20% of plasma radius) peaks by a factor as large as 20, due to neoclassical transport effects. As a result, the measured spectra are dominated by the emissions in this narrow region with small radial variations in the electron temperature ($T_e = \pm 200$ eV). Additionally, temporal variations of the central temperature allow us to scan different ionization stages. Similarly as in the EBIT spectra (see [19,20,22,42] and next subsection), the distinct shift of the center of gravity of the QC as well as the appearance of single lines can be observed. However, the additional broad structure appearing around 6 nm, which is typical of the W QC measured in tokamak plasmas, does not show up in the EBIT spectra [22]. This obvious difference is possibly related to the higher electron densities in the tokamak compared to the EBIT, which could enhance many weak spectral lines as indicated in the calculations presented in [19].

Accumulation discharges were also used to deduce the temperature dependence of the fractional abundances of W^{27+}–W^{46+} (as an example, results for W^{43+} and W^{44+} are presented in Figure 11.5). Here, one uses the fact that the emissivity of a VUV spectral line only weakly depends on the plasma temperature, but is strongly related to the abundance of the related ion stage, because the excitation rate is almost constant over the temperature range where the ion exists [22,43]. By varying T_e in the accumulation zone, the temperature dependence of the fractional abundance can be deduced. When comparing the results of the measurements to calculations using ionization/recombination rates from ADPAK [39], where the ionization rates were corrected for excitation autoionization [17], a strong temperature shift in the equilibrium is observed. Using the latest ionization rates from configuration average distorted wave (CADW) calculations, Loch et al. [44] shift the calculated fractional abundances to too high temperatures. Finally, the solid lines represent calculations where the recombination rates were adjusted (modified ADPAK) to reproduce the experimental findings. Due to the high actual importance of W in fusion plasmas, all the necessary rates are presently revisited and recalculated [18,43] within the framework of the ADAS database [45].

Similar to the increased amount of information in the VUV spectral range, a lot of new data have been produced in the SXR spectral range [18,23,43,46–49]. Here, usually transitions with $\Delta n \geq 1$ are observed, making atomic physics calculations much more reliable. A comprehensive compilation of transitions within W^{27+}–W^{46+} from *ah initio* calculations with the Hebrew University Lawrence Livermore Atomic structure Code package (HULLAC) is found

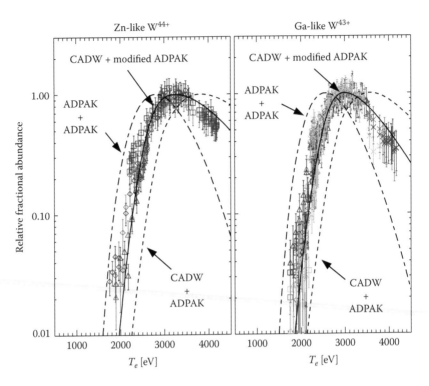

FIGURE 11.5
Measured fractional abundances of Zn-like W^{44+} and Ga-like W^{43+} are compared to various theoretical curves (details in text).

in [50]. Generally, the wavelength as well as the strength of the single spectral lines compare well to the measurements from the ASDEX Upgrade tokamak, where SXR spectral lines originating from W^{39+}–W^{49+} have been identified in the spectral range from 0.4 to 2.0 nm (see Figure 11.4a). Again, the assignment of the transitions is supported by measurements at the LLNL and Berlin EBITs, where electron beam energies up to 4.7 keV were used to record spectra in the 0.5–0.6 nm range (see [51] and Figure 11.6).

The measurements both from the VUV and the SXR spectral regions are routinely used to derive W density profiles in ASDEX Upgrade discharges (see, e.g., [52] and Figure 11.7). Both features have been targeted with atomic data calculations [16,18,43,50,53,54] such that a fairly good understanding between the two spectra exists. The calculated wavelengths of the most prominent spectral lines emitted by W^{41+} to W^{48+} are given in Table 11.2, together with measurements at the ASDEX Upgrade Tokamak (λ_{tok}) and the Berlin (λ_{Berlin}) LLNL ($\lambda_{L/N}$ (L)) [51], and NIST ($\lambda_{L/N}$ (N)) [21] EBITs. The uncertainties of λ_{tok} and λ_{Berlin} are typically ± 0.001 nm in the SXR spectral range (<2 nm) and ± 0.005 nm in the VUV (>2 nm), respectively. The ionization potentials (IP) are taken from [55] and the theoretical wavelengths originate either from

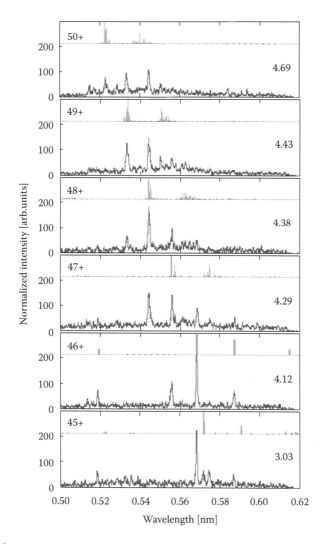

FIGURE 11.6

Measured and predicted soft x-ray spectra of tungsten ions. The darker line marks the experimental observation and is labeled (right) with the electron beam energy in units of (keV) at which EBIT is operated. The light gray histogram above each spectrum shows the calculated line intensity and position and is marked with the corresponding charge stage of the ion (left). (The theoretical data are taken from K. Fournier, *At. Data Nucl. Data Tables* 68, 1–48, 1998 and private communication. With permission.)

GRASP (G) [56] or from the Cowan code (C). Only lines which could unambiguously be identified and which do not show strong blending from other transitions are presented. The transitions are not only identified by their wavelength, but also by their spectral radiance, which was calculated within the

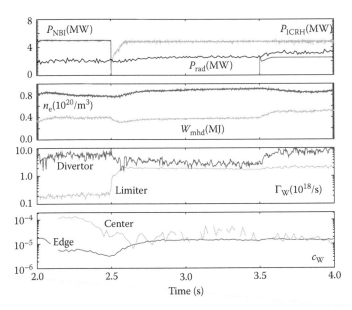

FIGURE 11.7
W-influx and W concentration in ASDEX Upgrade during discharge #23476. The two top graphs present the heating power (P_{NBI}, P_{ICRH}) and total radiation (P_{rad}) as well as the line-averaged density (n_e) and the stored energy (W_{mhd}) of the plasma. The third graph highlights the W-influx (Γ_W) from the limiters and the divertor and the bottom insert shows the deduced W concentration (c_W) at the plasma edge and the center.

collisional radiative model of ADAS. In the modeling, some of the initial levels are characterized by a dominant configuration in the *LS*-coupling scheme, others are better characterized in the *jj*-coupling scheme. In order to provide a consistent description for all levels, they are described by their configuration, their total angular momentum *J*, and the *A*-value for the transition to the corresponding lower level in Table 11.2. For most transitions, this is the ground state, which is then omitted for simplicity. In the few cases, where the lower level corresponds to an excited state, the energy and the configuration of this state are given separately.

For the diagnostic of W in ITER-like plasmas, higher charged tungsten ions have to be taken into account, because of the predicted central electron temperature of about 18 keV [57]. In Figure 11.4c, a modeling of the spectral emissions between 0.05 and 20 nm is presented (left-hand side of Figure 11.4c) for an ASDEX Upgrade plasma (top part) and an ITER plasma. On the right-hand side of Figure 11.4c, the fractional abundances of the ionization stages versus normalized plasma radius are presented, taking into account typical parameters for impurity transport [18]. The ionization stages investigated so far in ASDEX Upgrade and other fusion devices cover only the plasma boundary and the edge region. In the central region, ionization stages up to W[68+] are expected. The W ionization stages W[49+]–W[56+] exhibit only transitions which

TABLE 11.2

Experimentally Observed and Predicted Spectral lines of W Ions Recommended for Diagnostics of W in a Fusion Plasma

Ion	IP (eV)	λ_{calc}	λ_{tok}	λ_{Berlin}	$\lambda_{L/N}$	J_{low}	J_{up}	Configuration	A (s^{-1})
W^{41+}	1995	6.482 (G)	6.481	6.482	6.488 L	$\frac{3}{2}$	$\frac{5}{2}$	$4s4p^4$	1.7(4)
(As-like)		13.106 (G)	13.121			$\frac{3}{2}$	$\frac{5}{2}$	$4s^24p^3$	1.7(6)
		13.968 (G)	13.896		13.914 N	$\frac{3}{2}$	$\frac{3}{2}$	$4s^24p^3$	8.1(6)
W^{42+}	2149	0.5818 (C)	0.583			0	1	$3d^94s^24p^24f$	2.5(14)
(Ge-like)		0.5823 (C)				0	1	$3d^94s^24p^24f$	1.5(14)
		0.6010 (C)	0.601			0	1	$3d^94s^24p^24f$	1.3(14)
		4.685 (G)	4.718	4.697	4.719 L	0	1	$4s^24p4d$	2.0(12)
		6.115 (G)	6.123	6.130	6.130 L	0	1	$4s4p^3$	3.5(11)
		12.940 (G)	12.945		12.941 N	0	2	$4s^24p^2$	1.8(5)
a		13.029 (G)	12.912			1	2	$4s4p^3$	5.1(6)
b		13.690 (G)	13.475		13.495 N	2	2	$4s4p^3$	5.1(6)
W^{43+}	2210	0.5798 (G)	0.579			$\frac{1}{2}$	$\frac{1}{2}$	$3d^94s^24p4f$	3.7(14)
(Ga-like)		0.5801 (G)				$\frac{1}{2}$	$\frac{3}{2}$	$3d^94s^24p4f$	4.0(14)
		0.5988 (G)	0.598			$\frac{1}{2}$	$\frac{3}{2}$	$3d^94s^24p4f$	1.3(14)
		0.5989 (G)				$\frac{1}{2}$	$\frac{1}{2}$	$3d^94s^24p4f$	1.3(14)
		4.760 (G)	4.791	4.769	4.790 L	$\frac{1}{2}$	$\frac{3}{2}$	$4s^24d$	1.4(12)
		6.020 (G)	6.063	6.061	6.061 L	$\frac{1}{2}$	$\frac{1}{2}$	$4s4p^2$	7.8(11)
		6.119 (G)	6.135	6.129	6.133 L	$\frac{1}{2}$	$\frac{3}{2}$	$4s4p^2$	4.1(11)
		12.587 (G)	12.639		12.629 N	$\frac{1}{2}$	$\frac{3}{2}$	$4s^24p$	5.3(6)
		12.899 (G)	12.824		12.817 N	$\frac{1}{2}$	$\frac{1}{2}$	$4s4p^2$	3.0(10)
c		13.682(G)	13.534		13.481 N	$\frac{3}{2}$	$\frac{5}{2}$	$4s4p^2$	1.6(10)
W^{44+}	2355	0.5749 (G)	0.575			0	1	$3d^94s^24f$	4.0(14)
(Zn-like)		0.5938 (G)	0.595			0	1	$3d^94s^24f$	1.3(14)
		6.073 (G)	6.093	6.087	6.093 L	0	1	$4s4p$	7.2(11)
		13.230 (G)	13.287	13.275	13.288 N	0	1	$4s4p$	2.0(10)
W^{45+}	2414	0.5721 (G)	0.572	0.5716	0.5719 L	$\frac{1}{2}$	$\frac{1}{2}$	$3d^94s4f$	4.1(14)
(Cu-like)		0.5725 (G)		0.5723	0.5724 L	$\frac{1}{2}$	$\frac{1}{2}$	$3d^94s4f$	4.1(14)
		0.5911 (G)	0.591	0.6003		$\frac{1}{2}$	$\frac{1}{2}$	$3d^94s4f$	1.3(14)
		0.5912 (G)				$\frac{1}{2}$	$\frac{3}{2}$	$3d^94s4f$	1.3(14)
		0.7268 (G)	0.725			$\frac{1}{2}$	$\frac{1}{2}$	$3d^94s4p$	5.0(12)
		0.7273 (G)				$\frac{1}{2}$	$\frac{3}{2}$	$3d^94s4p$	4.7(12)
		6.217 (G)	6.232	6.232	6.234 L	$\frac{1}{2}$	$\frac{3}{2}$	$4p$	4.4(11)
		12.609 (G)	12.720	12.701	12.712 N	$\frac{1}{2}$	$\frac{1}{2}$	$4p$	4.8(10)

continued

TABLE 11.2 (continued)

Experimentally Observed and Predicted Spectral lines of W Ions Recommended for Diagnostics of W in a Fusion Plasma

Ion	IP (eV)	λ_{calc}	λ_{tok}	λ_{Berlin}	$\lambda_{L/N}$	J_{low}	J_{up}	Configuration	A (s^{-1})
W^{46+}	4057	0.5201 (C)		0.5188	0.5196 L	0	1	$3p^5 3d^{10} 4d$	8.7(13)
(Ni-like)		0.5687 (G)	0.569	0.5684	0.5690 L	0	1	$3d^9 4f$	4.1(14)
		0.5875 (G)	0.587	0.5872	0.5867 L	0	1	$3d^9 4f$	1.3(14)
		0.7035 (G)	0.702			0	1	$3d^9 4p$	1.4(13)
		0.7184 (G)	0.716			0	1	$3d^9 4p$	5.8(12)
		0.7944 (G)	0.793			0	2	$3d^9 4s$	6.0(9)
W^{47+}	4180	0.5550 (C)	0.556	0.5553	0.5564 L	$\frac{5}{2}$	$\frac{5}{2}$	$3d^8 4f$	2.3(14)
(Co-like)		0.5553 (C)		0.5558	0.5568 L	$\frac{5}{2}$	$\frac{7}{2}$	$3d^8 4f$	4.2(14)
		0.5740 (C)		0.5747	0.5748 L	$\frac{5}{2}$	$\frac{7}{2}$	$3d^8 4f$	1.1(14)
		0.5746 (C)				$\frac{5}{2}$	$\frac{5}{2}$	$3d^8 4f$	7.7(13)
W^{48+}	4309	0.5438 (C)	0.545	0.5443	0.5449 L	4	5	$3d^7 4f$	4.2(14)
(Fe-like)		0.5444 (C)			0.5452 L	4	4	$3d^7 4f$	1.9(14)
		0.5446 (C)			0.5457 L	4	3	$3d^7 4f$	2.0(14)
		0.5609 (C)		0.5610	0.5617 L	4	5	$3d^7 4f$	7.2(13)
d		0.5621 (C)				2	3	$3d^7 4f$	1.1(14)
		0.5622 (C)				4	4	$3d^7 4f$	3.7(13)
		0.5625 (C)				4	5	$3d^7 4f$	4.6(13)
		0.5631 (C)		0.5644	0.5638 L	4	4	$3d^7 4f$	3.8(13)

Note: IP: ionization potentials. λ_{calc}: predicted wavelengths from GRASP (G) or from the Cowan code (C). λ_{tok}, λ_{Berlin}, $\lambda_{L/N}$: experimental wavelengths from tokamak and EBIT measurements (all wavelengths in nm). J_{low}, J_{up}: total angular momentum of the lower and upper state, configuration: configuration of upper level, A: A-value, exponent given in parentheses. Transitions to excited states are denoted in first column (energy, configuration of lower level: a: 91.23 eV, $4s^2 4p^2$; b: 95.82 eV, $4s^2 4p^2$; c: 98.50 eV, $4s^2 4p$; d: 9.72 eV, $3d^8$).

lead to quasicontinuous structures extending from 0.4 to 0.7 and from 2.7 to 4.0 nm. The strongest spectral lines of the ions W^{57+} to W^{68+} are predicted in the wavelength ranges from 0.10 to 0.15 and 1.8 nm to 3.5 nm and some weaker spectral lines are predicted at 7–8 nm. All of these spectral regions seem to be well suited for the monitoring of these highly charged W ions, because they will not be blended by spectral lines from lower charge stages. It is to be noted that the simulations predict spectral radiances in ITER up to a factor of 10 larger than those predicted (and observed) in ASDEX Upgrade, providing a good perspective for spectroscopic W diagnostic in ITER. Table 11.3 provides a detailed calculated wavelength for transitions in W^{55+} to W^{66+} in the wavelength range of 0.12–0.15 nm.

TABLE 11.3

Same as Table 11.2 for W-ions with $55+ \leq q \leq 66+$

Ion	IP (eV)	λ_{calc}	λ_{Berlin}	J_{low}	J_{up}	Config. up st.	A-value
W^{55+}	5348	0.12313	0.12347	$\frac{3}{2}$	$\frac{1}{2}$	$2p^53s^23p^63d^2$	1.5(15)
(K-like)		0.12323		$\frac{3}{2}$	$\frac{3}{2}$	$2p^53s^23p^63d^2$	3.0(15)
		0.12356		$\frac{3}{2}$	$\frac{5}{2}$	$2p^53s^23p^63d^2$	5.1(14)
W^{56+}	5719	0.12274	0.12264	0	1	$2p^53s^23p^63d$	2.0(15)
(Ar-like)		0.14034	0.14029	0	1	$2p^53s^23p^63d$	2.4(15)
W^{57+}	5840	0.12234	0.12247	$\frac{3}{2}$	$\frac{3}{2}$	$2p^53s^23p^53d$	2.0(15)
(Cl-like)		0.12236		$\frac{3}{2}$	$\frac{5}{2}$	$2p^53s^23p^53d$	2.0(15)
		0.12246		$\frac{3}{2}$	$\frac{1}{2}$	$2p^53s^23p^53d$	1.9(15)
		0.13961	0.13975	$\frac{3}{2}$	$\frac{5}{2}$	$2p^53s^23p^53d$	7.7(14)
		0.13966		$\frac{3}{2}$	$\frac{1}{2}$	$2p^53s^23p^53d$	1.9(15)
		0.13978		$\frac{3}{2}$	$\frac{3}{2}$	$2p^53s^23p^53d$	2.4(15)
		0.14003		$\frac{3}{2}$	$\frac{5}{2}$	$2p^53s^23p^53d$	1.6(15)
W^{58+}	5970	0.12195	0.12207	2	2	$2p^53s^23p^43d$	9.1(14)
(S-like)		0.12197		2	3	$2p^53s^23p^43d$	2.0(15)
		0.12201		2	1	$2p^53s^23p^43d$	1.9(15)
		0.12205		2	2	$2p^53s^23p^43d$	1.1(15)
		0.13921	0.13927	2	1	$2p^53s^23p^43d$	2.2(15)
		0.13925		2	2	$2p^53s^23p^43d$	1.3(15)
		0.13941		2	3	$2p^53s^23p^43d$	1.1(15)
		0.13944		2	2	$2p^53s^23p^43d$	9.0(14)
W^{59+}	6093	0.12149	0.12161	$\frac{3}{2}$	$\frac{1}{2}$	$2p^53s^23p^33d$	2.1(15)
(P-like)		0.12152		$\frac{3}{2}$	$\frac{5}{2}$	$2p^53s^23p^33d$	2.0(15)
		0.12156		$\frac{3}{2}$	$\frac{3}{2}$	$2p^53s^23p^33d$	2.0(15)
		0.13867	0.13871	$\frac{3}{2}$	$\frac{1}{2}$	$2p^53s^23p^33d$	2.3(15)
		0.13874		$\frac{3}{2}$	$\frac{3}{2}$	$2p^53s^23p^33d$	2.3(15)
		0.13879		$\frac{3}{2}$	$\frac{5}{2}$	$2p^53s^23p^33d$	1.7(15)
W^{60+}	6596	0.12110	0.12118	0	1	$2p^53s^23p^23d$	2.1(15)
(Si-like)		0.13810	0.13824	0	1	$2p^53s^23p^23d$	2.5(15)
W^{61+}	6735	0.12023	0.12060	$\frac{1}{2}$	$\frac{3}{2}$	$2p^53s^23p3d$	6.0(14)
(Al-like)		0.12039		$\frac{1}{2}$	$\frac{1}{2}$	$2p^53s^23p3d$	2.0(15)
		0.12065		$\frac{1}{2}$	$\frac{3}{2}$	$2p^53s^23p3d$	1.6(15)
		0.13752	0.13760	$\frac{1}{2}$	$\frac{3}{2}$	$2p^53s^23p3d$	2.6(15)
		0.13755		$\frac{1}{2}$	$\frac{1}{2}$	$2p^53s^23p3d$	2.6(15)

continued

TABLE 11.3 (continued)

Same as Table 11.2 for W-ions with $55+ \leq q \leq 66+$

Ion	IP (eV)	λ_{calc}	λ_{Berlin}	J_{low}	J_{up}	Config. up st.	A-value
a		0.15121	0.15102	$\frac{3}{2}$	$\frac{1}{2}$	$2p^53s^23p^2$	4.9(13)
b		0.15141		$\frac{1}{2}$	$\frac{3}{2}$	$2p^53s^23p^2$	2.7(13)
W^{62+}	7000	0.11998	0.11994	0	1	$2p^53s^23d$	2.1(15)
(Mg-like)		0.13692	0.13697	0	1	$2p^53s^23d$	2.6(15)
c		0.15036	0.15058	1	0	$2p^53s^23p$	9.6(13)
W^{63+}	7130	0.11957	0.11961	$\frac{1}{2}$	$\frac{3}{2}$	$2p^53s3d$	2.1(15)
(Na-like)		0.11961		$\frac{1}{2}$	$\frac{1}{2}$	$2p^53s3d$	2.0(15)
		0.13638	0.13643	$\frac{1}{2}$	$\frac{1}{2}$	$2p^53s3d$	2.6(15)
		0.13643		$\frac{1}{2}$	$\frac{3}{2}$	$2p^53s3d$	2.7(15)
		0.15011	0.14986	$\frac{1}{2}$	$\frac{3}{2}$	$2p^53s3d$	9.6(13)
W^{64+}	15566	0.11919	0.11936	0	1	$2p^53d$	2.1(15)
(Ne-like)		0.13590	0.13593	0	1	$2p^53d$	2.7(15)
		0.14935	0.14967	0	1	$2p^53s$	1.3(14)
W^{65+}	15896	0.13341	0.13350	$\frac{3}{2}$	$\frac{5}{2}$	$2p^43d$	1.2(15)
(F-like)		0.13395	0.13397	$\frac{3}{2}$	$\frac{3}{2}$	$2p^43d$	1.9(15)
		0.13397		$\frac{3}{2}$	$\frac{5}{2}$	$2p^43d$	1.4(15)
W^{66+}	16252	0.13167	0.13172	2	3	$2p^33d$	2.2(15)
(O-like)							

Note: The experimentally observed wavelengths are extracted from Berlin EBIT measurements. Calculated wavelengths are from GRASP code. Transitions to excited states are denoted in first column (energy and configuration of lower level: a: 544.4 eV, $3s3p^2$; b: 555.1 eV, $3s3p^2$; c: 547.2 eV, $3s3p^2$).

11.3.3 EBIT Investigations Preparing the Diagnostic of W in Hot Fusion Plasma

An EBIT equipped with an appropriate spectroscopic equipment serves as an instrument dedicated to precisely measure the radiation of highly charged ions. The monoenergetic, strongly compressed electron beam of EBIT successively ionizes atoms up to a selected charge stage, excites the ions at any particular electron beam energy, and traps the ions for a long observation time. The Berlin EBIT can operate with electron beam energies up to 40 keV for production of tungsten ions in charge stages up to 72+ and electron currents of 150 mA resulting in electron densities of about 3×10^{18} m^{-3}. This ensures that electron excitation processes similar to fusion plasmas dominate radiative transitions. Due to the monoenergetic electron beam energy, no integration over a range of electron temperatures and densities occurs. The radiation emitted by the highly charged ions trapped in the Berlin EBIT is analyzed by high-resolution x-ray and EUV spectroscopy [58]. In the wavelength range 0.04–1.6 nm, x-ray lines are spectrally dispersed with a large area flat-crystal in Bragg geometry and registered by a position-sensitive proportional counter leading to a resolution of $\lambda/\Delta\lambda \geq 1000$. Diagnostics of the EUV emission in the wavelength region 3–100 nm is performed with a 2-m Schwob–Fraenkel grazing incidence spectrometer reaching a resolution of $\lambda/\Delta\lambda \sim 200$–4000 depending on grating and wavelength. The spectral measurements are calibrated using well-known reference lines from a standard database. Further, a solid-state detector monitors the x-radiation between 0.4 and 30 keV energy with large solid angle and high detection efficiency providing broadband spectra of the highly charged ion inventory of EBIT. The measurement procedure consists of incrementing the electron beam energy in small steps and recording spectra for each step. By this procedure, the variation of spectral features can be followed across a range of beam energies and ionization thresholds revealing which lines can be attributed to a particular charge stage. The lines are classified by comparison with theoretical predictions. As stated before, W charge stages from Zn-like W^{44+} up to Cr-like W^{50+} will exist in the edge of the main plasma of ITER. The dominant line radiation of ions with these charge stages is located in the soft x-ray spectral region around 0.5 nm [18]. This wavelength interval between 0.5 and 0.62 nm was investigated at the Berlin EBIT with the Bragg spectrometer using a flat ADP crystal. Figure 11.6 shows a series of spectra when increasing the electron beam energies from 3.03 to 4.69 keV. The measured line intensities vary as function of beam energy and enhance considerably as the population of the emitting ions is at its maximum. Stepping from 3.03 keV beam energy to 4.12 keV, the most abundant tungsten charge stage shifts to Ni-like W^{46+}. The spectrum is dominated by a line at 0.5685 ± 5 nm assigned to a $3d^{10}$–$3d^9$ $4f$ transition in Ni-like W^{46+}. In the 3.03-keV beam-energy spectrum additional to lines from Cu-like W^{45+} at 0.5716 nm, a minor contribution from Zn-like W^{44+} is observed with a line at 0.5747 nm. When progressively increasing the beam energy, the ion population raises to higher charge stages. The spectrum plotted for 4.29 keV beam energy shows a weakening of the 0.5685-nm Ni-like W line, which indicates the extinction of W^{46+} ions in the trap for beam energies above 4.3 keV. Increasing the beam energy further results in the successive production and excitation of Co-like W^{47+}, Fe-like W^{48+}, Mn-like W^{49+}, and finally Cr-like W^{50+} at 4.69 keV. These tungsten charge stages give rise to a multitude of lines within a narrow wavelength range. The line features from these ions incorporate contributions from a large number of $3d$–$4f$ transitions, since they are wider than expected for a single line. Above the plot of the measured spectra in Figure 11.6, the upper panel displays a calculated spectrum for each single tungsten charge stage. The simulated spectra are generated by *ab initio* calculations using HULLAC and relative populations for each level are obtained from collisional-radiative modeling [50]. From the comparison of the experimental and theoretical spectra shown in Figure 11.6, it is evident that the observations confirm the general pattern, intensity relation, and line position within each spectrum. For each tungsten charge stage, a small range of very closely spaced lines is predicted and measured as an unresolved single feature characterizing the spectrum. Similar studies of M-shell spectra of tungsten ions conducted at the Lawrence Livermore National Laboratory EBIT confirm the identifications [51].

As discussed in the previous section, it is predicted that tungsten will be ionized up to C-like W^{68+} in the central plasma of an ITER or a reactor. These W ions have an open or even empty $n = 3$ shell and when excited, emit strong lines from $n = 2$–3 transitions. Investigations on these ions and their radiation have been started at the Berlin EBIT ($\Delta n = 0$ transitions, located in the VUV spectral region, were targeted by recent measurements at the NIST EBIT [59,60]). L-shell spectra at electron beam energies between $E_{beam} = 9.5$ and 20.6 keV have been recorded to investigate the line intensity distribution as function of the ion charge stage abundance in EBIT. As an example, the spectra measured with the flat-crystal spectrometer at $E_{beam} = 15$ keV and $E_{beam} = 19$ keV are shown in Figures 11.8a and 11.8d. The spectra are assembled from two spectrometer settings each due to the limited wavelength range

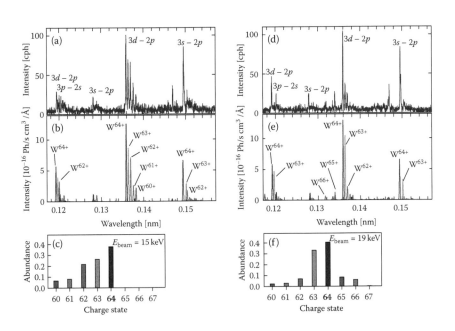

FIGURE 11.8

X-ray spectra of highly charged tungsten observed with a Bragg-crystal spectrometer at two electron beam energies of (a) 15 keV and (d) 19 keV. The charge stage abundance for the respective beam energy (c) and (f) is obtained from the measurement of the radiative recombination process with a solid-state detector. The theoretical intensity distribution is shown in (b) and (e) (details in text).

accepted by the detector and the narrow rocking curve of the LiF (220) crystal. At a beam energy of 15 keV, tungsten ions up to Ne-like W^{64+} are produced, because the energy required to ionize them to the next higher charge stage is 15.566 keV. Monitoring the x-ray emission from the trap with the broad-band solid-state detector, the line emission from directly excited transitions, and, characteristically for EBIT, the x-rays from the radiative recombination (RR) process appearing as distinct line features at energies larger than the electron beam energy are observed. The x-ray energy is well defined by the narrow electron-beam energy distribution of typically 50 eV full-width at half-maximum (FWHM) and the binding energy of the capture state of the recombining ion. We have exploited the information of the intensity distribution of radiative recombination to the $n = 3$ level to gain knowledge about the charge stage abundance trapped and excited in EBIT. For each ion charge stage, the energy of recombination to each nl subshell (l is the angular momentum state) is determined from atomic structure calculations and the predicted RR line is folded with the electron-beam energy distribution and the detector response. The theoretical RR intensity is calculated from the well-known RR cross-section and the detection efficiency. A fit of the theoretical distribution for all relevant ion stages to the experimentally observed RR spectrum yields the abundance of tungsten charge stages. The resulting charge stage distribution measured for 15 keV electron beam energy is plotted in Figure 11.8c. Four distinct line structures are observed, each splitting into a sequence of lines from different ion charge stages. The structures can be attributed to groups of $2s - 3p$, $2p - 3d$, and $2p - 3s$ transitions and a weaker line group of $2p - 3s$ transitions at 0.128 nm. An increase in the electron-beam energy to $E_{beam} = 19$ keV above the ionization potential of Ne-like W^{64+} raises the highly charged ion abundance up to $q = 67+$ as shown in Figure 11.8f. While the spectrum at a beam energy of 15 keV clearly shows emissions from tungsten ions up to $q = 64+$, but not from ions with larger q, the charge stage abundance for 19 keV beam energy is shifted to higher values. Nevertheless, at this energy the x-ray emission is still dominated by the very strong transitions of Na-like W^{63+} and Ne-like W^{64+} tungsten ions. As stated before, the classification of the lines has been supported by calculations using the computer codes from the ADAS [45], which include atomic structure calculations with the Cowan code and excitation cross-sections in coronal approximation. Additionally, the ion abundance extracted from the analysis of the radiative recombination measurement enters the calculation. Some of the theoretical lines are very closely spaced and are folded with a 0.00028-nm FWHM Gaussian to resemble the detector resolution. The wavelength values have been corrected by fully relativistic atomic structure calculations of GRASP [56] and are given in the fourth column of Table 11.3. The predicted x-ray intensity distribution shown in Figure 11.8b (resp. Figure 11.8e) demonstrates that each line within a group corresponds to transitions of a particular ion charge stage. To the shorter wavelength side, each of the transition groups ($n = 2s - 3p$, $2p - 3d$, $2p - 3s$) is headed by a line produced by the dominant Ne-like W^{64+}

ion. Lines from transitions of the next lower charge stage are shifted to larger wavelength values. The largest discrepancy is observed for the W^{64+} line measured at 0.1468 nm with larger intensity than the theory predicts for the $2p^6\,^1S_0 - 2p^5\,3p\,^3D_2$ electric quadrupole transition, which may not be adequately treated by the model as already observed for a similar transition in Ni-like W^{46+} [18].

11.3.4 W-Concentration Measurements in Tokamaks

In principle, the procedure described in Section 11.3.1 can be used to obtain W concentrations from the measurement of spectral lines. However, the interpretation of spectral emissions in the VUV to soft x-ray range is often not straightforward, because it is challenging to obtain an absolute intensity calibration and additionally, the atomic data may still contain considerable uncertainties for the absolute values. Therefore, a calibration procedure is presented in the following, which connects the spectral emissions of W to the total radiative losses they cause in the plasma. The latter value is then related to the W-concentration. Since the primary threat of W in a fusion plasma is its strong radiation, this procedure has the advantage that the concentration is directly connected to the total radiation without having the uncertainty in the theoretical and experimental absolute intensity of a spectral line. Vice versa, the procedure also allows a consistency check for the atomic data.

This calibration of the spectral measurements is obtained by injecting W in a plasma discharge by a laser ablation system (see, e.g., [61]). The increased W-impurity content (Δc_W) of the plasma leads to an enhancement of the plasma radiation (ΔP_{rad}) which is related to an increase in the intensity of specific spectral lines under investigation similar to the W-concentration, which is derived using the cooling factor L_W of W [39,40,62]. Prerequisites for this procedure are the detailed knowledge of the plasma parameters, an adequate diagnostic of the total radiation from the plasma (which is available at all major tokamaks) and the injection of only minor amounts of W, in order not to change the plasma background. Special care should be taken while designing the plasma discharge to yield a flat W concentration profile around the plasma radius at which the charge stages of interest exist. This region of the plasma is denoted in the following plasma volume of interest and the increase of the radiation in this region is then given as

$$\Delta P_{rad} = \int_{PVI} n_e^2 \Delta c_W L_W(T_e, n_e)\, dV$$

$$= \Delta c_W \int_{PVI} n_e^2 L_W\, dV \tag{11.12}$$

The various spectral emissions (i) of the W-ions with charge q exhibit for the same discharge an increase in intensity of ΔI_q^i measured on an LOS ℓ, similar to that described in Equation 11.10. Hence

$$\Delta I_q^i = \frac{h\nu}{4\pi F_c} \int_\ell \Delta c_W f_q(T_e, n_e) n_e^2 X_{eff}^i(T_e, n_e) B \, dl$$

$$= \Delta c_W \frac{h\nu}{4\pi F_c} \int_\ell f_q n_e^2 X_{eff}^i \, dl, \tag{11.13}$$

where F_c is a constant calibration factor, which accounts for the unknown sensitivity of the spectrometer-detector system and for inaccuracies in the absolute value of the X_{eff}^i. From the above equations, F_c can be deduced as

$$F_c = \frac{\Delta P_{rad}}{\Delta I_q^i} \frac{h\nu}{4\pi} \frac{\int_\ell f_q n_e^2 X_{eff}^i B \, dl}{\int_{PVI} n_e^2 L_W \, dV}. \tag{11.14}$$

For discharges other than the described calibration discharges, c_W can then be determined by

$$c_W = \frac{4\pi I_q^i F_c}{h\nu \int_\ell f_q n_e^2 X_{eff}^i B \, dl}. \tag{11.15}$$

In principle, the knowledge of X_{eff}^i and f_q and the measurements of n_e, T_e, and I_q^i are necessary to evaluate c_W at the plasma radius where the ionization stage q is abundant, however, the uncertainty in the absolute value of X_{eff}^i is removed from the problem in this way. Moreover, as stated already in Section 11.3.2, the relative behavior of f_q versus T_e is dominant compared to that of X_{eff}^i for highly charged W ions, and the latter is therefore approximately constant in the T_e range in which f_q is larger than 10^{-2} of its maximum value [22]. Thus, the following simplification of Equation 11.15 can be assumed for most of the transitions of highly charged W ions ($\Delta n \leq 1, q > 20+$):

$$c_W = \frac{4\pi I_q^i F_c^i}{\int_\ell f_q n_e^2 \, dl}, \tag{11.16}$$

where $F_c^i = 4\pi F_c/h\nu X_{eff}^i B$ and X_{eff}^i is evaluated at the plasma radius where f_q has its maximum. In practice, F_c^i is derived directly from Equation 11.14 and thus no detailed knowledge on X_{eff}^i is necessary.

The tungsten concentration in fusion plasmas depends on the W source rate at the plasma boundary and the impurity transport into and within the main plasma. However, due to very complex and only partly understood processes, a large variation of the central W concentration is observed even for a constant W source rate or—vice versa—an almost constant W concentration is found for strongly varying W sources. As an example, Figure 11.7 shows temporal evolution of some plasma parameters for a typical ASDEX Upgrade discharge (#23476) with different heating methods. The W-infiux is deduced from the WI line at 400.9 nm as described in Section 11.2. The W concentration is deduced from the QC emission at 5 nm and the spectral line at 0.794 nm emitted from Ni-like W^{46+}. The first gives the W concentration close to the plasma edge, whereas the latter represents the central concentration in typical ASDEX Upgrade discharges, as indicated in the upper part of Figure 11.4c. Already from these few time traces a lot of information of the qualitative behavior of W for different heating methods can be derived. During the first phase (until $t = 2.5$ s), the plasma is heated by neutral beam injection (NBI) only. During this phase, the W source at the limiter in the main chamber is more than a factor of 10 smaller than the divertor W source. At the same time, the W concentration is strongly peaked, as can be seen from the ratio of the central to edge W concentration, which is about 20. This strong peaking is explained by the so-called neoclassical inward drifts. At $t = 2.5$ s, the NBI is switched off and at the same time the ion cyclotron resonance heating (ICRH) is switched on. Immediately, the limiter source increases by a factor of 10. On a longer timescale—reflecting the "slow" particle transport within the plasma—the edge W concentration rises, but at the same time the central concentration decreases to a value close to that at the edge. This phase is characterized by dominant diffusive transport (the so-called anomalous transport) which tends to reduce W density gradients. At the same time it can be concluded that the main chamber (limiter) source has a much larger impact on the edge concentration than the divertor W source, since the sum of both barely changes during this phase. This is a strong hint for the superior behavior of the divertor, concerning its role as an impurity source. This fact is further illustrated in the third phase of the discharge from $t = 3.5$ s on. There, a part of the NBI is switched on again, resulting in an increased W source in the divertor, which is not reflected in the main chamber concentrations at all. This example highlights the importance of an as complete as possible W diagnostic to optimize plasma performance with external tools. Although ITER will operate at much higher central temperatures and the main heating will be provided by fast α-particles from the fusion reaction, a very similar behavior of W transport is expected from recent model calculations [63].

11.4 Summary and Conclusions

Emissions from neutral tungsten and from tungsten ions with low charge number allow to quantify the W-influx. In the main plasma, spectral lines from highly charged ions can be used to derive the local W density. Large progress on the tungsten spectroscopy in fusion devices has been made since the early investigations in tokamaks. Equally, the production and benchmarking of atomic data for the interpretation of the measured spectra has been considerably enforced. Under the premises of using spectroscopy as a quantitative tool, the excitation rates and in the case of nonresonant transitions, even the setup of a collisional radiative model is necessary. However, not only the rates within one ionization stage need a careful revisiting, but also the ionization and recombination rates as could be extracted from tokamak measurements and in EBIT investigations on the ionization equilibrium.

In present-day fusion devices, emissions from W-ions with charge stages below $q \approx 50+$ can be studied. The most intense spectral features of W for these plasma conditions are observed between 0.4 and 0.8 nm, around 5 nm and between 10 and 30 nm. The lines at 0.4–0.8 nm are emitted by ionization stages from Br-like W^{39+} to about Mn-like W^{49+}. Investigations in EBITs have confirmed the identification of the spectral lines, moreover—in some cases—they provided the only clue to unambiguously identify the transitions. The modeling in that wavelength range agrees well for the distribution of emissions versus the wavelength. The intensities of single spectral lines are predicted typically with deviations of around factor 2, while the wavelengths for transitions in the soft x-ray spectral range are predicted within 0.002 nm.

The experience gained in recent experiments and atomic physics calculations encourage the use of this tool to extrapolate from the current experimental findings to spectra, as they might be observed in future experiments. However, the extrapolation in electron temperature/ionization stages is quite far and currently no fusion device is at hand to bridge this gap. Calculations for the ITER standard scenario, where a central electron temperature $T_e \approx 18$ keV is envisaged, reveal that the above-described emissions will occur at the outer half of the plasma. New spectral features will be emitted from mid-radius to the plasma center by ionization stages of Cr-like W^{50+} to C-like W^{68+}. Important spectral ranges will be at 0.1–0.15 and 1.8–3.5 nm for ionization stages of Co-like W^{47+} to C-like W^{68+}. Predictions of spectral lines around 7–8 nm for the ionization stages above Cl-like W^{57+} are relatively weak, but also seem to be an interesting alternative in the VUV spectral range. In very recent EBIT investigations (this paper and [59,60]), most of the predicted lines could be identified. However, due to the different electron densities in the EBIT ($n_e \leq 10^{19}$ m^{-3}), in the center of fusion plasmas ($n_e \geq 10^{20}$ m^{-3}), and the monoenergetic electron beam (EBIT) instead of a Maxwellian distribution, investigations in fusion devices are still indispensable, to explore the composition of the spectra emitted from fusion plasmas. Besides these highest charge stages of W also the ionization stages with low charge numbers should be addressed in future experiments. At the moment, there is a huge gap in the observed ionization stages between the neutral or singly ionized tungsten at the very edge and the ions which emit the QC ($q \approx 30+$). The measurement of the charge stages in-between would allow one to investigate the behavior of the W-influx and the W transport within the edge in more detail. The final goal of the investigations will be the incorporation of all available information in a data structure such as ADAS to make them most commonly available in the fusion community.

In ITER and future fusion reactors, the measurement of the W concentration will be of utmost importance because of its potentially deleterious effect on the fusion power production due to its high cooling factor. In this context, W spectroscopy will be one of the key diagnostics to optimize fusion performance by providing input for control tools to minimize the W source and the W density in the plasma.

References

1. R. Aymar, P. Barabaschi, and Y. Shimomura, *Plasma Phys. Control. Fusion* 44, 519–565, 2002.
2. H. Bolt, V. Barabash, W. Krauss, J. Linke, R. Neu et al., *J. Nucl. Mater.* 329–333, 66–73, 2004.
3. J. Hosea, R. Goldston, and P. Colestock, *Nucl. Fusion* 25, 1155, 1985.
4. ORMAK-ISX Group, *Nucl. Fusion* 25, 1137–1143, 1985.
5. V. Arunasalam, C. Barnes, K. Bol, K. Brau, N. Bretz et al., Recent results from the PLT tokamak, In *Proceedings of the 8th Conference of EPS*, Prague, 1977, Vol. 2, Geneva: EPS, 1978, pp. 17–28.
6. E. Hinnov, K. Bol, D. Dimock, R. Hawryluk, D. Johnson et al., *Nucl. Fusion* 18, 1305, 1978.
7. R. Neu, R. Dux, A. Geier, O. Gruber, A. Kallenbach et al., *Fusion Eng. Design* 65(3), 367–374, 2003.
8. J. Wesson, *Tokamaks* (3rd edition). Oxford: Clarendon Press, 2003.
9. A. Herrmann and O. Gruber, *Fusion Science and Technology* 44(3), 569–577, 2003.
10. R. Neu, M. Balden, V. Bobkov, R. Dux, O. Gruber et al., *Plasma Phys. Control. Fusion* 49(12B), B59–B70, 2007.
11. U. Samm, *Fus. Sci. Technol.* 47, 73–75, 2005.
12. J. Pamela, G. Matthews, V. Philipps, and R. Kamendje, *J. Nucl. Mater.* 363–365, 1–11, 2007.
13. R. Neu, *Phys. Scripta T* 123, 33–44, 2006.
14. R. Isler, R. Neidigh, and R. Cowan, *Phys. Lett. A* 63, 295, 1977.
15. E. Hinnov and M. Mattioli, *Phys. Lett.* 66A, 109–111, 1978.
16. M. Finkenthal, L. Huang, S. Lippmann, H. Moos, P. Mandelbaum et al., *Phys. Lett. A* 127, 255, 1988.
17. K. Asmussen, K. B. Fournier, J. M. Laming, R. Neu, J. F. Seely et al., *Nucl. Fusion* 38(7), 967–986, 1998.
18. T. Pütterich, R. Neu, R. Dux, A. D. Whiteford, M. G. O'Mullane et al., *Plasma Phys. Control. Fusion* 50(8), 085016, 2008.
19. R. Radtke, C. Biedermann, J.-L. Schwob, P. Mandelbaum, and R. Doron, *Phys. Rev. A* 64, 012720, 2001.
20. S. B. Utter, P. Beiersdorfer, and E. Träbert, *Can. J. Phys.* 80, 1503–1515, 2002.
21. Y. Ralchenko, J. Reader, J. Pomeroy, J. Tan, and J. Gillaspy, *J. Phys. B* 40, 3861–3875, 2007.
22. T. Pütterich, R. Neu, C. Biedermann, R. Radtke, and ASDEX Upgrade Team, *J. Phys. B: At. Mol. Opt. Phys.* 38(16), 3071–3082, 2005.
23. Y. Ralchenko, *J. Phys. B* 40, F175–F180, 2007.
24. Y. Ralchenko, A. Kramida, J. Reader, and NIST ASD Team, *NIST Atomic Spectral Database*, vers. 3.1.5. Gaithersburg, MD: National Institute of Standards and Technology, 2008. http://physics.nist.gov/asd3.

25. I. Beigmann, A. Pospieszczyk, G. Sergienko, I. Y. Tolstikhina, and L. Vainshtein, *Plasma Phys. Control. Fusion* 49, 1833–1847, 2007.
26. K. Behringer, H. Summers, B. Denne, M. Forrest, and M. Stamp, *Plasma Phys. Control. Fusion* 31, 2059–2099, 1989.
27. A. Pospieszczyk, *Phys. Scripta T* 119, 71–82, 2005.
28. D. Naujoks, K. Asmussen, M. Bessenrodt-Weberpals, S. Deschka, R. Dux et al., *Nucl. Fusion* 36(6), 671–687, 1996.
29. A. Thoma, K. Asmussen, R. Dux, K. Krieger, A. Herrmann et al., *Plasma Phys. Control. Fusion* 39(9), 1487–1499, 1997.
30. A. Geier, H. Maier, R. Neu, K. Krieger, and ASDEX Upgrade Team, *Plasma Phys. Control. Fusion* 44(10), 2091–2100, 2002.
31. V. Shevelko and L. Vainshtein, *Atomic Physics for hot Plasmas*, Bristol, UK: Institute of Physics Publishing, 1993.
32. H. van Regemorter, *Astrophys. J.* 136, 906, 1962.
33. R. Neu, K. Asmussen, K. Krieger, A. Thoma, H.-S. Bosch et al., *Plasma Phys. Control. Fusion* 38, A165–A179, 1996.
34. R. Dux, V. Bobkova, A. Herrmanna, A. Janzera, A. Kallenbacha, et al., Plasma-wall interaction and plasma behaviour in the non-boronised all tungsten ASDEX Upgrade, *Proceedings of the 18th International Conference on Plasma-Surface Interactions in Controlled Fusion Device, Journal of Nuclear Materials*, Volumes 390–391, pp. 858–863, 2009.
35. K. Behringer, *Description of the impurity transport code STRAHL*. Technical Report JET-R(87)08. Abingdon, UK: JET Joint Undertaking, 1987.
36. R. Dux, *STRAHL Manual*. Report IPP 10/30, Garching: Max-Planck-Institut für Plasmaphysik, 2006.
37. R. Hulse, *Nucl. Technol. Fusion* 3, 259, 1983.
38. R. Dux, *Impurity Transport in Tokamak Plasmas*. Rep. IPP 10/27. Garching: Max-Planck-Institut für Plasmaphysik, 2004.
39. D. Post, R. Jensen, C. Tarter, W. Grasberger, and W. Lokke, *At. Data Nucl. Data Tables* 20, 397–439, 1977.
40. D. Post, J. Abdallah, R. Clark, and N. Putvinskaya, *Phys. Plasmas* 2, 2328–2336, 1995.
41. R. Dux, A. G. Peeters, A. Gude, A. Kallenbach, R. Neu et al., *Nucl. Fusion* 39(11), 1509–1522, 1999.
42. C. Biedermann, R. Radtke, J.-L. Schwob, P. Mandelbaum, R. Doron, T. Fuchs, and G. Fußmann, *Phys. Scripta T* 92, 85–88, 2001.
43. T. Pütterich, *Investigations on Spectroscopic Diagnostic of High-Z Elements in Fusion Plasmas*. Technical Report IPP 10/29, April 2006.PhD Thesis, Universität Augsburg. Garching, Germany: IPP, 2005.
44. S. D. Loch, J. A. Ludlow, M. S. Pindzola, A. D. Whiteford, and D. C. Griffin, *Phys. Rev. A* 72(5), 052716, 2005.
45. H. P. Summers, *The ADAS User Manual*, version 2.6. http://adas.phys.strath.ac.uk, 2004.
46. R. Neu, K. B. Fournier, D. Schlögl, and J. Rice, *J. Phys. B: At. Mol. Opt. Phys.* 30, 5057–5067, 1997 [Preprint in IPP 10/7, June 1997].
47. R. Neu, K. B. Fournier, D. Bolshukhin, and R. Dux, *Phys. Scripta T* 92, 307–310, 2001.
48. R. Neu, *Tungsten as plasma facing material in fusion devices*. Technical Report 10/25. Garching, Germany: IPP, 2003.

49. T. Pütterich, R. Neu, R. Dux, A. Kallenbach, V. Bobkov et al., Investigations on tungsten walls at ASDEX upgrade. In F. De Marco and G. Vlad (Eds.), *Europhysics Conference Abstracts (CD-ROM), Proceedings of the 33rd EPS Conference on Plasma Physics*, Roma, 2006, Vol. 301. Geneva: EPS, 2006, pp. O–2.004.
50. K. Fournier, *At. Data Nucl. Data Tables* 68, 1–48, 1998 and private communication.
51. P. Neill, C. Harris, A. Safronova, S. Hamasha, S. Hansen et al., *Can. J. Phys.* 82, 931–942, 2004.
52. R. Neu, R. Dux, A. Kallenbach, T. Pütterich, M. Balden et al., *Nucl. Fusion* 45(3), 209–218, 2005.
53. K. Fournier, R. Neu, D. Bolshukhin, A. Geier, and the ASDEX Upgrade Team, Soft X-ray emission spectra from highly charged tungsten ions as a quantitative diagnostic of fusion plasmas. *Bull. Am. Phys. Soc.* 46, 267, 2001.
54. V. Janauskas, S. Kucas, and R. Karazija, *J. Phys. B* 40, 2179–2188, 2007.
55. A. Kramida and T. Shirai, *J. Phys. Chem. Ref. Data* 35, 423, 2006.
56. I. P. Grant, C. T. Johnson, F. A. Parpia, and E. P. Plummer, *Comput. Phys. Commun.* 55, 425–456, 1989.
57. G. N. Pereverzev, C. Angioni, A. G. Peeters, and O. V. Zolotukhin, *Nucl. Fusion* 45(4), 221–225, 2005.
58. R. Radtke, C. Biedermann, G. Fussmann, J. Schwob, P. Mandelbaum et al., *IAEA At. Plasma-Mater. Interaction Data Fusion* 13, 45, 2007.
59. Y. Ralchenko, I. Draganic, J. Tan, J. Gillaspy, J. Pomeroy et al., *J. Phys. B* 41, 021003, 2008.
60. U. Feldman, J. Seely, E. Landi, and Y. Ralchenko, *Nucl. Fusion* 48, 045004, 2008.
61. R. Neu, K. Asmussen, R. Dux, P. N. Ignacz, M. Bessenrodt-Weberpals et al., Behaviour of laser ablated impurities in ASDEX upgrade discharges. In B. Keen, P. Stott, and J. Winter (Eds.), *Europhysics Conference Abstracts. Proceedings of the 22th EPS Conference on Controlled Fusion and Plasma Physics*, Bournemouth, 1995, Vol. 19C, Part I. Geneva: EPS, pp. 65–68, 1995.
62. T. Pütterich, R. Neu, R. Dux, A. D. Whiteford, M. G. O'Mullane et al., Calculation and experimental test of the cooling factor of tungsten, *Nucl. Fusion* 50, 025012, 2010.
63. C. Angioni and A. G. Peeters, *Phys. Rev. Lett.* 96(9), 095003-01–095003-03, 2006.

12

X-Ray Emission Spectroscopy and
Diagnostics of Nonequilibrium Fusion
and Laser-Produced Plasmas

F. B. Rosmej

CONTENTS

12.1 Introduction

The emission of light is one of the most fascinating phenomena in nature. Everybody feels the beauty while looking at the colors when the sun is going down, when a thunder illuminates the night, or when the emission of the aurora moves like magic in the dark heaven. And every day we are looking at something in order to read information from a computer screen, to drive not into but around an obstacle, and to look into the eyes of the child to understand that it tries to hide that it just burned off father's stamp collection in an unlucky physical experiment. In general terms, we all use light to obtain information, to diagnose something, to control or optimize a process, or to understand what is true and what is right. The basic science related to all these questions is called "spectroscopy," and questions such as "Why the heaven is blue?", "What is the temperature in the flame of a candle?", "What makes the sun burning so wonderful?" have been the historical origin of scientific activity. We also might ask what light is by itself? This is a difficult question: although we have some imagination what light is, it is difficult to say what it really is and the interesting reader is referred to an excellent textbook devoted to this question [1].

Today plasma spectroscopy has proven to be one of the most powerful methods to understand various physical phenomena. It provides essential information about basic parameters (e.g., temperature, density, chemical composition, velocities) and relevant physical processes (Why the aurora is green at low altitudes but red at larger ones? Why can we look with x-rays into the human body but not with visible light? . . .).

An essential point in modern research is to characterize a certain unknown phenomenon by spectroscopic methods independent from other theories. A simple example may illustrate this important circumstance: let us imagine there is a theory, which predicts why the sun is burning so nicely. This theory should then also predict the live time (after some time, all energy is burned off) and the temperature of the sun (because of the heat flow and energy balance). How to prove whether the theory is right or not? This would be an urgent problem to solve for human existence would the theory predict that the burn would stop in about 100 years. We can prove the theory by spectroscopic methods: we could analyze the light emission in various spectral ranges, identify the chemical composition of the sun and obtain, for example, the result that hydrogen-like iron is present. In order to strip 25 electrons from the iron ($Z_n = 26$), the temperature needs to be of the order of the ionization potential of H-like iron for this manifold ionization, that is, some keV (equivalent to some 10 million degree Celsius). If the theory would have predicted, for example, a few 1000 degree Celsius, it could therefore be identified as being wrong due to the large discrepancy of the independent spectroscopic analysis. This discrepancy would then also imply that the prediction of the theory when the sunburn stops might also be wrong.

The accessible parameter range of spectroscopy covers orders of magnitude in temperature and (especially) density, because practically all elements of particular, selected isoelectronic sequences can be used for diagnostic investigations. These elements can occur as intrinsic impurities or may be intentionally injected in small amounts (the so-called "tracer elements"). This makes plasma spectroscopy a very interdisciplinary science, and excellent reviews of traditional spectroscopic methods have been published in books and review articles [2–11].

The rapid development of powerful laser installations, intense heavy ion beams, and the fusion research (magnetic and inertial fusion) enable the creation of matter under extreme conditions never achieved in laboratories so far. An important feature of these extreme conditions is the nonequilibrium nature of the matter, for example, a solid is heated by a femtosecond (fs)-laser pulse and undergoes a transformation from a cold solid, warm dense matter, strongly coupled plasma to a highly ionized gas while time is elapsing. We might think about using time-dependent detectors to temporally resolve the light emission in the hope to have then resolved the problem. However, this is not so simple: there are serious technical obstacles and also basic physical principles to respect. A simple technical reason is that for 10 fs laser radiation interacting with matter, we do not have any x-ray streak camera available (the current technical limit is about 0.5 ps). A simple principal reason is that the atomic system from which light originates has a characteristic time constant which might be much longer than 10 fs. The atomic system is therefore "shocked" and any light emission is highly out of equilibrium invalidating any standard methods for diagnostics (even if the experimental observation is time-resolved).

More complex physical reasons are related to the fact that the term "nonequilibrium" concerns not only the time evolution, but also the statistical properties of the system. One important example (known for laser-produced plasmas and inertial fusion research) is the creation of suprathermal electrons (or hot electrons). They seriously alter the light emission in a nontrivial manner (a trivial change is an enhanced ionization due to hot electrons, whereas a nontrivial change is the qualitative distortion of ion charge stage distribution which in turn invalidates the application of any standard diagnostic methods). In consequence, traditional equilibrium methods are not applicable anymore. Moreover, theoretical analysis shows that often standard methods cannot be corrected or modified to be anymore useful like they have been at the beginning of their historical foundation.

What to do? A powerful class of light emission that enables to formulate spectroscopy of nonequilibrium, high-temperature plasmas concerns light emission from multiple excited states: the so-called dielectronic satellite and hollow ion (HI) atomic/ionic emission.

In high-energy-density physics, laser-produced plasmas, and fusion research, the light emission in the x-ray spectral range is of particular interest, because only the x-ray emission is, in general, able to exit the volume without essential photo-absorption. We are, therefore, looking for x-ray dielectronic satellites and HI x-ray emission to develop nonequilibrium spectroscopy. This is a challenging field of activity (research and application) that also involves fascinating topics in atomic physics of dense plasmas and nonequilibrium atomic kinetics.

12.2 Equilibrium Concepts

It is often not quite clear what is meant with the word "equilibrium." Does this mean that the plasma parameters do not change in time? Or does it mean that the particle statistics follows certain laws? Or something else? There is yet another reason to look firstly for the understanding of equilibrium methods when developing nonequilibrium methods: it turns out that "equilibrium ideas" are quite useful to develop advanced nonequilibrium methods. Let us, therefore, briefly summarize the most important concepts related to plasma spectroscopy.

12.2.1 Thermodynamic Equilibrium

The laws of thermodynamic equilibrium are derived in many standard textbooks, and a very illuminating and transparent presentation concerning classical and quantum particles can be found in [12]. We start from Boltzmann genius invention that the entropy S is related to the probability $P(N, N_1, N_2, N_3, \ldots)$ for distributing N particles over energy states with respective populations $N, N_1, N_2, N_3, \ldots$ according to

$$S = k \ln\{P(N, N_1, N_2, N_3, \ldots)\}, \tag{12.1}$$

where k is the Boltzmann constant. For an isolated system, thermodynamic equilibrium relations are obtained, when the entropy has an extreme:

$$\delta S = 0. \tag{12.2}$$

Depending on the probability function P, the distribution functions for different types of particles in thermodynamic equilibrium can be derived.

12.2.1.1 Planck Radiation

The Planck radiation is the radiation in thermodynamic equilibrium. It is obtained applying the P-function for photons to Equations 12.1 and 12.2:

$$U_\omega = \frac{\hbar\omega^3}{\pi^2 c^3} \frac{1}{e^{\hbar\omega/kT} - 1}, \tag{12.3}$$

where U_ω is the energy density (energy/volume/angular frequency), c the velocity of light, ω the photon angular frequency, $\hbar$ the Planck constant over 2π, and T the temperature of the system. The spectral intensity (energy/time/area/solid angle/angular frequency) is given by

$$B_\omega = \frac{c}{4\pi} U_\omega = \frac{\hbar\omega^3}{4\pi^3 c^2} \frac{1}{\exp(\hbar\omega/kT) - 1}. \tag{12.4}$$

The number density of photons N in a spectral interval of the Planck radiation field is given by

$$N(\hbar\omega) = 1.319 \times 10^{13} \frac{(\hbar\omega)^2}{\exp(\hbar\omega/kT) - 1} \ (\text{Number/cm}^3 \text{ eV}) \tag{12.5}$$

with $\hbar\omega$ und kT in eV.

12.2.1.2 Collisions and Opacity

Equation 12.2 is not very helpful in understanding under which conditions an arbitrary radiation field will be well approximated by the Planck radiation field in a specific spectral range. For a two-level atom, these conditions can be formulated for a certain frequency interval as follows:

a. The collisional depopulation rate of the upper level must be much larger than the radiative decay rate from the upper level.

b. The opacity for a certain frequency should be large enough (practically larger than about 5).

Numerical simulations of complex level systems show that these conditions also remain useful in understanding the evolution of the radiation field. We note that the radiation field according to Equations 12.3 through 12.5 is the radiation field in vacuum. This circumstance enters in the thermodynamic derivation (Equations 12.1 and 12.2) via the absence of any material properties in the probability function $P(N, N_1, N_2, N_3, \ldots)$.

12.2.2 Saha–Boltzmann Relation

The classical probability function $P(N, N_1, N_2, N_3, \ldots)$ for electrons is particularly important for spectroscopy as it permits to understand atomic- and ionic-level populations.

12.2.2.1 Boltzmann Relation: Local Thermodynamic Equilibrium and Partial Local Thermodynamic Equilibrium

Let us consider N energy levels with energies $E_1^Z, E_2^Z, \ldots, E_N^Z$ and statistical weights $g_1, g_2, \ldots, g_N$ of a certain atomic ionization stage Z. The atomic/ionic-level populations are then given by

$$\frac{n_j}{n_i} = \frac{g_j}{g_i} \exp\left\{ -\frac{E_i^Z - E_j^Z}{kT} \right\}, \tag{12.6}$$

whereas for free electrons, the Maxwell–Boltzmann energy distribution function $F(E)$ is obtained as follows:

$$F(E) = \frac{2\sqrt{E}}{\sqrt{\pi}} \frac{e^{-E/kT}}{(kT)^{3/2}}. \tag{12.7}$$

If all levels $1, \ldots, N$ are well described by the Boltzmann distribution 12.6 and also Equation 12.7 holds, one speaks about local thermodynamic equilibrium (LTE). If only atomic/ionic energy levels with large principal quantum numbers are well described by Equation 12.6, then one speaks about partial local thermodynamic equilibrium (PLTE). As collisional rates increase with principal quantum number but decrease with radiative decay rates, PLTE starts from the high lying levels and proceeds down to lower principal quantum numbers as density increases.

The condition to ensure PLTE for levels with principal quantum number n larger than a certain critical one, n_{crit}, is that collisions are much larger (typically a factor of 10) than relevant radiative decay rates. In the hydrogenic approximation, this condition can be formulated for a plasma consisting of electrons, ions, and atoms as follows:

$$n_{e,crit} \geq 6 \times 10^{19} \, Z^7 \frac{(n_{crit} - 1)^{2n_{crit}-2}}{n_{crit}^3 (n_{crit} + 1)^{2n_{crit}+2}} \left(\frac{T_e \, (\text{eV})}{Z^2 \, Ry} \right)^{1/2} (\text{cm}^{-3}), \tag{12.8}$$

where $n_{e,crit}$ is the critical electron density (cm^{-3}) above which Equation 12.6 holds for all levels with principal quantum number larger than n_{crit} (PLTE), T_e the electron temperature (eV), Z the ionic charge, and $Ry = 13.6\,eV$. For hydrogen ($Z = 1$), $n_{crit} = 1$, $T_e = 1\,eV$, and $n_{e,crit} \approx 1 \times 10^{18}\,cm^{-3}$, whereas for H-like molybdenum and $T_e = 2\,keV$, the critical density is very high: $n_{e,crit} \approx 2 \times 10^{29}\,cm^{-3}$. This example shows that it is not the absolute density, which is of importance in obtaining thermodynamic equilibrium conditions, but rather the relation between the collisional and radiative decay rates.

12.2.2.2 Boltzmann Plot

Equation 12.6 can be useful in obtaining the electron temperature from the so-called Rydberg series of emission lines, for example, $np \rightarrow 1s + h\nu$, $nd \rightarrow 2p + h\nu$, $1snp\ ^1P_1 \rightarrow 1s^2 + h\nu$, $1snd\ ^1L \rightarrow 1s2p\ ^1L + h\nu$, $1snd\ ^3L \rightarrow 1s2p\ ^3L + h\nu$, $1s^2nd \rightarrow 1s^22p + h\nu$, and so on. The local line intensity (energy/area/time/solid angle) of a Rydberg series is given by

$$I_n = \frac{1}{4\pi}\hbar\omega_n n_n A_n\,dx, \tag{12.9}$$

where n_n is the atomic/ionic-level population density of the level with principal quantum number n, A_n the radiative decay rate, and dx the length of the local line of sight. Let us now consider the intensity ratio between two emission lines from the Rydberg series:

$$\frac{I_n}{I_{n'}} = \frac{\omega_n}{\omega_{n'}} \frac{n_n}{n_{n'}} \frac{A_n}{A_{n'}}. \tag{12.10}$$

If Equation 12.6 is valid, then Equation 12.10 transforms readily into ($n > n'$)

$$\frac{I_n}{I_{n'}} = \frac{\omega_n}{\omega_{n'}} \frac{A_n}{A_{n'}} \frac{g_n}{g_{n'}} \exp\left\{-\frac{E_{n'} - E_n}{kT_e}\right\}. \tag{12.11}$$

In a logarithmic plot, that is, the so-called Boltzmann plot (y-axis: $\ln(I_n/I_{n'})$; x-axis: $(E_{n'} - E_n)$, usually one fixes the lower n'-value to consider a Rydberg series of transitions, for example, $1s2p\ ^3P \rightarrow 1snl\ ^3L$ in He-like ions), the slope $(-1/kT_e)$ of the linear function $(E_{n'} - E_n)$ provides directly the electron temperature:

$$\ln\left\{\frac{I_n}{I_{n'}}\right\} = \ln\left\{\frac{\omega_n g_n A_n}{\omega_{n'} g_{n'} A_{n'}}\right\} - \frac{E_{n'} - E_n}{kT_e}. \tag{12.12}$$

The Boltzmann plot also permits to estimate whether the PLTE condition holds: if the Boltzmann plot shows a linear function starting from some particular n' (equivalent to n_{crit} in Equation 12.8), then PLTE is probably achieved for quantum numbers higher than n. According to Equation 12.8, a simple order of magnitude estimate of the electron density can then be performed [13].

12.2.2.3 Saha Equilibrium

Let us now consider energy levels, which belong to different ionization stages. They are linked by the so-called Saha–Boltzmann equation

$$\frac{n_i^Z}{n_j^{Z+1}} = n_e \frac{g_i}{2g_j} \left(\frac{2\pi\hbar^2}{m_e kT_e} \right)^{3/2} \exp\left\{ \frac{\Delta E}{kT_e} \right\} \tag{12.13}$$

or

$$\frac{n_i^Z}{n_j^{Z+1}} = 1.656 \times 10^{-22} n_e \frac{g_i}{g_j} \frac{\exp\{\Delta E/kT_e\}}{(kT_e)^{3/2}}, \tag{12.14}$$

where ΔE is the ionization energy (positive) (eV) of level i with ionization stage Z with final level j in ionization stage $(Z+1)$, $\Delta E = E_i^Z - E_j^{Z+1}$ (see Figure 12.1), kT_e the electron temperature (eV), and n_e the electron density (cm^{-3}). The larger the density, the lower the ionization due to increased three-body recombination effect. The critical density above which Equation 12.13 is valid can also be estimated from Equation 12.8 (for level i with energy E_i^{Z+1} and level j with energy E_n^Z). For highly charged ions, this critical density is very large (see the example above) and therefore very difficult to achieve in laboratory experiments. However, the Saha–Boltzmann equation can be useful when considering levels where the ionization energy ΔE is very small (in this case, the principal quantum number n is large and the critical electron density according to Equation 12.8 is not so high). The level diagram of Figure 12.1 illustrates the LTE, PLTE, and Saha–Boltzmann relations.

Let us note that in thermodynamic equilibrium, the Planck radiation as well as the Boltzmann and Saha–Boltzmann relations (and also Equation 12.7) are valid. However, the conditions for the x-ray radiation field to approach the Planck radiation are very severe and only LTE or PLTE conditions (if any) might be achieved. Note that LTE does *not* mean that thermodynamic equilibrium exists locally; it means only that Equations 12.6, 12.7, and 12.13 are valid (or their quantum mechanical equivalent of spin 1/2 particles: Fermi–Dirac statistic).

12.2.3 Collisional-Radiative Equilibrium

If, in an equilibrium situation, collisional rates are not much larger than radiative decay rates, then we speak about a collisional-radiative equilibrium. Neither Planck nor Boltzmann nor Saha–Boltzmann relations are valid equations for all population densities. Level population densities have then to be obtained from the so-called atomic population kinetics:

$$n_j \sum_{k=1}^{N} W_{jk} = \sum_{i=1}^{N} n_i W_{ij}, \tag{12.15}$$

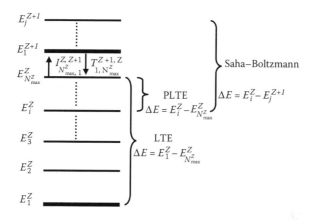

FIGURE 12.1
Schematic energy level diagram which illustrates LTE, PLTE, and Saha–Boltzmann relations. ΔE is the ionization energy of level i and ionization stage Z when the final state is j with charge state $Z + 1$. The Saha–Boltzmann relation connects different ionization stages, whereas the Boltzmann relation (LTE or PLTE) provides information about the population of levels within one ionization stage. As radiative decay rates decrease with principal quantum number ($A \propto 1/n^3$) but collisional rates increase with principal quantum number, the validity of the Boltzmann relation according to Equation 12.6 starts from high lying levels (PLTE) and then proceeds down to $n = 1$ as density increases. N_{max}^Z is the maximum number of levels in ionization stage Z.

where N is the number of levels included in the model (note that in this general notation also different ionization stages are included in the level system, i.e., the index j of a level characterizes the quantum numbers of a specific configuration in a definite ionization stage). The W-matrix for an optically thin plasma is given by

$$W_{ij} = A_{ij} + C_{ij} + I_{ij} + T_{ij} + R_{ij} + DC_{ij} + \Gamma_{ij} + \cdots, \qquad (12.16)$$

where A represents the spontaneous radiative emission rate, C denotes the collisional excitation/de-excitation rate, I the ionization rate, T the three-body recombination rate, R the radiative recombination rate, DC the dielectronic (radiationless electron) capture rate, and Γ represents the autoionization rate (the ellipsis indicates other processes not explicitly listed, e.g., charge exchange). The matrix elements (rate coefficients or cross-sections) for the inverse processes are obtained by the application of the principle of detailed balance or the principle of micro-reversibility. If a particular transition does not occur, the corresponding rate is set equal to zero. The set of population equations 12.15 is, sometimes, also called "rate equations." The level populations n_j are also called "non-LTE level populations," and the term "dynamic level populations" is often used in the framework of line profile calculations.

Equation 12.15 indicates equilibrium between depopulating (left-hand side of Equation 12.15) and populating (right-hand side of Equation 12.15) processes. This means that the level population densities are in equilibrium in the sense that they do not change in time. The question as to whether the level population densities are in statistical equilibrium (according to Equations 12.6 and 12.13) remains open. It also remains open, whether the radiation field or the particle distribution functions are in equilibrium. Other equations are requested to find an answer (to be discussed later on).

In the limit of high densities (i.e., when collisional rates are much larger than radiative decay rates), the atomic-level population kinetics 12.15 reproduces Equations 12.6 and 12.13. In complex atomic-level population kinetics comprising more than thousands of levels, Equations 12.6 and 12.13 allow a certain cross-check of the system. If the system does not correctly approach the high-density limit, the system is in error.

12.2.3.1 Non-LTE Line Ratios

Having once calculated the non-LTE level populations according to Equation 12.15, all combinations of line intensity ratios can be obtained:

$$\frac{I_{ji}}{I_{j'i'}} = \frac{\omega_{ji}}{\omega_{j'i'}} \frac{A_{ji}}{A_{j'i'}} \frac{n_j}{n_{j'}}. \tag{12.17}$$

Of particular interest are those intensity ratios which depend only on one plasma parameter. The ideal case of a temperature diagnostic is therefore given by

$$\frac{I_{ji}}{I_{j'i'}} = G_{jij'i'}(T_e), \tag{12.18}$$

whereas the ideal case of a density diagnostic is given by the relation

$$\frac{I_{kl}}{I_{k'l'}} = \gamma_{klk'l'}(n_e). \tag{12.19}$$

The functions G and γ are obtained from the solution of the system of rate Equations 12.15. Having measured these intensity ratios with appropriate line emissions, the application of Equations 12.18 and 12.19 provides readily temperature and density. However, the solution of Equations 12.15 shows that, in general, the intensity ratio depends both on temperature and density:

$$\frac{I_{ji}}{I_{j'i'}} = \chi_{jij'i'}(T_e, n_e). \tag{12.20}$$

One aim of spectroscopic research is to find line ratios whose dependence is close to those of the ideal equations 12.18 and 12.19. The difficulty in doing so lies in the fact that Equation 12.20 has multiple solutions, which means that for different sets of density and temperature, the same line intensity ratio is obtained. It is, therefore, necessary to employ several line ratios at the same time to avoid misleading parameter information from line ratios. For optically thin and stationary plasmas, powerful methods have been found and reviewed in [8,9]. We do not discuss these methods here in detail as they are often not very useful for nonequilibrium dense plasmas.

12.2.3.2 Radiation Transport

In high-density plasmas, radiation transport seriously influences the line intensities and also the energy balance. Excellent books have been devoted to these phenomena [14–16]. Line ratios are now given by the relation

$$\frac{I_{ji}}{I_{j'i'}} = \frac{\omega_{ji}}{\omega_{j'i'}} \frac{A_{ji}\Lambda_{ji}}{A_{j'i'}\Lambda_{j'i'}} \frac{n_j}{n_{j'}}, \tag{12.21}$$

where Λ_{ji} is the escape probability defined as the ratio of the number of photons leaving the plasma to the total local photon emissions. The escape probability itself depends on the population densities and on the geometry of the plasma volume. In low-density plasmas where collisional rates are much lower than the radiative ones, photo-absorption is directly followed by re-emission and the total intensity of the line remains unchanged (note that Λ_{ji} is smaller than 1, however, the upper population density is enhanced due to photo-absorption by a factor $1/\Lambda_{ji}$). The line profile, however, is seriously changed (for opacity broadening, see below). If collisions become important compared to radiative decay rates, depopulating collisions from the upper level decrease effectively the probability of re-emission. This results in an effective decrease in the line intensity.

Numerous different definitions of the escape probability are used in the literature: escape factor, escape probability, transmission factor, Bibermann–Holtstein factor, and so on. The important differences in the various definitions lie in the fact that whether nonlocal radiation transport effects are taken into account or not. For further discussions, the interesting reader is referred to the review articles of Irons [17–20].

The atomic population kinetics of an optically thick plasma becomes now a system of nonlinear equations:

$$n_j \sum_{k=1}^{N} \{W_{jk} + P_{ik}^{abs} + P_{ik}^{stim} + P_{ik}^{iz} + R_{ik}^{stim}\}$$

$$= \sum_{i=1}^{N} n_i \{W_{ij} + P_{ij}^{abs} + P_{ij}^{stim} + P_{ij}^{iz} + R_{ij}^{stim}\}, \tag{12.22}$$

where P^{abs} is the radiative absorption rate, P^{stim} the stimulated radiative emission rate, P^{iz} the photo-ionization rate, and R^{stim} denotes the stimulated radiative recombination rate. These rate coefficients depend also on the radiation field intensity. In the general case, when the intensity of the radiation field cannot be approximated by the Planck radiation B_ω (Equation 12.4), the radiation transport equation has to be solved simultaneously with the system of atomic rate equations 12.22.

Let us consider the one-dimensional transport equation without scattering for one ray:

$$\frac{\partial I_\omega}{\partial \tau_\omega} = -I_\omega + S_\omega, \tag{12.23}$$

where I_ω is the spectral radiation field intensity (energy/area/time/angular frequency/solid angle), S_ω the source function, and τ_ω the optical thickness for a photon with frequency ω. The optical thickness is given by

$$\tau_\omega(Z_0) = \int_0^{Z_0} \kappa_\omega \, dz, \tag{12.24}$$

where κ_ω is the absorption coefficient and Z_0 the considered path length of the photon in the plasma. Material properties can be included in the radiation transport equation taking into account the refractive index n_ω [21–25]. The importance of material properties can be estimated according to

$$n_\omega \approx 1 - \frac{\omega_p^2}{\omega^2}. \tag{12.25}$$

Rewriting the plasma frequency, we obtain

$$\omega_p = \sqrt{\frac{n_e e^2}{\varepsilon_0 m_e}} = 5.641 \times 10^4 \sqrt{n_e \, (\text{cm}^{-3})} \, (1/\text{s}). \tag{12.26}$$

In terms of the electron density, Equation 12.25 reads

$$n_\omega \approx 1 - 1.4 \times 10^{-21} \frac{n_e}{E_\omega^2}, \tag{12.27}$$

where E_ω is the photon energy (eV) and n_e the electron density (cm^{-3}). Equation 12.25 indicates that if the radiation frequency approaches the plasma frequency for a certain density, the further transport of radiation with this frequency (or lower) is impossible. This in turn modifies the radiation field and the atomic-level populations. We do not consider this effect here further; the interesting reader is referred to the excellent text books of Bekefi [21] and Pomraning [22].

If the source function is constant with respect to the z-coordinate, then Equation 12.23 has an analytical solution for the emerging intensity at the surface of the plasma volume (L is the total plasma length):

$$I_\omega(z = L) = S_\omega(1 - \exp(-\tau_\omega(z = L))). \tag{12.28}$$

Considering bound–bound transitions only, the source function for a radiative transition from level j to level i is given by (assuming complete frequency redistribution, i.e., local emission and absorption profiles are equal, ε_ω is the emission coefficient)

$$S_\omega = \frac{\varepsilon_\omega}{\kappa_\omega} = \frac{\hbar\omega^3}{4\pi^3 c^2} \frac{1}{n_i g_j / n_j g_i - 1}. \tag{12.29}$$

Equation 12.29 shows that if the plasma density is sufficiently large to ensure a Boltzmann population of the levels according to Equation 12.6, then the source function is equivalent to the Planck function B_ω (Equation 12.4). In this case, Equation 12.29 corresponds to Kirchhoff's law. From Equation 12.28, we see that opacity is a necessary condition to ensure that the emerging intensity reaches the Planck function: at about $\tau_\omega > 5$, the intensity I_ω is very close to the Planck intensity (near the certain frequency ω).

An important parameter in the photon transport theory is, therefore, the bound–bound opacity:

$$\tau_{\omega,ji}(z = L) = \int_0^L \kappa_{\omega,ji}\,dz = \int_0^L \frac{\pi^2 c^2}{\omega_{ji}^2} \frac{g_j}{g_i} A_{ji} n_i \phi_{ij}(\omega) \left\{1 - \frac{n_j g_i}{n_i g_j}\right\} dz, \tag{12.30}$$

where φ_{ij} is the absorption line profile. The importance of opacity can be estimated from the so-called line center optical thickness:

$$\tau_{0,ij} \approx \frac{2\pi^2 e^2 2\sqrt{\ln 2}}{m_e c \sqrt{\pi}\omega_{ji}} \frac{\omega_{ji}}{\text{FWHM}} f_{ij} n_i L_{\text{eff}} \{1 - e^{-\hbar\omega_{ji}/kT_e}\}$$

$$= 8.32 \times 10^{-15} \left(\frac{\omega_{ji}}{\text{FWHM}}\right) f_{ij} \frac{n_i}{(\text{m}^{-3})} \frac{L_{\text{eff}}}{(\text{m})} \{1 - e^{-\hbar\omega_{ji}/kT_e}\}, \tag{12.31}$$

where FWHM is the full width of the line profile at half maximum, ω_{ji} the transition frequency, f_{ij} the absorption oscillator strength, n_i the density of the absorbing ground state i (m^{-3}), and L_{eff} a characteristic plasma dimension (often close to the value of L but not always) relevant for photo-absorption (m). In the case of a Doppler line profile, the line center opacity is given by

$$\tau_{0,ij} \approx 1.08 \times 10^{-10} \frac{\lambda_{ji}}{(\text{m})} \sqrt{\frac{M\,(\text{a.m.u.})}{T_i\,(\text{eV})}} f_{ij} \frac{n_i}{(\text{m}^{-3})} \frac{L_{\text{eff}}}{(\text{m})} \{1 - e^{-\hbar\omega_{ji}/kT_e}\}, \tag{12.32}$$

where λ_{ji} is the wavelength of the bound–bound transition (m), T_i the ion temperature (eV), and M the atomic mass (a.m.u.).

If lines can be identified, line intensity ratios do depend now also on the source size L:

$$\frac{I_{ji}}{I_{j'i'}} = \eta_{jij'i'}(T_e, n_e, L). \tag{12.33}$$

Equation 12.33 has even more multiple solutions than Equation 12.20 and it is almost impossible to use a single line ratio to obtain a certain plasma parameter. Even if several line ratios are employed simultaneously, a unique conclusion for a certain set of parameters (T_e, n_e, L) remains difficult under practical circumstances. This is an important drawback of the line ratio method for diagnostics in optically thick plasmas. However, this situation can be considerably improved. Up to now, we do not have yet used the information of the line opacity broadening. This can be done by the analysis of the spectral distribution (see below).

A simple estimate of opacity broadening effects can be obtained from Equation 12.28 by setting $(1 - e^{-\tau_\omega}) = 1/2$. For a Doppler line profile (ω_0 is the line center frequency)

$$\phi^G(\omega) = \frac{1}{\sqrt{\pi}\Gamma^G} \exp\left(-\frac{(\omega - \omega_0)^2}{(\Gamma^G)^2}\right), \tag{12.34}$$

$$\mathrm{FWHM}^G = 2\sqrt{\ln 2}\,\Gamma^G, \tag{12.35}$$

$$\Gamma^G = \omega_0 \sqrt{\frac{2kT_i}{Mc^2}}, \tag{12.36}$$

we obtain

$$\Gamma^G(\tau_0) \approx \Gamma^G \sqrt{\ln(\tau_0 + e)}. \tag{12.37}$$

For a Lorentz profile

$$\phi^L(\omega) = \frac{\Gamma^L}{2\pi} \frac{1}{(\omega - \omega_0)^2 + (\Gamma^L/2)^2}, \tag{12.38}$$

$$\mathrm{FWHM}^L = \Gamma^L, \tag{12.39}$$

$$\Gamma_{ij}^L = \sum_k W_{ik} + \sum_l W_{jl}. \tag{12.40}$$

Compared to Equation 12.37, a quite different dependence on opacity is obtained:

$$\Gamma^L(\tau_0) \approx \sqrt{\frac{1}{\ln 2} - 1}\,\Gamma^L \sqrt{\tau_0 + \frac{\ln 2}{1 - \ln 2}}. \tag{12.41}$$

Due to the pronounced line wings of the Lorentz profile (compared to the Gauss profile), photons can more easily escape from the line wings, and the total line broadening for a fixed value of line center opacity is much more pronounced. In the more general case of a Voigt profile (convolution of the Gauss and Lorentz profile), the opacity broadening is found to be located between the two limiting expressions 12.37 and 12.41. Equations 12.37 and 12.41 are called "equivalent width" of a line that increases with increasing opacity. It indicates that even a single line can fill the total Planck curve with its equivalent width.

Equations 12.37 and 12.41 are very useful in order to understand the dependence of the source size on diagnostic ratios. In the ideal case, an optically thin and an optically thick emission line is employed to conclude directly for the line center opacity. An important example is the use of the resonance and intercombination line of He-like ions as they can be well separated with high-resolution spectroscopic methods. A more general method to use line intensity ratios along with the line profile information is the spectral simulation method to be discussed below.

12.2.3.3 Non-LTE Spectral Distributions

The non-LTE spectral distribution of optically thin plasmas is given by

$$I(\omega) = \frac{1}{4\pi} \sum_{i,j=1}^{N} \hbar\omega_{ji} n_j A_{ji} \phi_{ji}(\omega_{ji}, \omega, \alpha_{ji}), \qquad (12.42)$$

where φ_{ji} is the emission line profile and α_{ji} a set of plasma parameters (e.g., the ion temperature T_i, average charge Z, electron density n_e, ion density n_i, and so on). The non-LTE feature of the spectral distribution according to Equation 12.42 enters via the non-LTE populations n_j. Equation 12.42 is the most powerful method to interpret spectral emission in optically thin plasmas.

In optically thick plasmas, the spectral distribution is directly given by the solution of Equation 12.23. Note that usually numerous rays have to be taken into account to simulate a particular geometry of the source and the detector.

If single emission lines can be identified in a reasonable manner, an optically thick line profile Φ_{ji} for the transition $j \to i$ emerging from the volume can be defined [26] via

$$\Phi_{ji}(\omega) = \frac{\int_0^{\tau_\omega(z=L)} S_{\omega,ji} \exp(-\tau_{\omega,ij}) \, d\tau_{\omega,ij}}{\int_0^{\tau_\omega(z=L)} \int_0^\infty S_{\omega,ji} \exp(-\tau_{\omega,ij}) \, d\tau_{\omega,ij} \, d\omega}, \qquad (12.43)$$

where $S_{\omega,ji}$ is the space-dependent source line source function. The total spectral distribution is given by

$$I(\omega) = \frac{1}{4\pi} \sum_{i,j=1}^{N} \hbar\omega_{ji} n_j A_{ji} \Lambda_{ji} \Phi_{ji}(\omega). \qquad (12.44)$$

Equation 12.43 allows one to study the main effects on the line profile due to radiation transport effects, namely opacity broadening, inhomogeneous density distribution, and differential plasma motion. Moreover, Equations 12.43 and 12.44 are extremely useful for rapid simulations when employing analytical expressions for Λ_{ji} (e.g., via the usual escape factor technique) and the spatial distribution $n_j(z)$. Of particular interest is a parabolic density distribution according to

$$n(z) = n_0 \left(1 - \left(\frac{L/2 - z}{L_s} \right)^2 \right), \tag{12.45}$$

$$s = \frac{1}{2} \frac{L}{L_s} \tag{12.46}$$

(L_s is a scaling length and s is a dimensionless scaling length parameter) and a unique relative frequency shift

$$\Delta \omega = \omega_{ji} \frac{V}{c} \tag{12.47}$$

of emission and absorption profiles (V is the relative velocity of emitting and absorbing atoms, c the velocity of light), that is,

$$\phi_{ij} = \phi_{ij}(\omega_{ji} + \Delta \omega, \omega), \tag{12.48}$$

$$\phi_{ji} = \phi_{ji}(\omega_{ji}, \omega). \tag{12.49}$$

Equations 12.45 through 12.49 allow one to study almost all principal radiation transport effects on the line profile. Moreover, when employing Equations 12.45 through 12.49, the optically thick line profile (Equation 12.43) has an analytical solution that even can include line-overlapping effects:

$$\Phi_{ji}(\omega) = \frac{\tilde{\Phi}_{ji}(\omega)}{\int_0^\infty \tilde{\Phi}_{ji}(\omega)\, d\omega}, \tag{12.50}$$

$$\tilde{\Phi}_{ji}(\omega) = \frac{\tilde{\tau}_{\omega,ji}}{\tau_\omega} \left\{ 1 - \exp(-\tau_\omega) + \left(\frac{L_{\text{eff}}}{L_s} \right)^2 K(\tau_\omega) \right\}, \tag{12.51}$$

$$K(\tau_\omega) = (1 - \exp(-\tau_\omega)) \cdot \left(\frac{1}{4} + \frac{1}{\tau_\omega^2} \right) + \frac{1}{\tau_\omega^2}(1 + \exp(-\tau_\omega)), \tag{12.52}$$

$$\tau_\omega = \sum_{ji} \kappa_{0,ij} L_{\text{eff}} \phi_{ij}(\omega_{ij} + \Delta \omega, \omega, \alpha_{ij}), \tag{12.53}$$

$$\tilde{\tau}_{\omega,ji} = \kappa_{0,ij} L_{\text{eff}} \phi_{ji}(\omega_{ji}, \omega, \alpha_{ji}), \tag{12.54}$$

$$\kappa_{0,ij} = \frac{\pi^2 c^2}{\omega_{ji}^2} \frac{g_j}{g_i} A_{ji} n_i \left\{ 1 - \frac{n_j\, g_i}{n_i\, g_j} \right\}. \tag{12.55}$$

The function $K(\tau)$ describes inhomogeneity effects originating from the upper-level populations. As can be seen from Equation 12.53, even line overlapping effects can be included via the opacity τ_ω that originates from all possible line transitions. Employing Equations 12.50 through 12.55, even time-dependent simulations of the spectral distribution of differentially moving optically thick plasmas can be performed with reasonable numerical burden. Note that the remaining integral in Equation 12.50 is just the sum of all frequency points of a line transition which has already been calculated for the spectral distribution of this transition.

Equation 12.51 shows that

$$\tilde{\Phi}_{ji}(\omega) \to \frac{\phi_{ji}(\omega_{ji}, \omega, \alpha_{ji})}{\phi_{ij}(\omega_{ji} + \Delta\omega, \omega, \alpha_{ij})}\{1 - \exp(-\tau_{\omega,ji})\} \tag{12.56}$$

if the scaling length L_s is much larger than the effective photon path length L_{eff} and line overlapping effects are neglected. If also no differential plasma motion is encountered ($V = 0$, i.e., $\Delta\omega = 0$), we obtain the well-known result for a constant source function and complete frequency redistribution:

$$\tilde{\Phi}_{ji}(\omega) \to \{1 - \exp(-\tau_{\omega,ji})\}. \tag{12.57}$$

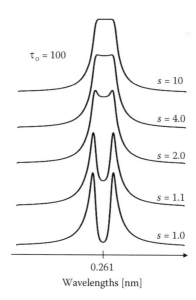

FIGURE 12.2
Effects of inhomogeneous plasma density on the optically thick line profile of the He-like resonance line of titanium at $\lambda_0 = 0.261$ nm, $kT_e = kT_i = 2000$ eV, $n_e = 10^{22}$ cm^{-3} and a line center optical thickness of $\tau_0 = 100$. A Voigt profile has been assumed as a local emission profile and a convolution with an apparatus profile with $\lambda/\delta\lambda = 5000$ has been made. A homogenous plasma is represented by a scaling length parameter $s \to \infty$, maximal nonhomogeneity, and strong self-reversal is obtained for $s = 1$.

Figure 12.2 shows the effects of inhomogeneous plasma density employing the parabolic expression according to Equations 12.45 and 12.46 and the analytical expressions according to Equations 12.50 through 12.55. For $s < 2$, strong self-reversal effects are seen. Figure 12.3 shows the effects of differential plasma motion. The line profile is strongly asymmetric due to the relative frequency shift of emission and absorption profiles. If two or more emission lines are very close to each other (means that their spectral separation is less than about $\Delta \omega$ from Equation 12.47), then the photon from one line might be strongly absorbed by another transition leading to "asymmetric repumping" and a corresponding strong distortion of line ratios [27]. The inclusion of these effects in the escape factor approach is difficult. Figures 12.2 and 12.3 demonstrate strong modifications of the line profiles in dense, optically thick plasmas, and subtle analysis of Stark broadening effects on the line profile appears to be difficult.

If free–free and bound–free absorption (the so-called continuum opacity) are important, Equation 12.44 is no longer useful to approximate the spectral distribution because photons that are emitted from a spectral line may be redistributed with high probability elsewhere in the continuum. Therefore, photons from a well-defined bound–bound emission are lost for the particular line emission when they are re-emitted far away from the line profile. Whether continuum absorption is important or not can be estimated from the following

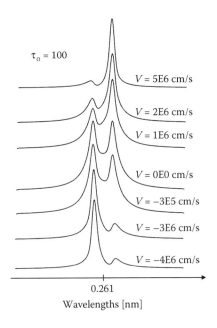

$\tau_0 = 100$

$V = 5E6$ cm/s

$V = 2E6$ cm/s

$V = 1E6$ cm/s

$V = 0E0$ cm/s

$V = -3E5$ cm/s

$V = -3E6$ cm/s

$V = -4E6$ cm/s

0.261

Wavelengths [nm]

FIGURE 12.3
Influence of differential plasma motion in inhomogeneous plasmas on the optically thick line profile for different velocities V; other parameters are as in Figure 12.2.

expressions:

$$\tau_{ff} \approx 2.4 \times 10^{-37} \bar{g}_{ff} \frac{n_e^2 Z_{eff}}{\sqrt{kT_e} E_\omega^3} \left(1 - \exp\left[-\frac{\hbar\omega}{kT_e}\right]\right) L_{eff}, \qquad (12.58)$$

$$\tau_{fb} \approx 2.9 \times 10^{-17} \bar{g}_{fb} \frac{E_i^{5/2} n_i}{Z_{eff} E_\omega^3} \left(1 - \exp\left[-\frac{\hbar\omega}{kT_e}\right]\right) L_{eff}, \qquad (12.59)$$

where τ_{ff} is the free–free opacity, τ_{fb} the bound–free opacity, E_ω the photon energy (eV), E_i the ionization potential (eV), kT_e the electron temperature (eV), n_e the electron density (cm^{-3}), n_i the population density from which photoionization proceeds (cm^{-3}), Z_{eff} the effective charge of the plasma, L_{eff} an effective photon path lengths (cm), $\bar{g}_{ff}$ and $\bar{g}_{fb}$ are the Gaunt factors for the free–free transitions (Bremsstrahlung) and the free–bound transitions (radiative recombination radiation). In order to operate in a meaningful manner with diagnostic line ratios, the continuum opacities τ_{ff} and τ_{fb} should be less than about 1.

12.3 Nonequilibrium Phenomena

As discussed in Section 12.1, nonequilibrium phenomena are not only connected with nonstationary phenomena, but also with nonequilibrium statistics. Below, we discuss several effects that are important for plasma spectroscopy. We begin with nonstationary phenomena.

12.3.1 Transient Plasma Evolution and Characteristic Time Scales of Atomic Systems

The basic equations to study transient plasma evolution on atomic systems are given by a set of time-dependent nonlinear system of differential equations of the type:

$$\frac{dn_j}{dt} = \frac{\partial n_j}{\partial t} + \nabla(\Gamma_j) = -n_j \sum_{k=1}^{N} \left\{ W_{jk} + P_{ik}^{abs} + P_{ik}^{stim} + P_{ik}^{iz} + R_{ik}^{stim} \right\}$$

$$+ \sum_{i=1}^{N} n_i \left\{ W_{ij} + P_{ij}^{abs} + P_{ij}^{stim} + P_{ij}^{iz} + R_{ij}^{stim} \right\}. \qquad (12.60)$$

The partial derivative is important to study the response of the atomic system due to time-dependent plasma parameters, whereas the second term is usually important to describe the transport of particles (to be considered below) with the particle flux Γ_j. Equation 12.60 readily indicates that time-dependent plasma parameters, such as, for example, $T_e(t)$ and $n_e(t)$ enter directly into the rate coefficients (W-matrix, photon rates P, see Equations 12.16 and 12.22), however, they do not translate directly into a corresponding change of the atomic populations $n_j(t)$. Equation 12.60 can therefore, be considered as a response function of the atomic system to external perturbations [28].

The response properties can be studied with a two-level atom which connect either different ionization stages (Figure 12.4a) or one radiating level inside one ionization stage (Figure 12.4b). Let us begin with the time-dependent evolution of the level n_z (Figure 12.4a):

$$\frac{\partial n_Z}{\partial t} = -n_Z I_{Z,Z+1} + n_{Z+1}(T_{Z+1,Z} + R_{Z+1,Z} + D_{Z+1,Z}), \qquad (12.61)$$

where I is the ionization rate, T the three-body recombination rate, R the radiative recombination rate, and D the dielectronic recombination rate. For the two-level atom, the normalization condition reads

$$n_Z + n_{Z+1} = 1, \qquad (12.62)$$

which means that the probability to find the atom either in the state Z or in the state $Z + 1$ is equal to 1. Let us consider a rapid heating process where all population is found in the ground level at $t = 0 : n_Z(t = 0) = 1$. The analytical solution of the differential equation 12.61 together with Equation 12.62 is then

$$n_Z(t) = (1 - \gamma)e^{-t/\tau_Z} + \gamma, \qquad (12.63)$$

$$\gamma = \frac{T_{Z+1,Z} + R_{Z+1,Z} + D_{Z+1,Z}}{I_{Z,Z+1} + T_{Z+1,Z} + R_{Z+1,Z} + D_{Z+1,Z}}, \qquad (12.64)$$

$$\tau_{Z,Z+1} = \frac{1}{I_{Z,Z+1} + T_{Z+1,Z} + R_{Z+1,Z} + D_{Z+1,Z}}. \qquad (12.65)$$

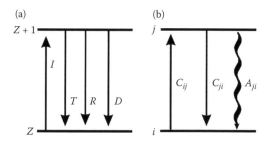

FIGURE 12.4
(a) Two-level atom connecting atomic levels from two different ionization stages; (b) two-level atom (one ionization stage) of a radiating system.

Equations 12.63 and 12.64 recover the initial condition and also the equilibrium condition at very long times (which could directly be obtained from Equations 12.61 and 12.62 setting $\partial/\partial t = 0$). As shown by Equation 12.65, the heating process that populates the level n_{Z+1} has a characteristic time scale, namely τ_Z. Therefore, even sudden heating processes do not lead to a sudden response of the atomic level populations. Moreover, the time scale for the ionization is not given by the inversion of the ionization rate itself but rather by the sum of the ionization and all recombination process. The physical reason is that equilibrium requests not only the equilibrium of the state which is ionized, but also the equilibrium of those which are populated by ionization. From these levels, however, recombination processes originate. Numerical calculations show [28,29] that for the K-shell of highly charged ions, Equation 12.65 can be estimated by $n_e \tau_{K\text{-shell}} \approx 10^{12}$ cm^{-3} s.

As Equation 12.65 demonstrates, each ionization stage and each element has, in principle, its own characteristic time scale. Therefore, in transient plasmas, the use of the radiation emission from tracer elements of different nuclear charges Z_n to establish intensity line ratios which are then employed for the determination of temperature and density in a stationary collisional radiative calculation according to Equations 12.15 and 12.21 is questionable (even if the same atomic transitions of the isoelectronic sequence are employed). Moreover, the use of different types of tracer elements requests to ensure that the observation volume contains a certain amount of tracer "A" and tracer "B", an additional experimental difficult. In a similar manner, even the use of resonance lines from different ionization stages but one tracer element (e.g., H-like Lyman-alpha and He-like Helium-alpha transitions to determine the temperature) in transient plasmas may provide misleading conclusion when the stationary set of Equations 12.15 and 12.22 is employed.

We now consider the transient evolution of photon emission according to Figure 12.4b. The relevant set of differential equations then reads

$$\frac{\partial n_j}{\partial t} = -n_j(A_{ji} + C_{ji}) + n_i C_{ij}, \tag{12.66}$$

$$n_i + n_j = 1, \tag{12.67}$$

which means that the probability to find the atom either in the state i or in the state j is equal to 1. Let us consider a rapid cooling processes with $n_j(t = 0) = 1$. The analytical solution of the differential Equation 12.66 together with Equation 12.67 is then

$$n_j(t) = (1 - \delta)e^{-t/\tau_{ji}} + \delta, \tag{12.68}$$

$$\delta = \frac{C_{ij}}{A_{ji} + C_{ji} + C_{ij}}, \tag{12.69}$$

$$\tau_{ji} = \frac{1}{A_{ji} + C_{ji} + C_{ij}}. \tag{12.70}$$

Equations 12.68 and 12.69 recover the initial condition and also the equilibrium condition at very long times (which could directly be obtained from Equations 12.66 and 12.67 by setting $\partial/\partial t = 0$). Equation 12.70 shows that photon emission is not instantaneous and at very small densities (when $A_{ji} \gg C_{ji} + C_{ij}$), the photon relaxation time is finite, namely $\tau_{ji} \approx 1/A_{ji}$. This is quite different from Equation 12.65: the lower the particle density, the longer the relaxation time: that means, for example, stellar clouds with particle densities of about $10^6 \, \mathrm{cm}^{-3}$ may need days to move to atomic equilibrium.

Equation 12.70 also shows that the population of levels which radiatively decay is strongly density-dependent if the collisional processes are at least of the order of the radiative decay rate. This phenomenon may lead to a so-called "mixing of relaxation times": in a multilevel system, a metastable level can "feed" a resonance emission for a long time via collisions. This phenomenon is shown in Figure 12.5 via a multilevel collisional radiative simulation for argon carried out with the MARIA code [29–31]. The shortest relaxation time is those of the resonance line $\tau(W) \approx 9 \times 10^{-15} \, \mathrm{s}$ (indicated by the arrow at the first step in Figure 12.5a). The next step is due to a collisional coupling between the levels $1s2p \, ^1P_1$ and $1s2s \, ^1S_0$. The relaxation time of the $1s2s \, ^1S_0$ level is determined by the two-photon decay $\tau(2E1) \approx 3 \times 10^{-9} \, \mathrm{s}$ as well as by collisions (Equation 12.70). At particle densities of $n_e = 10^{21} \, \mathrm{cm}^{-3}$, the relaxation time of the $1s2s \, ^1S_0$ level is determined by collisions [rate coefficient $C(1s2s \, ^1S_0 - 1s2p \, ^1P_1) \approx 2 \times 10^{-9} \, \mathrm{cm}^3 \, \mathrm{s}^{-1}$]. The effective relaxation time is therefore about $\tau(1s2s \, ^1S_0) \approx 4 \times 10^{-13} \, \mathrm{s}$ as indicated by the arrow "$1s2s \, ^1S_0$" (giving rise to a second step at about $t = 10^{-13}$–$10^{-12} \, \mathrm{s}$). The last step is due to the establishment of ionization equilibrium: the recombination rate from the H-like to He-like ions at $kT_e = 500 \, \mathrm{eV}$ is about $R \approx 4 \times 10^{-12} \, \mathrm{cm}^3 \, \mathrm{s}^{-1}$, giving a relaxation time of about $\tau(1s \, ^2S_{1/2}) \approx 3 \times 10^{-10} \, \mathrm{s}$. This is indicated by the arrow "$1s \, ^2S_{1/2}$" [note that $\tau(1s \, ^2S_{1/2}) \, n_e \approx 3 \times 10^{11} \, \mathrm{cm}^{-3} \, \mathrm{s}$].

In conclusion, in transient dense plasma evolution, collisional processes do not lead only to a transfer of population, but also to a mixing of relaxation times. This may result in some cases to a considerable prolongation of the radiation emission. Let us consider the intercombination line of He-like argon ions as an example: for argon, the radiative relaxation time is $\tau(Y = 1s^2 - 1s2p \, ^3P_1) \approx 6 \times 10^{-13} \, \mathrm{s}$, however, the fine structure $1s2l \, ^3L$ is also the origin of magnetic multipole transitions with very long relaxation times: $\tau(Z = 1s^2 - 1s2s \, ^3S_1) \approx 2 \times 10^{-7} \, \mathrm{s}$, $\tau(X = 1s^2 - 1s2p \, ^3P_2) \approx 3 \times 10^{-9} \, \mathrm{s}$. It is, therefore, possible that the intercombination line emission has a collisionally enhanced relaxation time by about seven orders of magnitude compared to the radiative relaxation time of the Y-line itself (indicated by the vertical arrow "$1s2l \, ^3L$" in Figure 12.5a). This can lead to very long-lasting intercombination line emission in cooling plasmas (e.g., in laser-produced plasmas, Z-pinch disruptions). Figure 12.5a demonstrates that the intercombination line intensity in the time interval of about 10^{-13}–$10^{-9} \, \mathrm{s}$ is much stronger than those of the resonance line. The traditional interpretation of intense intercombinaton line emission

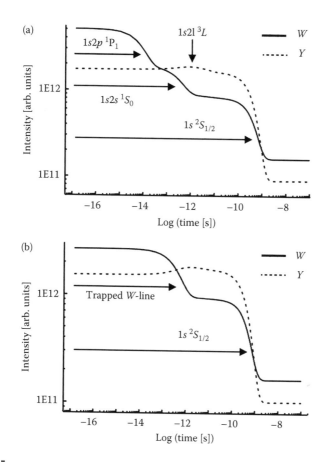

FIGURE 12.5
(a) Transient line emission of the He-like resonance line $W = 1s^2 - 1s2p\,^1P_1$ and the intercombination line $Y = 1s^2 - 1s2p\,^3P_1$ of argon: rapid plasma cooling from 2000 to 500 eV, $n_e = 10^{21}\,cm^{-3}$; (b) transient line emission in an optical thick plasma, $L_{eff} = 100\,\mu m$; other parameters are as in (a).

in recombining plasmas due to radiative recombination [8] is therefore not so obvious.

A "feeding" of the intercombination line emission due to charge exchange recombination processes in dense recombining laser-produced plasmas has therefore been identified on the basis of different space-resolved satellite x-ray emission lines [32]. We note that inner-shell ionization ($1s^22l + e \rightarrow 1s2l^{1,3}L + 2e$) may only explain at maximum three times larger intensities of the intercombinaton line compared to the resonance line (due to the statistical weights of the singlet and triplet system), whereas the collisional mixing of relaxation times may also explain order of magnitude different intensities.

Finally, we note that in optically thick plasmas, the scattering of photons (emission and re-emission) leads to a "delay" in photon escape. This is demonstrated in Figure 12.5b. Due to the large opacity of the He-like resonance line, their escape is prolonged by orders of magnitude.

It should be emphasized that the characteristic time scales according to Equations 12.65 and 12.70 are fundamental properties which determine the response times of the atomic system. Therefore, time-dependent measurements, for example, by means of a streak camera might not be able to solve all problems concerning "transient plasma evolution."

12.3.2 Particle Transport

The magnetic confinement of the fusion plasma is one of the most important issues, and intensive efforts have, therefore, been devoted to the understanding of the particle transport. However, the physical processes that underlie plasma transport in toroidally confined plasmas are not so well understood. The plasma transport induced by Coulomb collisions (the so-called classical or neoclassical transport) is often much less than what is actually observed [33,34] and thus the transport is called anomalous.

Methods that determine the particle transport *independent* of theoretical plasma models are, therefore, of fundamental importance in the magnetic fusion research. Spectroscopic methods have turned out to be very effective. One of the most powerful methods is based on the space- and time-resolved observation of the line emission from impurity ions [33–35]. The emission is simulated from an atomic physics model like, for example,

$$\frac{\partial n_Z}{\partial t} + \nabla(\vec{\Gamma}_Z) = -n_Z(I_{Z,Z+1} + T_{Z,Z-1} + R_{Z,Z-1} + D_{Z,Z-1})$$

$$+ n_{Z-1}(I_{Z-1,Z}) + n_{Z+1}(T_{Z+1,Z} + R_{Z+1,Z} + D_{Z+1,Z}), \qquad (12.71)$$

where $\vec{\Gamma}_Z$ is the particle flux (Z indicates the charge of the ion). With given temperature and density profiles, one tries to match the experimental observations by a best fit of $\vec{\Gamma}_Z$. For these purposes, it turned out to be convenient to split the flux into a diffusive and convective term according to $\vec{\Gamma}_Z = -D_Z \nabla n_Z + \vec{V}_Z n_Z$, where D_Z is the diffusion coefficient (note that D_Z is the diffusion coefficient, whereas $D_{Z,Z-1}$ is the dielectronic recombination rate coefficient connecting the charge states Z and $Z-1$) and $\vec{V}_Z$ is the convective velocity. These parameters are then varied in a numerical procedure in order to best fit the spectral emission data. The importance of this type of analysis lies in the fact that it provides a plasma simulation-independent information (independent from, e.g., turbulence models) for the diffusion coefficient and the convective velocity [33–35].

Under real experimental conditions of magnetically confined fusion plasmas, the impurity ions do interact with the plasma background H/D via charge exchange. This in turn leads to a change in the radial charge state distribution of the impurity ions, an effect which has a large impact on the analysis and the interpretation of possible particle transport: diffusion in space (particle transport) and diffusion in charge states (charge exchange) are of similar nature in the framework of the traditional particle transport analysis (via diffusion coefficients D and convective velocities V) [36,37]. This can easily be seen from the more generalized equation

$$\frac{\partial n_Z}{\partial t} + \nabla(\vec{\Gamma}_Z) = -n_Z(I_{Z,Z+1} + T_{Z,Z-1} + R_{Z,Z-1} + D_{Z,Z-1} + Cx_{Z,Z-1})$$
$$+ n_{Z-1}(I_{Z-1,Z} + Cx_{Z-1,Z}) + n_{Z+1}(T_{Z+1,Z} + R_{Z+1,Z}$$
$$+ D_{Z+1,Z} + Cx_{Z+1,Z}), \tag{12.72}$$

where $Cx_{Z,Z-1}$, and so on indicate possible charge exchange processes between the radiating test element (e.g., intrinsic impurities) and other species (namely hydrogen, deuterium, tritium, and helium). Let us assume that the partial derivative is zero and integrate the set of Equation 12.72 over space. The integration over space transforms the diffusion term into the so-called "tau approximation." Note that the tau approximation is a rather powerful method of particle transport analysis which even permits to study details of the line emission not only of resonance lines, but also from forbidden lines too [36,37]. In the "tau approximation," Equation 12.72 takes the form

$$n_Z(I_{Z,Z+1} + T_{Z,Z-1} + R_{Z,Z-1} + D_{Z,Z-1}) + n_Z\left(Cx_{Z,Z-1} + \frac{1}{\tau_{Z,Z+1}}\right)$$
$$= n_{Z-1}(I_{Z-1,Z}) + n_{Z-1}\left(Cx_{Z-1,Z} + \frac{1}{\tau_{Z-1,Z}}\right)$$
$$+ n_{Z+1}(T_{Z+1,Z} + R_{Z+1,Z} + D_{Z+1,Z}) + n_{Z+1}\left(Cx_{Z+1,Z} + \frac{1}{\tau_{Z+1,Z}}\right), \tag{12.73}$$

where $\tau_{Z,Z+1}$, and so on, are the respective diffusion times. It is clearly seen that diffusion/transport (represented by the tau terms in Equation 12.73) are of the same origin as charge exchange processes (Cx terms in Equation 12.73). It is, therefore, difficult to characterize the particle transport on the basis of Equation 12.71: if the charge exchange is a free parameter as well as diffusion D_Z and convective velocity V_Z, their significance is not so evident as charge exchange (diffusion in charge states) and particle transport (diffusion in space) are overlapping effects.

In order to circumvent this difficulty, a self-consistent analysis has been proposed [38] to eliminate the free parameters for the charge exchange: the coupling is a self-consisted excited states coupling of the tracer (impurity) kinetics to the plasma background (H, D, T) via atomic physics processes (charge exchange). The matrix coupling elements $M_{ji}(H, D, T, X)$ can schematically be written as

$$M_{ji}(H, D, T, X) = n_j^{H,D,T} n_i^X \langle \sigma_{ji}^{Cx} V_{rel} \rangle \qquad (12.74)$$

where H, D, and T indicate hydrogen, deuterium, and tritium and X is the spectroscopic tracer element (e.g., He, an intrinsic impurity or any other element intentionally introduced for diagnostic purposes), $n_j^{H,D,T}$ the population density of the elements (H, D, T) in state j, n_i^X the population density of the tracer element in state i, σ_{ji}^{Cx} the charge exchange cross-section from state j to state i between the elements (H, D, T) and X, V_{rel} the relative particle velocity, and the brackets indicate an average over the particle energy distribution functions. As the coupling matrix elements according to Equation 12.74 contain the product of different population densities, the system of equations (H, D, T) and (X) is nonlinear (even in the optically thin plasma approximation). The self-consistent numerical simulation of multi-ion, multilevel (LSJ-split) non-LTE atomic kinetic systems coupled by charge exchange processes via the excited states coupling matrix (Equation 12.74) are now realized in the "SOPHIA code."

The great advantage in using the coupling matrix according to Equation 12.74 lies in the fact that the selection rules for the charge exchange processes are respected: charge transfer from excited states is *directly* coupled to excited states. Therefore, the population flow due to charge exchange is consistently treated without any free parameter along with the population flow of usual collisional radiative processes.

We note that the excited states coupling also avoids critical divergences which arise from the strong scaling of the charge exchange cross-sections with principal quantum number n : $\sigma^{Cx} \propto n^4$. Moreover, under typical conditions of ITER, the excited-state populations of hydrogen increase rapidly due to the increasing statistical weights resulting altogether in an effective divergence $\propto n^6$. This charge exchange-driven divergence is therefore much more pronounced than the well known divergence of the partition sum. Table 12.1 demonstrates this effect: excited states driven change exchange processes become even more important than the ground state for about $n > 15$ and at $n = 20$ all charge exchange flow is driven by excited states rather than ground states. Therefore, any level cut-off is highly critical, and numerical simulations are nonpredictive. In this respect, effective rate coefficients as, for example, proposed in [39] have to be employed with caution. In the self-consistent excited states coupling approach [38], however, no critical level cut-off is present (or necessary) because charge exchange and collisions are treated on a unique footing: a large charge exchange flow in highly excited states is directly redistributed by collisions between the excited states before radiative decay can populate the ground states. Figure 12.6 demonstrates this effect schematically. The left-hand side shows the thermal limit (see Equation 12.8) above which PLTE holds. The diagram on the right-hand side shows the case, when charge exchange flow (indicated by arrows) populates the levels: the thermal limit according to Equation 12.8 is changed. However,

TABLE 12.1

Effective Charge Exchange Contributions from Excited States of Hydrogen Obtained from the SOPHIA-Code: Population Density n_H Multiplied by the 4th Power of the Principal Quantum Number

Niveau	$n_H * n^4$	$n_H * n^4/n_H(1s)$
1s	1.16D−03	1.00D+00
$n = 2$	1.24D−06	1.07D−03
$n = 3$	7.18D−07	6.21D−04
$n = 4$	3.28D−06	2.84D−03
$n = 5$	1.17D−05	1.01D−02
$n = 10$	6.66D−04	5.77D−01
$n = 15$	7.40D−03	6.41D+00
$n = 20$	4.12D−02	3.57D+01
$n = 25$	1.57D−01	1.36D+02

Note: The second column indicates the absolute fraction, whereas the third column indicates the relative importance with respect to the hydrogen ground state 1s. The plasma parameters are $kT_e = 3\,eV$, $n_e = 10^{13}\,cm^{-3}$. The populations n_H are normalized according to $\Sigma n_H = 1$.

the new thermal limit $n_{thermal}^{Cx}$ is still below those quantum numbers $n_f^{Cx}(div.)$, where the charge exchange flow diverges. Therefore, the excited states coupling allows an instantaneous ionization (indicated by arrow I). As $I \gg A$, no

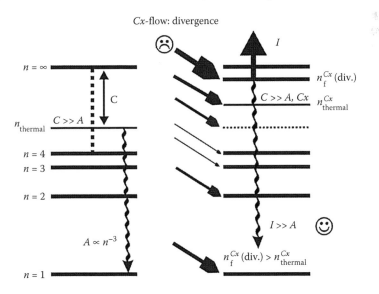

FIGURE 12.6
Schematic energy level diagram of the self-consistent excited states coupling realized in the SOPHIA-code. Free parameters for the charge exchange processes and flow divergences are avoided (see text).

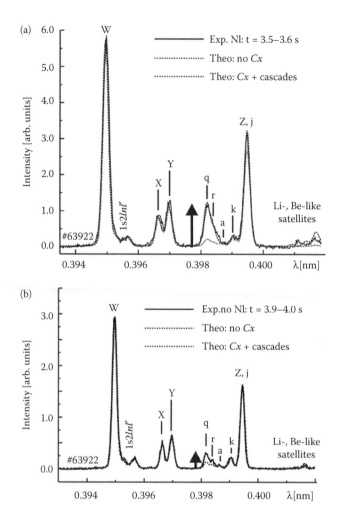

FIGURE 12.7

(a) Time-resolved soft x-ray impurity spectrum of gas puff-injected argon during neutral beam injection with 1.2 MW. The high spectral resolution enables the distinct observation of the He-like lines $W = 1s^2 - 1s2p\ ^1P_1$, $X = 1s^2 - 1s2p\ ^3P_2$, $Y = 1s^2 - 1s2p\ ^3P_1$, and $Z = 1s^2 - 1s2s\ ^3S_1$ and numerous Li-like satellites indicated as q, r, a, k, and $1s2lnl'$. A simulation which also includes the simultaneous calculation of the ion abundances shows nevertheless a very poor agreement for the q- and r-satellites, whereas the self-consistent MARIA charge exchange simulations provides excellent agreement in all spectral features (see arrow): $kT_e = 1700\,\text{eV}$, $f(H) = 1.7 \times 10^{-5}$. (b) Similar to (a), however, the NI is switched off, $kT_e = 1300\,\text{eV}$, $f(H) = 6 \times 10^{-6}$. (c) Experimental peak intensity ratio of the q- and k-satellites in dependence of the neutral beam injection power. It can be clearly seen that with increasing NI-power, the ratio continuously rises (solid and open squares) and relaxes to a common level (solid and open circles) after switching off the NI-injection. The line of sight for the x-ray emission crosses the injection direction for NI1 (solid symbols), however for NI2, no geometrical crossing occurs (open symbols). As can be seen, however, open and sold circles coincide within the error bars. These results suggest that the neutral beam is rapidly thermalized creating an enhanced neutral background. (d) MARIA simulations of the integrated intensity ratio $I(q)/I(k)$ and $I(k)/I(W)$ in dependence of the neutral beam fraction f. A strong dependence of the $I(q)/I(k)$-ratio is obtained, whereas the satellite-resonance line ratio $I(k)/I(W)$ stays almost constant (multiplied by a factor of 10 for better presentation): $kT_e = 1700\,\text{eV}$, $n_e = 2 \times 10^{13}\,\text{cm}^{-3}$.

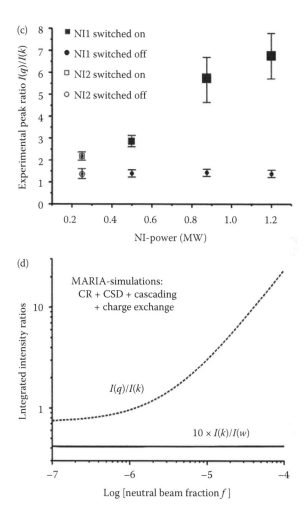

FIGURE 12.7
Continued

effective radiative decay can feed the lower levels from the divergent flow and the excited-state coupled system is naturally stabilized. In consequence, the divergent charge exchange flows do not lead to a divergent population of the atomic levels. This means that first, no artificial (and therefore uncertain) level cut-off is needed to stabilize the system and, second, the number of levels included in the simulations are not very critical if principal quantum numbers are included which are a few times larger than n_{thermal}^{Cx}.

It is worth noting that the foregoing discussion of Equation 12.72 can be extended to suprathermal electrons, because the rate coefficients I, T, R, and D are treated on a unique footing along with the charge exchange and the particle transport [40]. It is, therefore, advisable not only to improve the standard rate coefficients I, T, R and D, but charge exchange cross-sections from and to excited states too.

It is evident from Equation 12.72 and the related discussion that supplementary information is needed to provide a unique characterization. Therefore, the He-like lines W, X, Y, Z, the He-beta resonance line (W3 := $1s3p\ ^1P_1$ – $1s^2\ ^1S_0$) and intercombination line (Y3 := $1s3p\ ^3P_1$ – $1s^2\ ^1S_0$) as well as the Li-like satellites $1s2l2l'$ – $1s^2 2l''$ of highly charged ions have been proposed and successfully applied to experimental measurements [36,37,41] in order to distinguish particle transport and charge exchange effects. This turned out to be possible because the response functions of these different emission lines to charge exchange-driven cascading, particle transport, inner-shell excitation, and inner-shell ionization processes are quite different. Figure 12.7a and b shows two examples: the argon impurity x-ray emission (solid curve) during neutral beam injection (NI) (Figure 12.7a) and after switching off the NI (Figure 12.7b), observed at the TEXTOR tokamak [36,41].

Figure 12.7c demonstrates a strong rise of the q-satellite intensity with increasing power of the NI independent of whether the neutral beam injection is crossing the line of sight of the spectrometer or not. This observation suggests that NI is rapidly thermalized and that the enhanced neutral background (and the corresponding charge exchange processes) leads to the observed intensity variations. Figure 12.7d shows the MARIA simulations of the q- and k-satellite ratio in dependence of the neutral beam fraction $f = n_{H/D}/n_e$. A strong dependence of the ratio from the neutral beam fraction is obtained, indicating that charge exchange might be at the origin of the observations. Figure 12.7a shows the spectral simulations without and with charge exchange effects. Excellent agreement with the observations (solid curve in Figure 12.7a) and the MARIA charge exchange simulations (dashed curve in Figure 12.7a) are seen for $f = 1.7 \times 10^{-5}$ ($\Delta t = 3.5$–3.6 s). It is important to note that not only the q-satellite, but also all forbidden lines (X, Y, Z) are in almost perfect agreement with the simulations. Simulations not taking into account charge exchange and multilevel cascading processes (dashed curve in Figure 12.7a) result in a considerably underestimation of the intensities for the q-satellite and also for the forbidden line intensities X, Y, and Z.

After switching off the NI at $t = 3.7$ s, the neutral background is still enhanced as the simulations without charge exchange show (Figure 12.7b). The simulation method also permits us to determine the enhanced neutral background after neutral beam injection: $f = 6 \times 10^{-6}$ ($\Delta t = 3.9$–4.0 s). Due to the large sensitivity of this method, even the neutral background in purely ohmic discharges could be successfully determined [36].

As Figure 12.7 demonstrates, excellent agreement between theory and experiment is obtained. Surprisingly, just the opposite has been stated in [42] with a nontransparent discussion of the Z-line. It should be noted that the Z-line intensity is strongly plasma parameter-dependent due to its cascade sensitivity and inner-shell ionization population channel. Therefore, the surprises announced in [42] that their observations show quite different Z-line intensities when compared to [36] is irrelevant as the plasma parameters in [42] and [36,41] are quite different.

Further serious discrepancies between simulations and experimental data for the W3 and Y3 argon line emission in a well-diagnosed tokamak have been reported [43]. However, the statements made turned out also to be in error and correctly performed multilevel multi-ion stage simulations [44,45] carried out with the MARIA code [29–31] demonstrated excellent agreement with the data.

In conclusion, the high precision of the proposed methods and simulations [36–38] has been confirmed in a series of experiments with and without neutral beam injection at tokamaks [36,41] as well as in several plasma simulator experiments [46,47]. These methods serve now as a basis for advanced investigations in the magnetic fusion research (e.g., [46–49]).

Charge exchange processes turned out to be equally important in dense hot plasmas (laser-produced plasmas, Z-pinches), and numerous international projects have been devoted to these interesting phenomena. The interested reader is referred to [32,50–54].

12.3.3 Inhomogeneous Spatial Distributions

Efforts are made to create homogenous dense plasmas under extreme conditions to provide samples and emission properties, which can be directly compared with theory. Unfortunately, dense hot laboratory plasmas show almost always-large variations of the plasma parameters over space and, in consequence, a large variation of the spectral emission. As spatial parameter grid reconstructions are difficult, x-ray spectroscopy with spatial resolution is frequently applied to obtain supplementary information. The spatial resolution can be realized either with a slit mounted at a suitable distance between the source and the x-ray crystal or by means of curved x-ray crystals. The most commonly used curved crystal arrangements are the Johann geometry, the Johannson geometry, and two-dimensional curved crystals, and a review of these methods is given in [8]. It is worth emphasizing two particular methods which turned out to be extremely useful for dense plasma research:

1. The vertical Johann geometry [55] which is extremely suitable for line profile investigations (spatial resolution of some μm with simultaneous extremely high spectral resolution of about $\lambda/\delta\lambda \approx 6000$, the spectral range, however, is rather small, permitting only to observe, for example, the H-like aluminum Lyman α line and corresponding satellite transitions [56]). Due to the appearance of double-sided spectra, the geometry can provide line shift measurements without reference lines. Due to the high sensitivity of this method, even a spin-dependent line shift in He-like ions (W and W-lines) has recently been reported in dense laser-produced plasmas [57,58]. These results have

initiated first attempts to investigate the density dependence of the exchange energy [59].

2. The spherical x-ray crystals [8,60,61] do provide simultaneously high spectral ($\lambda/\delta\lambda \approx$ 1000–6000, depending on the large varieties of possible geometries) and spatial resolution (about $10\,\mu m$), large spectral windows in wavelengths (permitting, e.g., to observe all the K_α-satellite series until He-like $He_\alpha = W$ for aluminum [62]) and space (up to cm with $10\,\mu m$ resolution), and the possibility for x-ray microscopic applications (two-dimensional x-ray imaging). The two-dimensional imaging permits us, for example, to determine the angle of plasma jet diffusion in laser-produced plasma jets (see Figure 12.8).

12.3.4 Fast Ions

An important application of space-resolved spectroscopy is the experimental determination of the energy distribution function of fast ions in dense laser-produced plasmas measured via line shifts of spectrally highly resolved resonance lines. These shifts are induced by the directional Doppler shift. Figure 12.9a demonstrates the basic principle of these measurements. Figure 12.9b shows space-resolved line shift measurements (x-ray images

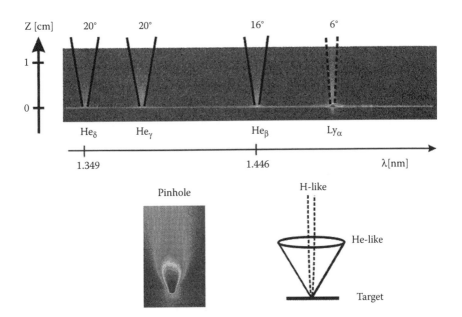

FIGURE 12.8
Angular distribution of the x-ray emission of H-like and He-like fluorine lines in a dense laser-produced plasma (laser energy of 50 J, pulse duration of 15 ns, and laser wavelength of $1\,\mu m$). The two-dimensional x-ray image was obtained with a spherical mica crystal. It can be seen, that the diffusion angle of H-like ions ($6°$) is considerably lower than for He-like ions ($20°$).

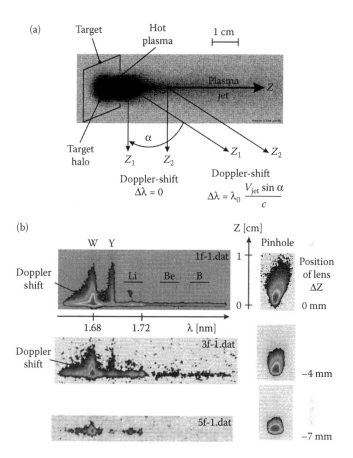

FIGURE 12.9
(a) Experimental scheme of space-resolved Doppler shift measurements, (b) of spectrally highly resolved resonance line emission (W) of He-like fluorine ions, the higher the laser intensity, the more pronounced are the "blue wings" of the He-like resonance line W, (c) Doppler-shifted line wings induced by fast ion motion observed at different angles α, and (d) the intensity in the line wing provides directly the energy distribution function of the fast ions. Due to the inherent line identification, the energy distribution function is measured in dependence of the charge state of the ion, which provides critical data for plasma simulations.

obtained from a spherical crystal) for different laser irradiation conditions (different focus conditions due to different positions of the focusing lens, laser pulse duration of 15 ns, laser wavelength of 1.06 μm, laser energy of 10–60 J). Also seen from Figure 12.9b is a correlation of the spatial extension of the spectrally resolved plasma jets (Z-axis, direction of the expanding plasma) and the x-ray pinhole measurements. Figure 12.9c shows the line profile of the H-like fluorine Ly_α and the strongly pronounced "blue" line wing. The Doppler-shifted position which corresponds to an ion energy of $E = 3.6\,\mathrm{MeV}$ is indicated. Figure 12.9d shows a fit of the blue line wing to obtain the energy

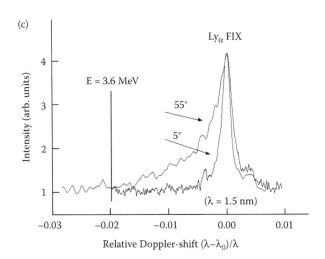

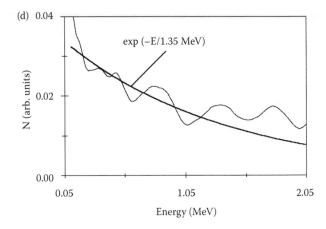

FIGURE 12.9
Continued

distribution function. A fit with a monodirectional Maxwellian function with a "hot ion temperature" of 1.35 MeV leads to a reasonable description for the energy distribution function. These types of measurements provide critical information to test kinetic plasma simulations, and numerous discrepancies to standard theories have been discovered (e.g. [63–65] and references therein).

Fast ion velocities in dense plasmas lead to differential shifts of emission and absorption coefficients which may lead to considerable modification of optically thick lines shapes (see Figures 12.2 and 12.3). For this reason, analysis of the energy distribution function as discussed in Figure 12.9c and d has been limited to the line wings, which are optically thin.

Fast ion velocities may also contribute directly to excitation and ionization processes via the W-matrix in the atomic population kinetics (Equations

12.15, 12.16, 12.22, and 12.60). Important examples are the redistribution of population in the fine structure of H-like ions ($2s$ $^2S_{1/2}$, $2p$ $^2P_{1/2}$, $2p$ $^2P_{3/2}$) due to a collisional coupling of the $2s$ $^2S_{1/2}$, $2p$ $^2P_{1/2}$. Their energy difference is very small (Lamb shift), and heavy particle collisions are therefore effective. Other examples are the collisional ionization of hydrogen levels with large principal quantum numbers by protons, which turned out to be an important effect in magnetic fusion research to analyze charge exchange processes [37].

An estimate of the importance of collisional cross-sections induced by fast ions can be made (in some cases) via the classical approach: if the ion velocity V_{ion} is much smaller than the effective Bohr velocity V_n for an electron with principal quantum number n, then the ion-induced cross-section might be negligible:

$$\sigma \approx 0: \quad V_{ion} \ll V_n = V_0 \frac{Z_{eff}}{n}, \tag{12.75}$$

where V_0 is the Bohr velocity (2.19×10^8 cm/s) and Z_{eff} the effective charge of the target ion where the atomic transitions are induced. More detailed formulas which can be easily applied for practical purposes are proposed in the excellent book of Sobelman et al. [66]. Useful formulas are also summarized in this book in order to calculate other collisional processes.

12.3.5 Suprathermal Electrons

As discussed in Section 12.1, nonequilibrium phenomena are not only connected with nonstationary phenomena, but also with nonequilibrium particle statistics. Nonequilibrium electron statistics (or non-Maxwellian energy distribution functions) are at the center of attention in worldwide research because they play an exceptionally important role for the fusion research:

a. In the indirect drive scheme (hohlraum) of internal fusion, hot electron production may lead to a preheat of the deuterium–tritium capsule and subsequent compression above solid density to obtain ignition is impossible.

b. In the direct drive scheme (fast ignition), hot electrons are created purposely by a PetaWatt laser via "hole boring" to initiate burn in the DT capsule (under less stringent parameter conditions). In the magnetic fusion research, suprathermal electrons connected with the plasma disruption will lead to serious wall destruction.

c. In atomic physics, suprathermal electrons lead to strong changes of almost all radiative properties: ion charge state distributions, radiative losses, line intensities, and line ratios.

The sensitivity of the radiation emission and, in particular, the sensitivity of the spectral distribution to non-Maxwellian electrons provide the basis for the development of advanced spectroscopic methods and diagnostics.

We note that a further driving force for the worldwide spectroscopic efforts is connected with the circumstance that, first, high-resolution x-ray spectroscopy (high spatial and high spectral resolution) provides a unique description in terms of plasma parameters (by the use of properly chosen tracer elements and atomic transitions) and, second, Thomson scattering (optical or x-ray) is not sensitive to suprathermal electrons.

As in the case of fast ion velocity analysis, space-resolved x-ray spectroscopy allows the characterization of hot electrons [62]. Figure 12.10 shows an example of dense laser-produced plasma experiments. Figure 12.10a shows the principal scheme of the experiment and Figure 12.10b shows the space-resolved x-ray emission (space resolved along the target surface in the x-direction). The image shows a strong K_α emission from the cold target. This indicates the existence of hot electrons (accelerated due to the laser–plasma interaction) which returned to the target due to strong plasma magnetic fields. When hitting the target surface, the K_α emission is then induced like in a x-ray tube. As can be seen, the x-ray image provides direct information about the spatial extension of the returning electrons.

Due to the simultaneously achieved high spectral resolution of the x-ray image, a detailed non-Maxwellian analysis of the radiative properties can be made. As demonstrated with non-Maxwellian spectral simulations employing the MARIA code [29–31], the sensitivity to suprathermal electrons is very high permitting to detect hot electron fractions as low as 10^{-4} [62]. It should

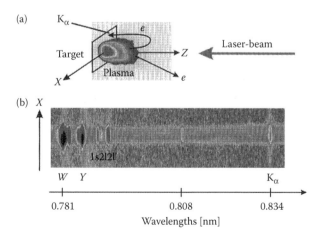

FIGURE 12.10
(a) Schematic scheme to detect hot electrons in laser target interaction experiments; (b) space-resolved x-ray spectroscopy provide direct information about the spatial area of the suprathermal "return-electrons" which illuminate the target like an x-ray tube.

be emphasized that the space-resolved non-Maxwellian spectral analysis presented in [62] concerns the hot electron fraction which exists *inside* the dense plasma (usually difficult to access otherwise).

12.3.6 Nonstatistical Line Shapes

High-resolution spectroscopy allows a detailed analysis of line shapes: an additional important method to characterize the plasma. The traditional use of line shapes employs the so-called "statistical lines shapes" which means that the atomic levels which are important for the electric field interaction with the atom are assumed to be in statistical equilibrium.

Line shapes, however, are also dependent on the nonequilibrium nature of dense plasmas providing additional diagnostic sensitivities (although much more difficult to realize). Interesting investigations have recently been devoted to the analysis of turbulence effects on line shapes: fluctuating densities and temperatures lead to a variation of line intensities and the fluctuation function (which characterizes the turbulence) changes the line shapes in a particular manner [67–69]. Whether line of sight integration as well as temporal integration effects (see discussions above) will allow a diagnostic applications under real experimental conditions to extract the fluctuation function [28,47,67–69] from the line shapes only remains to be demonstrated in future investigations.

As discussed above in connection with radiation transport effects, the opacity broadening may lead to serious changes of resonance line shapes (see Figures 12.2 and 12.3). It might therefore be useful to employ line profiles for the analysis which are optically thin, like, for example, the intercombination lines (their line center opacity is low due to low oscillator strengths; Equation 12.32). Despite the low oscillator strengths for intercombination lines, the line intensity can be nevertheless of the same order as the resonance line intensity (see, e.g., Figures 12.9b and 12.10) due to photo-absorption of the resonance line and nonstatistical effects in the upper-level populations (e.g., the singlet/triplet system of He-like ions).

Figure 12.11 demonstrates the effect of the so-called "dynamical line shapes" for the He-like resonance and intercombination lines of aluminum when the nonstatistical populations of the 1s2l levels are taken into account. The line shape calculations have been performed with the PPP code [70,71], the dynamical properties of the level populations have been calculated with the MARIA code [29–31] employing a relativistic atomic structure (LSJ-split), multipole transitions, cascading and ionization balance shifts. Figure 12.11 demonstrates the case for He-like aluminum (spectral range of the W- and Y-line) at an electron density of $n_e = 10^{21}$ cm^{-3} and an electron temperature of $kT_e = 100$ eV. The simulations show that the intercombination line shape (Y) is essentially modified: intensity and line wings are enhanced by about an order of magnitude providing a larger diagnostic potential as believed in the framework of the statistical line shape approach. The two smaller peaks near

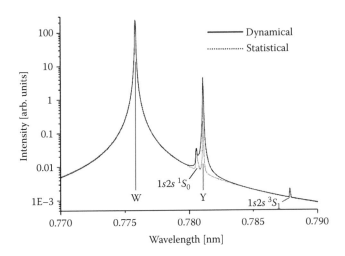

FIGURE 12.11
Statistical and dynamical line shapes of He-like aluminum, $n_e = 10^{21}\,\mathrm{cm^{-3}}$, $kT_e = 100\,\mathrm{eV}$. The simulations of the dynamical level populations have been performed with the MARIA code including a relativistic split level structure (LSJ), multipole transitions as well as a ionization balance calculation. Stark profile calculations have been carried out with the PPP code.

0.781 and 0.788 nm are due to Stark-induced transitions from the $1s2s\,^1S_0$ and $1s2s\,^3S_1$ levels, respectively.

12.4 Dielectronic Satellites

Although the technology of space-resolved spectroscopy advances the analysis of the radiation emission considerably, the principal difficult remains: there is always a line of sight and a corresponding spatial integration of the radiation emission along this line of sight. Spatial resolution as discussed above can provide spatial resolution only for coordinates perpendicular to the line of sight. Therefore, even when applying two-dimensional x-ray imaging methods (see, e.g., Figure 12.8) the principal effect of the line of sight integration remains. It should be noted that Abel inversion methods which are widely used in plasma tube experiments, magnetic fusion research, and so on are difficult to implement for dense plasmas experiments: the emission from different coordinates corresponds often to different emission times and therefore quite different regimes (with corresponding different excitation channels) are responsible for the line emission. An example is the radiation emission of the resonance line emission in dense laser-produced plasmas: during the heating processes, collisional excitation from the ground state strongly populates the upper levels and drives correspondingly intensive line emission. When the laser is turned off, the plasma recombines and the upper levels are populated by radiative recombination at much lower temperatures and densities. Interesting, high-density regimes are, therefore, masked by low-density recombining regimes. Moreover, the intensity driven by the low-density recombining regime can drive even more intense line emission than the high-density plasma heating phase itself (a pitfall for Stark profile interpretations). This is easily seen in Figure 12.9b: a massive CF_2 target is irradiated with a nanosecond high-energy-density laser and the line emission is observed with a spatial resolution in the direction of the expanding plasma. It can clearly be seen that even at cm-distances far from the target (corresponding to the recombining regime), the intercombination line emission is as intense as the resonance line emission and of similar intensity near the target surface (corresponding to the plasma heating regime).

Proposing the use of a streak camera sounds good to distinguish the heating and recombination regime, but may lead to another pitfall: we lose somewhat the space resolution. We can continue with endless discussions of technical improvements of the spectroscopic equipment, but it looks like that we somehow turn in a circle and do not reach our dream: a temporally resolved spectral photon distribution of useful atomic transitions (useful in the sense that they are sensitive either to temperature, density, hot electrons fractions, ion velocities, charge state distributions, chemical composition, etc.) originating from a single point entering the detector without interaction of the surrounding dense plasma.

Although technical developments are extremely important, their developments alone are not sufficient to make our dream come true. And this is because of general physical principles: the limited response function of the atomic system (Equations 12.65 and 12.70), the difficulty to have simultaneously high spectral, spatial, and temporal resolution as well as line of sight integration effects. Let us also add the remark of the limited hope for a 1 fs x-ray streak camera in nearest future.

We may be depressed at this point and the review would finish here would there be not yet other properties of the radiation emission of revolutionary power: the existence of a particular class of atomic x-ray transitions which can provide inherent high time resolution (some 10 fs), spatial resolution (e.g., limited to the regions of highest density and temperature), limited photoabsorption effects (almost negligible even in superdense plasmas), limited line of sight integration effects and a sensitivity to select different plasma regimes (heating or recombination).

This powerful class of x-ray transitions originate from multiple excited states and are called "dielectronic satellites" and "hollow ion (HI) transitions." Combined with modern spectroscopic techniques (high spectral resolution, high spatial resolution, high time resolution, and high luminosity), unique and powerful characterizations of dense hot plasmas under extreme conditions are possible. Therefore, large theoretical and experimental efforts are devoted to the investigation of the radiation emission from multiple excited states, and the interest for basic plasma research and applications is continuously growing. Even analytical approaches have been developed [30,31] to simulate the most complex transitions of multiple excited states (namely those originating from HIs) with spectroscopic precision. This opens up the possibility for diagnostic applications of high-resolution spectroscopy with complex simulations which have been thought to be impossible before. For further discussions employing relevant equations, simulations, and experiments, the interesting reader is referred to [72–75].

12.4.1 Electron Temperature

Gabriel [76] has introduced the dielectronic satellite transitions as a sensitive method to determine the electron temperature. In low-density plasmas, this method approaches the ideal picture of a temperature diagnostic according to Equation 12.18. This method is still applicable in high-density plasmas and is one of the most powerful methods for electron temperature determination. Let us therefore consider the basic principles via an example: the dielectronic satellites $2l2l'$ near the Lyman-alpha line of H-like ions. Figure 12.12 shows the relevant energy level diagram. As the He-like states $2l2l'$ are located above the ionization limit, a nonradiative decay to the H-like ground state (autoionization) is possible:

$$\text{Autoionization:} \quad 2l2l' \rightarrow 1s + e. \tag{12.76}$$

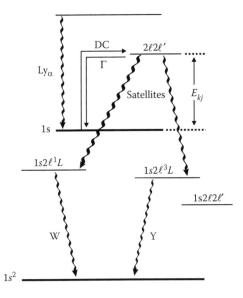

FIGURE 12.12
Energy level diagram of the He-like double excited states $2l2l'$; E_{kj} is the dielectronic capture energy for the satellite transitions.

By first quantum mechanical principles, the reverse process, the so-called "dielectronic capture" must exist:

$$\text{Dielectronic capture:} \quad 1s + e \rightarrow 2l2l'. \tag{12.77}$$

The radiative decay reads

$$\text{Radiative decay:} \quad 2l2l' \rightarrow 1s2l + \hbar\omega_{\text{satellite}}. \tag{12.78}$$

The emitted photon is called a "satellite." The satellite transition is of similar nature like the resonance transition $\text{Lyman}_\alpha = 2p \rightarrow 1s + \hbar\omega_{\text{Ly}_\alpha}$ except the circumstance that an additional electron is present in the quantum shell $n = 2$, the so-called "spectator electron." As the spectator electron screens the nuclear charge, the satellite transitions are essentially located on the long-wavelength side of the corresponding resonance line. However, due to intermediate coupling effects and configuration interaction, satellites on the short-wavelengths side also are emitted, the so-called "blue satellites" (see Figure 12.13). The interested reader is referred to [77] for further details.

As the number of possible angular momentum couplings increases rapidly with the number of electrons, usually numerous satellite transitions are located near the resonance line (which often cannot be resolved spectrally even with high-resolution methods). Figure 12.13 shows an example of the Lyman_α satellite transitions obtained in a dense laser-produced magnesium plasma. The experiment also shows higher-order satellites: the spectator electrons are located in quantum shells $n > 2$ (configurations $2lnl'$).

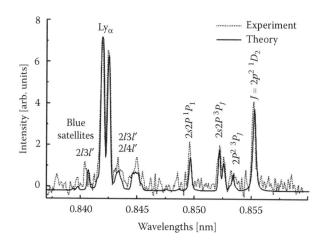

FIGURE 12.13
Soft x-ray spectrum of Ly_α and satellite transitions of a dense laser-produced magnesium plasma (50 J, 15 ns, 1.064 μm). Spectral simulations have been carried out with the MARIA code: $kT_e = 210$ eV, $n_e = 3 \times 10^{20}$ cm^{-3}, $L_{\text{eff}} = 500$ μm, $s = 1.3$.

Let us now proceed to the genius idea of Gabriel [76] to obtain the electron temperature from satellite transitions. In a low-density plasma, the intensity of the resonance line is given by

$$I_{k',j'i'}^{\text{res}} = n_e n_{k'} \frac{A_{j'i'}}{\sum_{l'} A_{j'l'}} \langle C_{k'j'} \rangle, \tag{12.79}$$

where n_e is the free electron density, $n_{k'}$ the ground-state density from which electron collisional excitation proceeds (k' is the 1s level in our example), $A_{j'i'}$ is the transition probability of the resonance transition $j' \rightarrow i'$ (the sum over A in the denominator accounts for possible branching ratio effects), and $\langle C_{k'j'} \rangle$ is the electron collisional excitation rate coefficient from level k' to level j'. The intensity of a satellite transition with a large autoionizing rate (and negligible collisional channel) is given by

$$I_{k,ji}^{\text{sat}} = n_e n_k \frac{A_{ji}}{\sum_l A_{jl} + \sum_m \Gamma_{jm}} \langle D_{kj} \rangle, \tag{12.80}$$

where A_{ji} is the transition probability of the particular satellite transition and $\langle D_{kj} \rangle$ the dielectronic capture rate coefficient from level k to level j. The sums over the radiative decay rates and autoionizing rates account for possible branching ratio effects (in our simple example, only $m = k$ exist, a particular upper level $2l2l'$ may have more than one radiative decay possibilities $j \rightarrow l$). We note that already for the He$_\beta$ satellites, numerous autoionizing channels exist which are very important in dense plasmas due to excited states coupling effects, the interested reader is referred to [78] for a detailed discussion. As both intensities (Equations 12.79 and 12.80) are proportional to the electron density n_e and to the same ground-state density ($k' = k$), the intensity ratio is a function of the electron temperature only, because the rate coefficients $\langle C \rangle$ and $\langle D \rangle$ depend only on the electron temperature but not on the density:

$$\frac{I_{k,ji}^{\text{sat}}}{I_{k'j'i'}^{\text{res}}} = G(T_e). \tag{12.81}$$

The dielectronic capture rate is an analytical function and given by

$$\langle D_{kj} \rangle = \alpha \Gamma_{jk} \frac{g_j}{g_k} \frac{\exp(-E_{kj}/kT_e)}{(kT_e)^{3/2}}, \tag{12.82}$$

where $\alpha = 1.656 \times 10^{-22} \text{ cm}^3 \text{ s}^{-1}$, g_j and g_k are the statistical weights of the states j and k, Γ_{jk} is the autoionizing rate (in s^{-1}), E_{kj} the dielectronic capture energy (in eV) (see also Figure 12.12) and kT_e the electron temperature (eV). The intensity of a satellite transition can therefore be written as

$$I_{k,ji}^{\text{sat}} = \alpha n_e n_k \frac{Q_{k,ji}}{g_k} \frac{\exp(-E_{kj}/kT_e)}{(kT_e)^{3/2}}, \tag{12.83}$$

where $Q_{k,ji}$ is the so-called "dielectronic satellite intensity factor" and given by

$$Q_{k,ji} = \frac{g_j A_{ji} \Gamma_{jk}}{\sum_l A_{jl} + \sum_m \Gamma_{jm}}. \tag{12.84}$$

The calculation of the dielectronic satellite intensity factors $Q_{k,ji}$ requests rather complicated multiconfiguration relativistic atomic structure calculations which have to include intermediate coupling effects as well as configuration interaction.

For simplicity of applications, we therefore provide an analytical set of all necessary formulas for the most important cases to apply the temperature diagnostic (Equations 12.79 through 12.81) via dielectronic satellite transitions near Ly$_\alpha$ and He$_\alpha$ of highly charged ions. For the dielectronic satellite intensity factor, the following formula is proposed:

$$Q = 10^{10} s^{-1} \frac{C_1 (Z_n - C_2)^4}{C_3 Z_n^{C_4} + 1}. \tag{12.85}$$

Table 12.2 provides the fitting parameters for the *J*-satellite near Ly$_\alpha$ as well as for the *k*-satellite and the *j*-satellite near He$_\alpha$ for all elements with nuclear charge $6 < Z_n < 30$. We note that the *k*- and *j*-satellites are treated separately, as line overlapping may request their separate analysis (see Figure 12.7a and b for a line overlap of the *j*-satellite and the Z-line). Note that $g_k = 2$ for the Ly$_\alpha$ satellites and $g_k = 1$ for the He$_\alpha$ satellites in Equation 12.82. The dielectronic capture energies can be approximated by

$$E_{kj} \approx \delta(Z_n + \sigma)^2 \, Ry. \tag{12.86}$$

For the Ly$_\alpha$ satellites $2l2l'$, $\delta = 0.5$, $\sigma \approx 0.5$, for the He$_\alpha$ satellites $1s2l2l'$, $\delta = 0.5$, $\sigma \approx 0.1$, and $Ry = 13.6\,eV$. The electron collisional excitation rate coefficients have been calculated with the Coulomb–Born exchange method including intermediate coupling effects and effective potentials (using the ATOM code

TABLE 12.2

Z_n-Scaled Fitting Parameters of Dielectronic Satellite Intensity Factors Q According to Equation 12.85 (the Range of Validity is $6 < Z_n < 30$)

Satellite	C_1	C_2	C_3	C_4	Max. Error (%)
$J = 2p^2 \, {}^1D_2 - 1s2p \, {}^1P_1$	5.6696E−1	1.4374E−8	5.8934E0	2.2017E−2	1.5
$j = 1s^1 2p^2 \, {}^2D_{5/2} - 1s^2 2p \, {}^2P_{3/2}$	3.4708E−1	1.5569E−7	4.9939E0	8.6347E−1	1.5
$k = 1s^1 2p^2 \, {}^2D_{3/2} - 1s^2 2p \, {}^2P_{3/2}$	2.4072E−1	6.7212E−9	5.9468E0	1.1362E0	3

[79,80]) and fitted into a simple Z- and β-scaled expression:

$$\langle C, T_e \rangle \approx \frac{10^{-8} \text{ cm}^3 \text{ s}^{-1}}{Z^3} \left(\frac{E_u}{E_l} \right)^{3/2} \sqrt{\beta} A \frac{\beta + 1 + D}{\beta + \chi} \exp \left\{ -\frac{E_l - E_u}{kT_e} \right\}, \quad (12.87)$$

$$\beta = \frac{Z^2 \, Ry}{kT_e}, \quad (12.88)$$

where Z is the spectroscopic symbol ($Z = Z_n + 1 - N$, where Z_n is the nuclear charge and N the number of bound electrons), the fitting parameters A, χ, and D are given in Table 12.3. E_l and E_u are the ionization energies of lower and upper states, respectively. If not particularly available, they can be approximated by the simple expression

$$E_j \approx \delta (Z_n - \sigma)^2 \, Ry. \quad (12.89)$$

For the $1s$ level, $\delta = 1$, $\sigma \approx -0.05$, for the $2p$ levels $\delta = 0.25$, $\sigma \approx -0.05$, for the $1s^2$ level, $\delta = 1$, $\sigma \approx 0.6$ and for the $1s2p\ ^1P_1$ level $\delta = 0.25$, $\sigma \approx 1$.

Higher-order satellites, namely $2lnl'$ and $1s2lnl'$ provide further possibilities for plasma diagnostics even if single transitions are not resolved. A rather tricky variant of electron temperature measurement which employs only satellite transitions has been proposed in [81]:

$$\frac{I_n^{\text{sat}}}{I_2^{\text{sat}}} \approx \frac{Q_n}{Q_2} \exp \left\{ -\frac{(Z_n - 0.6)^2 \, Ry}{4kT_e} \left(1 - \frac{4}{n^2} \right) \right\}, \quad (12.90)$$

where Q_n and Q_2 are the total dielectronic satellite intensity factors for the $2lnl' \rightarrow 1snl'$ and $2l2l' \rightarrow 1s2l'$ transitions, respectively. The considerable advantage of this method is that it is even applicable, when the resonance line is absent due to high photo-absorption or due to very low electron temperatures, a typical situation in dense strongly coupled plasmas. The interesting reader is referred to [78,81–84] for further information.

We note that another important excitation channel for satellite transitions is via electron collisional excitation from inner-shells. Concerning the above-discussed example of satellite transitions near Ly_α, this excitation channel reads

$$\text{Inner-shell excitation:} \quad 1s2l + e \rightarrow 2l2l' + e. \quad (12.91)$$

TABLE 12.3

Fitting Parameters for Z- and β-scaled Electron Collisional Excitation Rates of H-like Ly_α and He-like He_α ($1/32 < \beta < 32$)

Transition	A	χ	D	Max. Error (%)
$\text{Lyman}_\alpha = 1s\ ^2S_{1/2} - 2p\ ^2P_{1/2,3/2}$	24.1	0.145	−0.120	4
$\text{He}_\alpha = 1s^2\ ^1S_0 - 1s2p\ ^1P_1$	24.3	0.198	1.06	6

This excitation channel is important for satellite transitions with low autoionizing rates but high radiative decay rates. It drives satellite intensities, which allow an advanced characterization of the plasma, for example, the determination of charge exchange effects in tokamaks as discussed in connection with Figure 12.7 (increase in the collisionally excited Li-like qr-satellites) or the characterization of suprathermal electrons (to be discussed below). For electron temperature measurements, the inner-shell excitation channel should be avoided.

Figure 12.14a shows the MARIA simulations of the spectral distribution near Ly$_\alpha$. Dielectronic satellites $2l2l'$ as well as $2l3l'$ satellites are included in the simulations for a dense plasma: $n_e = 10^{21}$ cm^{-3}. Several $2l3l'$ satellites are located at the blue wavelength side of Ly$_\alpha$. For these particular transitions, LS-coupling effects are as important as the screening effect originating from the spectator electron. As can be seen, numerous satellites are located at

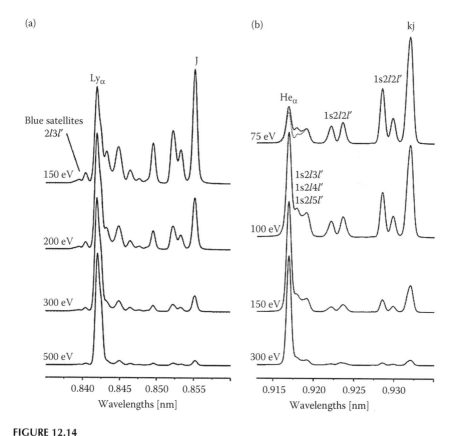

FIGURE 12.14
Electron temperature sensitivity of the spectral distribution of H-, He- and Li-like ions of Mg in dense plasmas, $n_e = 10^{21}$ cm^{-3}. (a) Ly$_\alpha$ and dielectronic satellites $2lnl'$ and (b) He$_\alpha$ and dielectronic satellites $1s2lnl'$. The dashed line shows a simulation where only $1s2l2l'$- and $1s2l3l'$-satellites are included.

the blue wavelengths wing of the resonance line, so-called "blue satellites" [77], which have indeed been observed in experiments (see Figure 12.13). Figure 12.14b shows the MARIA simulations of the spectral distribution near He_α, dielectronic satellites $1s2l2l'$, $1s2l3l'$, $1s2l4l'$, and $1s2l5l'$ are included in the simulations. In all cases (Figure 12.14a and b), a strong sensitivity to electron temperature is seen.

Figure 12.13 shows the fitting of the experimental spectrum obtained in a dense laser-produced plasma experiment taking into account also opacity effects (important only for the Ly_α line): $kT_e = 210\,eV$, $n_e = 3 \times 10^{20}\,cm^{-3}$. An ion temperature of $kT_i = 100\,eV$ is assumed and a convolution with an apparatus function $\lambda/\delta\lambda = 5000$ has been made. We note that the opacity broadening of Ly_α has been used to stabilize the fitting procedure: $L_{eff} = 500\,\mu m$ and $s = 1.3$ (see Equation 12.46) have been obtained. In this case, $\tau_w(Ly_\alpha) = 6/12$, $\tau_0(2l2l') \approx 2 \times 10^{-2}$. Figures 12.13 and 12.14 demonstrate that even in high-density plasmas, the temperature diagnostic via dielectronic satellite transitions works very well.

12.4.2 Ionization Temperature

Gabriel has also introduced the "ionization temperature T_Z" to plasma spectroscopy in order to characterize ionizing and recombining plasmas [76]. In general terms, the ionization temperature is the temperature used to solve Equation 12.71 for a certain density setting the left-hand side to zero (stationary and nondiffusive). This provides a certain set of ionic populations n_Z. If in an experiment, the electron temperature is known (e.g., by means of the dielectronic satellite method described above), however the ratio of the determined ionic populations n_{Z+1}/n_Z is smaller than it would correspond to the solution of Equation 12.71 (the left-hand side is zero), the plasma is called ionizing, if n_{Z+1}/n_Z is larger, the plasma is called recombining. The physical picture behind this is as follows: let us assume a rapid increase in the electron temperature which results in a subsequent plasma heating (e.g., a massive target is irradiated by a laser). Due to the slow relaxation time according to Equation 12.65, the ionic populations need a considerable time to adopt their populations to the corresponding electron temperature. In the initial phase, the ionic populations are lagging behind the electron temperature and the plasma is called ionizing. Only after a rather long time (order of $\tau_{Z,Z+1}$), the ionic populations correspond to the electron temperature. The simulations of Figure 12.5a and b provide detailed insights to this example. At an electron density of $10^{21}\,cm^{-3}$, only after 1 ns the ionic populations have been stabilized. It is important to note that absolute time is not important for the rapidity of the ionization, but the inverse of the rates which are density-dependent (see Equation 12.65). In more general terms, the ionic populations have stabilized after $t > 10^{12}\,cm^3\,s/n_e$ for the K-shell of highly charged ions.

Let us now assume that the electron temperature is rapidly switched off. Also in this case, the ionic population need the time according to Equation

12.65 to decrease the plasma ionization. The plasma is therefore called recombining because higher charge states disappear successively until the ionic populations correspond to the decreased electron temperature.

In the original work of Gabriel [76], the radiation emission of the Li-like $1s2l2l'$-satellite transitions which had strong inner-shell excitation channels but low dielectronic capture (e.g., the qr-satellites) and strong dielectronic capture but low inner-shell excitation channel (e.g., the jk-satellites) have been used to determine the ionic populations of the Li- and He-like ions (note that the dielectronic capture channel for the Li-like $1s2l2l'$-satellites is connected to the He-like ground-state $1s^2\,^1S_0$, whereas the inner-shell excitation channel is connected to the Li-like states $1s^22l$). In the work [85], satellite transitions near Ly$_\alpha$ have been employed to characterize the plasma regime. Also other emission lines can be used in order to characterize the ionizing/recombining nature of a plasma. The use of Rydberg line emission is another important example: in recombining plasmas, the Rydberg series emission is enhanced, whereas in ionizing plasma, high n-members of the Rydberg series are barely visible.

The long time (Equation 12.65) to establish equilibrium in the ionic populations does not permit to employ standard temperature diagnostics which are based on the intensity ratio of resonance lines originating from different ionization stages, for example, the line intensity ratio of the H-like Ly$_\alpha$ and the He-like He$_\alpha$ (if the time scale of characteristic changes of plasma parameters is much shorter). The error of this method is connected with the fact that theoretical line intensity ratios are calculated from the stationary equations 12.15 and 12.21 which ignores that T_Z and T_e might be different.

12.4.3 Relaxation Times of Satellite Transitions

For the temperature diagnostic based on dielectronic satellite transitions (as discussed above), the obstacle of the long relaxation times according to Equation 12.65 does *not* exist, because the employed line ratios concern only one ionization stage which then cancels in the line ratio method. Therefore, independent of any plasma regime (stationary, ionizing, recombining), the dielectronic satellite method allows to access the electron temperature and this is yet another reason why Gabriel's idea to employ satellite intensities for the temperature diagnostic is really a genius one.

Moreover, the response time of satellite transitions is much faster than for resonance lines according to Equation 12.70. The reason is connected with the large autoionizing rate which has a characteristic time scale of the order of some 10-fs for L-shell electrons. For atomic transitions of multiple excite states, Equation 12.70 has therefore to be modified as

$$\tau_{ji} = \frac{1}{A_{ji} + C_{ji} + C_{ij} + \sum_k \Gamma_{jk}}. \tag{12.92}$$

This means that satellite transitions respond on a time scale of about some 10-fs irrespective of any population mixing by collisional processes. As the dielectronic capture population channel is proportional to the exponential temperature dependence (see Equation 12.83), low electron temperatures are practically cut off because the dielectronic capture energy (e.g., Equation 12.86) is very large for highly charged ions:

$$I_{ji}^{sat}(\text{high } \Gamma_{kj}) \propto \frac{\exp(-E_{kj}/kT_e)}{(kT_e)^{3/2}}. \tag{12.93}$$

In consequence, satellite transitions inherently cut of the low-density, low-temperature recombining regime, an extremely important property in high-density plasma research. This effect can clearly be seen from Figure 12.9b: the satellite transitions (indicated at Li, Be, B) are confined near the target surface, whereas the He-like resonance and intercombinations lines (W and Y, respectively) exist also far from the target surface.

12.4.4 Spatially Confined Emission Areas of Satellite Emission

Inspection of the dielectronic capture channel and the correspondingly induced satellite line intensity (Equation 12.80) shows that the intensity is proportional to the square of the electron density (because the ground-state n_k is proportional to the electron density):

$$I^{sat}(\text{high } \Gamma) \propto n_e^2. \tag{12.94}$$

Together with Equation 12.93, the emission is therefore confined to high-density high-temperature plasma areas. This effect is clearly seen in Figure 12.10: satellite transitions are visible just around the laser spot size. Line of sight integration effects are therefore minimized, as Equations 12.93 and 12.94 act like a "local emission source."

For He$_\beta$ 1s3l3l'-satellite transitions, an even stronger density dependence is expected. In high-density plasmas, their dominant excitation channel is dielectronic capture from the 1s2l-states and even density dependences up to $\propto n_e^3$ are possible. Figure 12.15 shows this effect on a space-resolved x-ray image of Si. In the spectral range around the He-like He$_\beta$-line, the 1s3l3l'-satellites are much more confined to the target surface (the Z-direction is the direction of the expanding plasma; see Figure 12.9a) than the 1s2l3l'-satellites. The interesting reader is referred to [78,83] for a more detailed discussion.

There is yet another wonderful property of satellite transitions which minimizes line of sight integration effects with respect to photon–plasma interaction: their line center opacity (see Equation 12.32) is small because the absorbing ground states, for example, the 2l2l' satellite transitions are the excited states 1s2l and not the atomic ground-state 1s² (like it is the case for the He-like resonance line). As the population ratio $n(1s2l)/1s^2$ is rather small

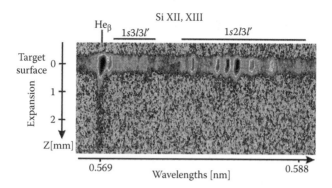

FIGURE 12.15

X-ray image of the He_β-line and satellite transitions. The $1s3l3l'$-satellites are strong confined near the target surface, whereas the $1s2l3l'$-satellites and in particular the He_β-line show strong expansion effects (strong emission far from the target surface).

even in high-density plasmas (the maximum upper limit can be estimated with Equation 12.6). This results in a corresponding very low line center opacity of the satellite transitions.

We note that radiation transport effects in satellite transitions have been observed for Li-like $1s2l2l'$ transitions. This, however, is an exceptional case because their absorbing ground states coincide with the atomic ground and first excited states of the Li-like ions, namely the $1s^2 2l$ configuration. The interesting reader is referred to [50,86,87] for a detailed discussion. Also these obstacles can be avoided: employing higher-order satellite transitions from multiple excited states, other multiple excited configurations or even transitions from HIs.

At this point, the reader may feel the inherent power of satellite transitions with respect to the critical discussion at the beginning of this paragraph (time resolution, inhomogeneity, line of sight integration effects, opacity, etc.). Let us therefore continue to explore further the revolutionary power of this class of transitions for plasma spectroscopy.

12.4.5 Electron Density

In dense plasmas, where electron collisions between the autoionizing levels become of increasing importance (compared to the radiative decay rates and autoionizing rates), population is effectively transferred between the autoionizing levels of a particular configuration (e.g., the $2l2l'$ and $1s2l2l'$ configurations). These mixing effects result in characteristic changes of the satellite spectral distribution (total contour). Figure 12.16 shows the principal mechanism for the $1s2l3l'$ configurations (indicated numerical values are for Si). In low-density plasmas, only those autoionizing levels are strongly populated which have a high autoionizing rate (because in this case, the

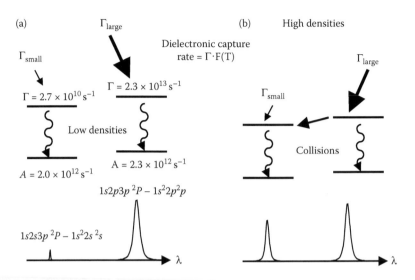

FIGURE 12.16
The principle mechanism of density diagnostics via electron collisional induced pouplation redistribution effects (collisional mixing, (b)). The dielectronic capture (a) is proportional to the autoionizing rate Y and to a function $F(T)$ which depends only on the electron temperature T.

dielectronic capture rate is large). This results in a high intensity of satellite transitions which do have high autoionizing rates and high radiative decay rates. In contrast, satellite transitions with high radiative decay rates but low autoionizing rates have small intensities (because the dielectronic capture is small). In high-density plasmas, population can be transferred from highly populated levels to low populated ones, resulting in a density-dependent change of satellite line intensity. These characteristic changes of the spectral distribution can then be used for density diagnostics. The great advantage of this type of density diagnostic is that essentially optically thin lines are employed which are confined to the highest density areas only (see discussion above). The principles of this type of density diagnostic has first been explored in [88] together with the analysis of the $2l2l'$-satellites near Ly$_\alpha$ and has later been extended to the $1s2l2l'$-satellites [89] and also to non-Maxwellian plasmas by employing Be-satellite transitions too [90].

Figure 12.17a and b shows the mixing effect on the satellite transitions near Ly$_\alpha$ and He$_\alpha$ of highly charged Mg ions. The simulations have been carried out with the MARIA code taking into account an extended level structure: LSJ-split levels of different ionization stages for ground, single, and multiple excited states have simultaneously been included. Figure 12.17a shows the density effect on the Ly$_\alpha$ satellites. Strong density effects are indicated by arrows. Not only the $2l2l'$-satellites show strong density effects near $\lambda \approx 0.853$ nm but the $2l3l'$-satellites near $\lambda \approx 0.847$ nm too. The density sensitivity of the $2l3l'$-satellites starts for lower densities, because the collisional rates between the $2l3l'$-configurations are in general larger than those for the

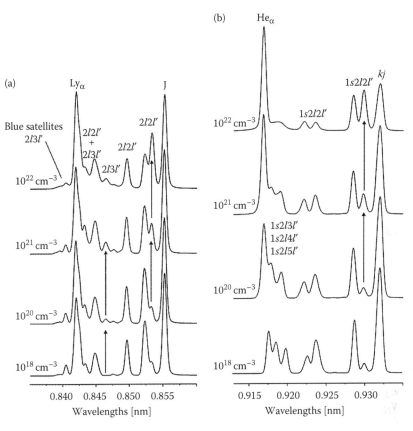

FIGURE 12.17
Electron density sensitivity of the spectral distribution of H-, He- and Li-like ions of Mg in dense plasmas. (a) Ly$_\alpha$ and dielectronic satellites $2lnl'$, $kT_e = 150$ eV, the arrows indicate collisional-induced mixing effects for the $2l3l'$-satellites as well as for the $2l2l'$-satellites, (b) He$_\alpha$ and dielectronic satellites $1s2lnl'$, $kT_e = 100$ eV, strong mixing effects are seen for the $1s2l2l'$-satellites.

$2l2l'$-configuration (collisional rates $C(2lnl' - 2lnl'')$) increase with principal quantum number n), whereas corresponding radiative rates ($A \propto 1/n^3$) and nonradiative rates (autoionization rate $\Gamma \propto 1/n^3$) are smaller. Figure 12.17b shows similar effects for the satellite transitions near He$_\alpha$. Strong density effects are visible near $\lambda \approx 0.930$ nm. Higher-order satellite transitions originating from the $1s2l3l'$-, $1s2l4l'$- and $1s2l5l'$-configurations have been included in the simulations, however, due to their large line overlap, density effects are not strongly pronounced under real experimental conditions.

The density effect of the $1s2l2l'$-satellites have been employed to diagnose the electron density of a plasma created by a vacuum spark: electron densities near 10^{23} cm^{-3} have been stated [91] being in large discrepancies with other measurements [92,93]. An analysis of the Be-like satellite transitions has shown [94] that in transient dense plasmas (such as the vacuum spark), satellite transitions of the type $K^1L^2M^1 \rightarrow K^2L^1M^1 + h\nu$ have a strong overlap with the density-sensitive spectral interval of the $1s2l2l'$-satellites (see Figure 12.17b). Neglecting these Be-like satellites (as done in [91]) may result therefore in a misleading high-density interpretation of the $1s2l2l'$-satellites.

Collisional mixing effects turned out to be very useful also for the satellite transitions $1s2l3l' \rightarrow 1s^22l' + h\nu$ near the He_β line [77]: their density sensitivity is located in a convenient interval of about 10^{19}–10^{22} cm^{-3} (for Al) which is difficult to access otherwise (e.g., by mixing effects of the $1s2l2l'$-satellites or by Stark broadening).

In very-high-density plasmas (near solid density), the Stark broadening analysis of satellites is very useful and has firstly been demonstrated for the $2l2l'$- and $1s2l2l'$-satellites [95]. Figure 12.18 shows the Stark broadening simulations for the $2l2l'$-satellites of Mg carried out with the PPP-code assuming a statistical population between of the autoionizing levels. It can clearly be seen that strong density sensitivities are obtained only for densities $n_e > 10^{22}$ cm^{-3}.

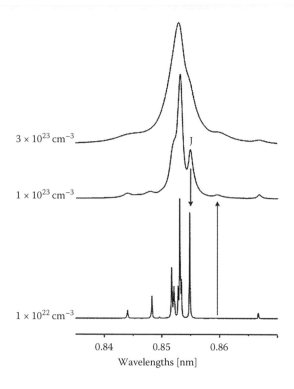

FIGURE 12.18
Stark broadening PPP simulations of the $2l2l'$-satellite transitions of He-like Mg, $kT_e = 100$ eV. Doppler broadening and apparatus function convolution is switched off and a statistical population of the $2l2l'$-levels have been assumed in order to demonstrate the microfield effects on the spectral distribution only. There are two distinct effects indicated by arrows: an intensity decrease of the J-satellite intensity and an increase of a satellite on the red side of the J-satellite.

In order to access lower electron densities, Stark broadening analysis of the satellite transitions near He_β has been carried out in [96,97]. Moreover, it was demonstrated that even Rydberg-satellite transitions of the type $1s2\,ln\,l' \rightarrow 1s^2 2l' + h\nu$ can be recorded in dense laser-produced plasma experiments with high spectral and spatial resolution [98] which has stimulated Stark broadening calculations even of Rydberg-satellite transitions [84].

12.4.6 Hot Electrons

12.4.6.1 Non-Maxwellian Atomic Population Kinetics

Non-Maxwellian energy distributions of the particles enter into the radiation emission via the transition matrix W of Equation 12.16. The rate coefficients for the process X ($X \neq T$) has to be determined via an integral over the arbitrary energy distribution function $F(E)$:

$$X_{ij} = \int_{E_0}^{\infty} dE\, \sigma_{ij}^X(E) V(E) F(E), \tag{12.95}$$

where σ_{ij}^X is the cross-section for the process X for the transition $i \rightarrow j$, V the relative velocity of the colliding particles and E_0 the threshold energy (if no threshold exist, then $E_0 = 0$). For the three-body recombination coefficient, one has to take care of the fact that the energy distribution function of simultaneously two particles ("1" and "2") has to be taken into account:

$$T_{ij} = \frac{\pi^2 \hbar^3}{m_e^2} \frac{g_i}{g_j} \int_0^\infty dE_1 \int_0^\infty dE_2 \frac{E}{\sqrt{E_1 E_2}} \sigma_{ij}^I(E, E_1) F(E_1) F(E_2), \tag{12.96}$$

where $\sigma_{ij}^I(E, E_1)$ is the double differential ionization cross-section for the transition $i \rightarrow j$, g_i and g_j the statistical weights of the level i and j, respectively, and

$$E = \Delta E_{ij} + E_1 + E_2. \tag{12.97}$$

where ΔE_{ij} is the ionization energy. For further reading on the coupling of non-Maxwellian distribution functions to atomic kinetics, the interesting reader is referred to [28].

Let us consider the following approximation to the non-Maxwellian energy distribution function $F(E)$ for suprathermal electrons:

$$F(E, T_1, T_2) = (1 - f) F_M(E, T_1) + f F_M(E, T_2), \tag{12.98}$$

where $F_M(E, T_1)$ and $F_M(E, T_2)$ are Maxwellian energy distribution functions with the parameters T_1 and T_2 (note that T_1 and T_2 are not temperatures in a thermodynamic sense) and f is the fraction of electrons which are described by the energy distribution function $F_M(E, T_2)$.

Many experiments with hot dense plasmas have shown (e.g., dense laser-produced plasmas, dense pinch plasmas) that Equation 12.98 is a reasonable approximation to the measured distribution function (e.g., obtained by means of the Bremsstrahlung). Moreover, T_2 is often much larger than T_1. In this case, it is convenient to speak of a "bulk" electron temperature T_{bulk} for T_1 and to interpret T_2 as a hot electron temperature T_{hot}, f is the hot electron fraction defined by

$$f = \frac{n_e(\text{hot})}{n_e(\text{bulk}) + n_e(\text{hot})},$$ (12.99)

where $n_e(\text{bulk})$ is the density of the "bulk" electrons and $n_e(\text{hot})$ those of the hot or suprathermal electrons.

The introduction of the "bulk" and "hot" electron temperature permits us to understand the basic effects of suprathermal electrons on the radiation emission, the spectral distribution, and line intensity ratios. Figure 12.19 shows the intensity line ratios of the He-like J-satellite and the H-like Ly$_\alpha$ of argon for different fractions of hot electrons: $kT_{hot} = 20\,\text{keV}$ and $n_e(\text{hot}) + n_e(\text{bulk}) = 10^{22}\,\text{cm}^{-3}$. The case $f = 0$ corresponds to a Maxwellian plasma, and the numerical simulations are close to those described by the analytical equations 12.79 through 12.89. Deviations from the analytical results for very low temperatures are due to collisional radiative effects. The monotonic dependence on the electron temperature indicates a strong sensitivity for electron temperature measurements. Hot electron fractions, however, lead to a nonmonotonic

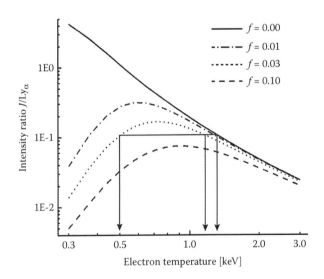

FIGURE 12.19
MARIA simulations of the intensity ratios of the J-satellite ($J = 2p^2\ {}^1D_2-1s2p\ {}^1P_1$) and H-like Ly$_\alpha$ of Ar in a dense non-Maxwellian plasma, for different fractions of hot electron density, $n_e = 10^{22}\,\text{cm}^{-3}$, $kT_{hot} = 20\,\text{keV}$.

behavior. This is connected with the different asymptotic behavior of the cross-sections for the collisional excitation and the dielectronic capture: for the collisional excitation, all electrons whose energy is larger than the excitation energy contribute to the excitation, the dielectronic capture, however, is a resonance processes and only those electrons in the continuum contribute to the cross-section which match the resonance energy (E_{kj} in Equation 12.86). Therefore, hot electrons contribute strongly to the collisional excitation of the resonance line but only little to the dielectronic capture of the satellite lines. As for low bulk electron temperatures, the hot electron-induced collisional excitation is large compared to those of the bulk electrons, the intensity ratio is much lower than for a Maxwellian plasma. For high bulk electron temperatures, the hot electrons do not contribute much compared to the bulk electrons and the curves for $f > 0$ approach those for $f = 0$. The arrows show the principal difficult for diagnostics: the same line ratio can be obtained for three different sets of parameters: (1) $kT_{bulk} = 0.5$ keV, $f = 3\%$, (2) $kT_{bulk} = 1.17$ keV, $f = 3\%$, and (3) $kT_{bulk} = 1.24$ keV, $f = 0$. Of particular importance is the difference between the solutions (1) and (3): low bulk temperature and hot electrons versus a high electron temperature without hot electrons. This example indicates the general difficult to interpret the measurements: the difficult is not connected with the particular selection of the resonance line and the satellite transitions but is based on the general asymptotic dependence of the cross-sections (resonance process and threshold process). Therefore, all line ratios of any resonance line and its satellite transitions are affected in the same manner.

Figure 12.20 shows the line ratios of the H-like Ly_α and He-like He_α of argon: $n_e = 10^{22}$ cm^{-3} and $kT_{hot} = 20$ keV. In a Maxwellian plasma ($f = 0$), the strong monotonic dependence is very convenient for temperature diagnostics. The presence of hot electrons, however, rises considerably the intensity ratio for lower bulk electron temperatures due to increased ionization induced by hot electrons: $1s^2 + e(hot) \rightarrow 1s + 2e$. The arrows indicate an example of the principal difficult for diagnostics due to multiple solutions for the same line intensity ratio: (1) $kT_{bulk} = 400$ eV, $f_{hot} = 0.1$, and (2) $kT_{bulk} = 930$ eV, $f_{hot} = 0$. Also this example shows that the neglect of hot electrons leads to a considerable overestimation of the bulk electron temperature.

It should be noted that the intensity ratio of Figure 12.20 poses other difficulties in transient plasmas: in ionizing plasmas, the ionic populations are lagging behind the electron temperature and therefore the Ly_α emission is lower than it would correspond to the given electron temperature. In other words, using the stationary line intensity ratio (i.e., neglecting the ionizing nature of the plasma) underestimates the electron temperature. In recombining plasmas, the electron temperature is overestimated because the Ly_α intensity is too large for the given electron temperature. The line intensity ratio of Ly_α and He_α is therefore more indicative for the ionization temperature and not for the electron temperature as often erroneously stated.

Figure 12.21 shows the line intensity ratio of the He-like intercombination line $1s2p\ ^3P_1 \rightarrow 1s^2\ ^1S_0 + h\nu$ and the He-like resonance line $1s2p\ ^1P_1 \rightarrow 1s^2$

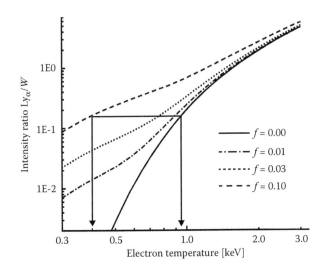

FIGURE 12.20
MARIA simulations of the intensity ratios of the H-like Ly_α-line and the He-like He_α-line (W) of Ar in a dense non-Maxwellian plasma for different fractions of hot electron density, $n_e = 10^{22}\,\mathrm{cm}^{-3}$, $kT_{hot} = 20\,\mathrm{keV}$.

$^1S_0 + h\nu$, $kT_{bulk} = 600\,\mathrm{eV}$. The continuous decrease in the line intensity ratio with increasing density is due to an effective collisional population transfer from the triplet system to the singlet system. As the intercombination transition has a rather low transition probability compared to the resonance line (spin forbidden transition in the LS-coupling scheme), the population of

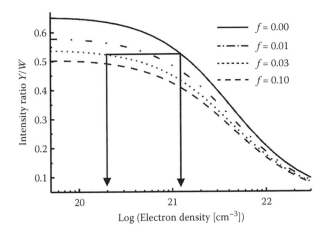

FIGURE 12.21
MARIA simulations of the intensity ratios of the He-like intercombination line Y and the He-like resonance line He_α-line (W) of Ar in a dense non-Maxwellian plasma for different fractions of hot electron density, $kT_{bulk} = 600\,\mathrm{eV}$, $kT_{hot} = 20\,\mathrm{keV}$.

the triplet system is much larger than those of the singlet system. Therefore, the transfer of population from the singlet to the triplet system is smaller than opposite resulting in an effective population transfer from the triplet to the singlet levels. This in turn results in a decrease in the line intensity ratio. As Figure 12.21 indicates the density sensitivity of this line intensity ratio is strong and can therefore be used as an electron density diagnostic.

Hot electrons, however, result in an overall decrease in the intensity ratio for all electron densities. This effect is due to the different asymptotic dependence of direct and exchange excitation cross-sections:

$$\sigma_{direct} \propto \frac{\ln E}{E}, \tag{12.100}$$

$$\sigma_{exchange} \propto \frac{1}{E^3}. \tag{12.101}$$

In a pure LS-coupling scheme, the collisional excitation from the ground state to the triplet levels is carried only by the exchange part of the cross-section. Due to the strong decrease in the exchange cross-sections with increasing impact energy (compared to the direct cross-section), the hot electron-induced excitation for the singlet levels is much larger than for the triplet levels. This results in a strong decrease in the line ratio and in turn to a large overestimation of the electron density when hot electrons are neglected. We note that this effect is not connected with the particular line ratio of the Y and W line: it is a consequence of the general asymptotic dependence of the cross-sections according to Equations 12.100 and 12.101. Therefore, all line ratios which are based on resonance and intercombination line transitions are perturbed in a similar manner. The arrows in Figure 12.21 (calculated for $kT_{bulk} = 600$ eV, $kT_{hot} = 20$ keV) indicate an example of the principal difficult of density diagnostics due to multiple solutions for the same line intensity ratio: (1) $n_e = 2 \times 10^{20}$ cm^{-3}, $f_{hot} = 3\%$, and (2) $n_e = 1.2 \times 10^{21}$ cm^{-3}, $f_{hot} = 0.0$. This example shows that the neglect of hot electrons leads to a considerable overestimation (order of magnitude) of the electron density.

Let us consider the influence of intermediate coupling effects. For the transitions $1s^2 + e \rightarrow 1s2l$ $^{1,3}L + e$, the excitation cross-sections in intermediate coupling σ_{IC} can be written [66,79] as follows:

$$\sigma_{IC} = Q_d \sigma_{direct} + Q_e \sigma_{exchange}. \tag{12.102}$$

For example, for the excitation of the resonance line W (cross-section $1s^2$ $^1S_0 + e \rightarrow 1s2p^1P_1 + e$) and the intercombination line Y (cross-section $1s^2$ $^1S_0 + e \rightarrow 1s2p^3P_1 + e$), we have in the pure LS-coupling scheme: $Q_d^{LS}(W) = 2$, $Q_e^{LS}(W) = 0.5$, $Q_d^{LS}(Y) = 0$, and $Q_e^{LS}(Y) = 0.5$. For argon, the angular factors are only slightly different in the intermediate coupling scheme: $Q_d^{IC}(W) = 1.9684$, $Q_e^{LS}(W) = 0.5$, $Q_d^{IC}(Y) = 0.0316$, and $Q_e^{LS}(Y) = 0.5$. The intermediate coupling angular factors Q indicate that the triplet levels have a small admixture of the direct cross-sections part. This admixture is, however, of importance for rather heavy elements, for example, for molybdenum we have $Q_d^{IC}(W) = 1.515$, $Q_e^{LS}(W) = 0.5$, $Q_d^{LS}(Y) = 0.485$, and $Q_e^{LS}(Y) = 0.5$. In conclusion, even in the intermediate coupling scheme (note that the above simulations, Figure 12.21, include intermediate coupling effects), the discussion concerning the asymptotic behavior of the excitation cross-sections (Equations 12.100 and 12.101) remains valid.

What is the general conclusion from Figures 12.19 through 12.21? The standard line ratios are excellent methods for density and temperature diagnostics in stationary Maxwellian plasmas. However, for plasmas containing hot electrons, the development of other methods is mandatory. Of primary importance are the stable determination of the bulk electron temperature and the hot electron fraction.

12.4.6.2 Bulk Electron Temperature

Let us consider a plasma whose electron energy distribution function is given by Equations 12.98 and 12.99. The rate coefficients for the processes X are then given by

$$\langle X \rangle = (1 - f)\langle X, T_{\text{bulk}} \rangle + f \langle X, T_{\text{hot}} \rangle. \tag{12.103}$$

For the three-body recombination rate coefficient $\langle TR \rangle$, the expression is more complicated due to the need for simultaneously two energy distribution functions of the continuum electrons:

$$\langle TR \rangle = (1 - f)^2 \langle TR, T_{\text{bulk}} \rangle + f^2 \langle TR, T_{\text{hot}} \rangle + 2f(1 - f)\langle TR, T_{\text{bulk}}, T_{\text{hot}} \rangle. \tag{12.104}$$

The first term describes the usual three-body recombination at a temperature T_{bulk} ($\langle TR, T_{\text{bulk}} \rangle$ being the three-body recombination rate coefficient at temperature T_{bulk}), the second one those at a temperature T_{hot} ($\langle TR, T_{\text{hot}} \rangle$ being the three-body recombination rate coefficient at temperature T_{hot}). The last term is a mixed term which requests the integration over the double differential cross-section (see Equation 12.96). This term cannot be expressed by simple combinations of usual three-body coefficients with T_{bulk} and/or T_{hot} like the first and second terms of Equation 12.104. This is very inconvenient for numerical simulations because the double integration in a multilevel multi-ion-stage atomic system requests considerable computational resources. Numerical simulations carried out with the MARIA code indicate that the "mixed" term of Equation 12.104 can be roughly approximated by

$$\langle TR, T_{\text{bulk}}, T_{\text{hot}} \rangle \approx 0.95 \sqrt{\langle TR, T_{\text{bulk}} \rangle \langle TR, T_{\text{hot}} \rangle} \left(\frac{T_{\text{bulk}}}{T_{\text{hot}}} \right)^{0.1}, \tag{12.105}$$

where $T_{\text{bulk}} < T_{\text{hot}}$. Equation 12.105 leads to a reasonable approximation of Equation 12.104 for most of the practical applications.

We consider now a new method to determine the electron bulk temperature in dense non-Maxwellian plasmas [13] which is based on the analysis of the x-ray line emission of a He-like satellite $2l2l' \to 1s2l + h\nu$ and the He-like Rydberg series $1snp\,^1P_1 \to 1s^2\,^1S_0 + h\nu$ of highly charged ions. The intensity of a He-like $2l2l'$-satellite with high radiative and high autoionizing rate is given by

$$I_{k,ji}^{\text{sat}} = \hbar\omega_j n_e n_k (k = 1s)\big\{(1 - f)\langle \text{DR}, T_{\text{bulk}}\rangle + f\langle \text{DR}, T_{\text{hot}}\rangle\big\}. \tag{12.106}$$

The rate coefficient of the dielectronic recombination is given by

$$\langle \text{DR}, T \rangle = \alpha \frac{Q_{k,ji}}{g_k} \frac{\exp(-E_{kj}/kT)}{(kT)^{3/2}}. \tag{12.107}$$

If the electron density is sufficiently high to ensure a balance between the levels $1s$ and $1snp\,^1P_1$ via collisional ionization and three-body recombination, we can determine the population density of the $1snp\,^1P_1$-level analytically:

$$n_e n(1snp^1P_1)\big\{(1 - f)\langle I, T_{\text{bulk}}\rangle + f\langle I, T_{\text{hot}}\rangle\big\}$$
$$\approx n_e^2 n(1s)\big\{(1 - f)^2\langle \text{TR}, T_{\text{bulk}}\rangle + f^2\langle \text{TR}, T_{\text{hot}}\rangle + 2f(1 - f)\langle \text{TR}, T_{\text{bulk}}, T_{\text{hot}}\rangle\big\}. \tag{12.108}$$

From Equations 12.107 and 12.108, the intensity ratio of the dielectronically captured satellite and the He-like Rydberg series is given by (note that the term $\langle \text{TR}, T_{\text{hot}}\rangle$ is rather small due to the scaling relation $\langle \text{TR}, T\rangle \propto 1/T^{9/2}$):

$$\frac{I_{k,ji}^{\text{sat}}}{I_n} = \frac{\omega_j}{\omega_n} \frac{n_e n(1s)\big\{(1 - f)\langle \text{DR}, T_{\text{bulk}}\rangle + f\langle \text{DR}, T_{\text{hot}}\rangle\big\} \times \big\{(1 - f)\langle I, T_{\text{bulk}}\rangle + f\langle I, T_{\text{hot}}\rangle\big\}}{A_n n_e n(1s)\big\{(1 - f)^2\langle \text{TR}, T_{\text{bulk}}\rangle + f^2\langle \text{TR}, T_{\text{hot}}\rangle + 2f(1 - f)\langle \text{TR}, T_{\text{bulk}}, T_{\text{hot}}\rangle\big\}}, \tag{12.109}$$

where A_n is the spontaneous transition probability of the Rydberg series $1snp\,^1P_1 \to 1s^2\,^1S_0$ and ω_j and ω_n are the transition frequencies of the satellite and Rydberg transitions, respectively. Developing Equation 12.109 in the series of f gives

$$\frac{I_{k,ji}^{\text{sat}}}{I_n} = \frac{\langle \text{DR}, T_{\text{bulk}}\rangle \langle I, T_{\text{bulk}}\rangle}{A_n \langle \text{TR}, T_{\text{bulk}}\rangle}\{1 + f \cdot G_1 + \cdots\} \tag{12.110}$$

with

$$G_1 = \frac{\langle I, T_{\text{hot}}\rangle}{\langle I, T_{\text{bulk}}\rangle} + \frac{\langle \text{DR}, T_{\text{hot}}\rangle}{\langle \text{DR}, T_{\text{bulk}}\rangle} - 2\frac{\langle \text{TR}, T_{\text{bulk}}, T_{\text{hot}}\rangle}{\langle \text{TR}, T_{\text{bulk}}\rangle}. \tag{12.111}$$

With

$$\langle DR, T_{\text{bulk}} \rangle = \frac{1}{g(1s)} \frac{(2\pi)^{3/2} \hbar^3}{2(m_e k T_{\text{bulk}})^{3/2}} Q_{k,ji} \exp\left(-\frac{E_{kj}}{kT_{\text{bulk}}}\right) \tag{12.112}$$

and

$$\langle TR, T_{\text{bulk}} \rangle = \langle I, T_{\text{bulk}} \rangle \frac{g_n}{2g(1s)} \frac{(2\pi)^{3/2} \hbar^3}{(m_e k T_{\text{bulk}})^{3/2}} \exp\left(\frac{E_n}{kT_{\text{bulk}}}\right), \tag{12.113}$$

the intensity ratio of Equation 12.109 can be written as

$$\frac{I_{k,ji}^{\text{sat}}}{I_n} = \frac{\omega_j}{\omega_n} \frac{Q_{k,ji}}{A_n g_n} \exp\left(-\frac{E_{kj} + E_n}{kT_{\text{bulk}}}\right) \cdot \{1 + f \cdot G_1 + \cdots\}. \tag{12.114}$$

Calculations show that for almost all practical cases, $G_1 < 1$. This indicates that the zeroth-order approximation of Equation 12.110, namely

$$\left.\frac{I_{k,ji}^{\text{sat}}}{I_n}\right|_{\text{zeroth-order}} = \frac{\omega_j}{\omega_n} \frac{Q_{k,ji}}{A_n g_n} \exp\left(-\frac{E_{kj} + E_n}{kT_{\text{bulk}}}\right) \tag{12.115}$$

is of extraordinary importance: Equation 12.115 represents the ideal case of a bulk electron temperature diagnostic in non-Maxwellian plasmas (it is important to note that the zeroth-order approximation does not depend on the hot electron fraction but still describes the line ratios very accurately: that is why this approximation is of great importance to determine the bulk-electron temperature in non-Maxwellian plasmas). An extremely useful He-like satellite transition is the J-satellite $(2p^2\ ^1D_2 \rightarrow 1s2p\ ^1P_1 + hv)$ because of its large autoionizing rate, high radiative decay rate, and its easy experimental registration (well separated from other satellite transitions, see Figure 12.13). A further advantage of Equation 12.115 is that it contains only a few atomic data. These data can be expressed in an analytical manner with high precision. For A_n, we propose the following expression:

$$A_n(1/s) = (A_0 - \alpha Z^2) \frac{n(n-1)^{2n-2}}{(n+1)^{2n+2}} Z^4, \tag{12.116}$$

$$Z = Z_n - \sigma. \tag{12.117}$$

Note that the first factor in Equation 12.116 takes into account intermediate coupling effects which strongly depend on the nuclear charge. With $A_0 = 4.826 \times 10^{11}$, $\alpha = 7.873 \times 10^7$, and $\sigma = 0.8469$, a precision better than 6% is reached for all n and $Z_n = 6$–32. The dielectronic satellite intensity factor for the *J*-satellite has already been described by Equation 12.85. The energies in Equation 12.115 can be approximated by

$$E_{1s,J} + E_n = 13.6 \, \text{ev} \left[\frac{3}{4} Z_n^2 + \frac{3}{4} (Z_n - 0.4)^2 - (Z_n - 0.5)^2 \cdot \left(1 - \frac{1}{n^2} \right) \right].$$

(12.118)

The precision is better than 2%. For example, for $Z_n = 18$ and $n = 4$, the exact data are: $A_4 = 1.21 \times 10^{13} \, \text{s}^{-1}$, $Q_J = 4.30 \times 10^{14} \, \text{s}^{-1}$, $E_{1s,J} + E_4 = 2.55 \, \text{keV}$, whereas the analytical formulas proposed above provide $A_4 = 1.19 \times 10^{13} \, \text{s}^{-1}$, $Q_J = 4.36 \times 10^{14} \, \text{s}^{-1}$, $E_{1s,J} + E_{i,4} = 2.56 \, \text{keV}$. This indicates the high precision of the simple analytical expressions.

Figure 12.22 shows the numerical simulations of the intensity ratio of the He-like *J*-satellite line and the Rydberg series for $n = 2$–7 (solid lines) in dependence of the hot electron fraction f. The bulk electron temperature is $kT_{\text{bulk}} = 600 \, \text{eV}$, the electron density is $n_e = 10^{22} \, \text{cm}^{-3}$, and the hot electron temperature is $kT_{\text{hot}} = 20 \, \text{keV}$. Also shown in Figure 12.22 are the intensity ratios of Ly_α / He_α and J/Ly_α which has been discussed in connection with Figures 12.19 and 12.20. Figure 12.22 shows a large stability of the intensity ratios $J/1snp \, {}^1P_1$ even for very large hot electron fractions up to 10%. For such large hot electron fractions, the standard line intensity ratios are already off by an order of magnitude (which means that they are meaningless in the original sense of a temperature diagnostics). Figure 12.22 also indicates that for higher Rydberg series transitions, the intensity ratio is only very weakly dependent on the hot electron fraction because the thermalization threshold to ensure PLTE has already been reached. This is demonstrated schematically in Figure 12.23 which also explains the basic characteristics of the bulk-electron

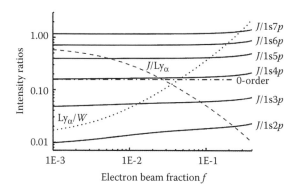

FIGURE 12.22
MARIA simulations of the intensity ratios Ly_α / He_α, J/Ly_α, and $J/1snp \, {}^1P_1$ of Ar in a dense non-Maxwellian plasma, $n_e = 10^{22} \, \text{cm}^{-3}$, $kT_{\text{bulk}} = 600 \, \text{eV}$, $kT_{\text{hot}} = 20 \, \text{keV}$. Also indicated is the analytical zeroth-order approximation for the ratio $J/1s4p \, {}^1P_1$ to determine the bulk-electron temperature in non-Maxwellian plasmas according to Equation 12.115.

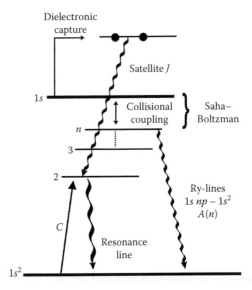

FIGURE 12.23
Energy level diagram showing the collisional coupling of Rydberg states $1snl$ and the dielectronic capture of the $2l2l'$-states to the H-like ground state $1s\,^2S_{1/2}$.

temperature diagnostic with the help of an energy level diagram. Figure 12.22 also presents the results of the zeroth-order approximation (Equation 12.115) for the $1s4p\,^1P_1$-line. Excellent agreement is seen between the numerical non-Maxwellian simulations and the zeroth-order analytical approximation. This confirms the great importance of the zeroth-order approximation to determine the bulk-electron temperature.

This advanced method of bulk-electron temperature diagnostics has been employed to characterize dense pinching plasmas [13,99] and also in dense laser-produced plasmas [100]. As demonstrated in [13], the observation of the stabilization of the bulk-electron temperature with increasing principal quantum number of the Rydberg series permits also to estimate the electron density according to Equation 12.8.

12.4.6.3 Hot Electron Fraction

Having once determined the bulk electron temperature, the intensity ratios of the He-like $2l2l'$-satellites with the Ly_α line can be used for the determination of the hot electron fraction if resonance line intensity and satellite intensity are well approximated by

$$I^{res}_{k',j'i'} \approx n_e n_{k'} \frac{A_{j'i'}}{\sum_{l'} A_{j'l'}} \langle C_{k'j'} \rangle \qquad (12.119)$$

and

$$I_{k,ji}^{\text{sat}} \approx \alpha n_e n_k \frac{Q_{k,ji}}{g_k} \frac{\exp(-E_{kj}/kT_e)}{(kT_e)^{3/2}}. \tag{12.120}$$

If the energy distribution function is described by Equations 12.98 and 12.99 and the rate coefficients by Equations 12.103 and 12.104, the fraction of hot electrons is given by

$$f_{\text{hot}} \approx \frac{1}{1 + (R\langle \text{CR}, T_{\text{hot}}\rangle - \langle \text{DR}, T_{\text{hot}}\rangle)/(\langle \text{DR}, T_{\text{bulk}}\rangle - R\langle \text{CR}, T_{\text{bulk}}\rangle)}, \tag{12.121}$$

where the line intensity ratio R is given by

$$R = \frac{I_{k,ji}^{\text{sat}}}{I_{k',j'i'}^{\text{res}}}. \tag{12.122}$$

The relevant rate coefficients of Equation 12.121 are given by

$$\langle \text{CR}, T \rangle = \frac{A_{j'i'}}{\sum_{l'} A_{j'l'}} \langle C_{k'j'}, T \rangle, \tag{12.123}$$

$$\langle \text{DR}, T \rangle = \alpha \frac{Q_{k,ji}}{g_k} \frac{\exp(-E_{kj}/kT)}{(kT)^{3/2}}. \tag{12.124}$$

In order to best fulfill the parameter range of validity of Equation 12.120, the use of the He-like J-satellite (see also discussion above) is recommended. Equations 12.85 through 12.89 provide the necessary data for the Ly_α- and He_α-satellites and their resonance lines.

In high-density plasmas, the range of validity of Equations 12.119 through 12.121 might be limited and other methods have to be employed [29].

In order to develop other methods, let us consider the ionic fractions in dense plasmas with and without fractions of hot electrons. Figure 12.24 shows the MARIA simulations for $n_e = 10^{21}\,\text{cm}^{-3}$, $L_{\text{eff}} = 300\,\mu\text{m}$, and $f = 0.0$ (solid curves) and $f = 0.09$ (dashed curves), $kT_{\text{hot}} = 20\,\text{keV}$. The comparison between the solid and the dashed curves shows that a qualitative deformation of the ionic fractions has taken place. The arrow indicates a strong rise of the H-like abundance for lower bulk-electron temperatures, whereas other ionic fractions (He-, Li-, Be-like) are much less influenced. This qualitative deformation can in turn be used for the determination of the hot electron fraction by visualizing the various fractions via corresponding x-ray line emissions from the H-, He-, Li-, Be-, B-, C-like ions [29].

Experimentally, it is difficult to observe simultaneously the line emission from H- and He-like ions (K-shell emission) and those of Li-, Be-, B-, C-like ions (L-shell emission) due to the strongly different spectral ranges (requesting different types of spectrometers and their relative intensity calibration). However, by means of inner-shell satellite transitions $1s2s^n2p^m \rightarrow 1s^22s^n2p^{m-1} + h\nu$,

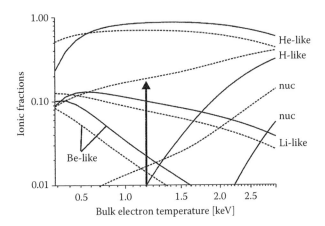

FIGURE 12.24
MARIA simulations of the ionic fractions of titanium in dense non-Maxwellian plasmas for different fractions f of hot electron density, $n_e = 10^{21}$ cm^{-3}, $L_{eff} = 300 \mu$m, $f = 0.0$ (solid lines), $f = 0.09$ (dashed lines), $kT_{hot} = 20$ keV.

this drawback can be circumvented: the wavelength interval of all line transitions is located in a similar spectral range (K-shell) and all transitions are therefore simultaneously observable with one type of spectrometer.

Figure 12.25 shows the soft x-ray emission spectrum of titanium in a dense optically thick plasma, $n_e = 10^{21}$ cm^{-3}, $L_{eff} = 300 \mu$m. Figure 12.25a shows the MARIA simulation for $f = 0.09$, $kT_{bulk} = 800$ eV, $kT_{hot} = 20$ keV, Figure 12.25b those for $f = 0.0$, $kT_e = 2.3$ keV, and Figure 12.25c those for $f = 0.0$, $kT_e = 1$ keV. As Figure 12.25c demonstrates, the Ly$_\alpha$ emission is practically absent for low electron temperatures, whereas the K$_\alpha$-satellite series is strongly pronounced. At high electron temperatures (Figure 12.25b), the Ly$_\alpha$-emission is strong, however, the K$_\alpha$-satellites series of Be- B- and C-like ions is practically absent. This reflects the general behavior of the ionic charge state distribution (Figure 12.24): the ionic populations of highly charged ions (nuc, H-like) are never at the same time as large as those for low charged ions (Be-, B-, C-like). In non-Maxwellian plasmas, however, large fractions of highly and low charged ions (see arrow in Figure 12.24) can exist simultaneously. This circumstance is related to the fact that for hot electrons, the exponential factor in the expression for the rate coefficients is close to 1 because kT_{hot} is larger than the threshold energy (excitation, ionization). Therefore, the shell structure is not anymore strongly reflected in the distribution of ionic populations. The "admixture" of a considerable fraction of H-like ions simultaneously with Li-, Be-like ions is connected with the following: the ionization rate for the bulk electrons is in strong competition with the hot electron-induced rate. However, the ionization of the He- and H-like ions is essentially carried by the hot electrons as the rate coefficients for the bulk electrons is exponentially small. Therefore, hot electrons lead only to a minor decrease of the low charged ions

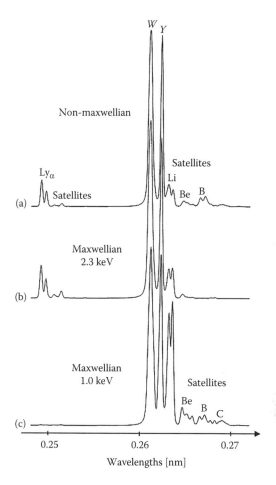

FIGURE 12.25
MARIA simulations of the spectral distribution of titanium in dense optically thick non-Maxwellian plasmas, $n_e = 10^{21}$ cm^{-3}, $L_{eff} = 300\,\mu$m: (a) $f = 0.09$, $kT_{bulk} = 800$ eV, $kT_{hot} = 20$ keV; (b) $f = 0.0$, $kT_e = 2.3$ keV; (c) $f = 0.0$, $kT_e = 1$ keV.

(Be-, B-, C-like) but to a strong increase of the H-like ions with corresponding simultaneously strong Ly$_\alpha$ emission and K$_\alpha$-satellite series emission.

The characteristic distortion of the ion charge stage distribution can in turn be used for the determination of the hot electron fraction [29]. This method has successfully been applied in laser-driven inertial fusion experiments to determine the time- and space-resolved hot electron fraction in the NOVA-hohlraums [101]. Moreover, the time resolution of the spectroscopically determined hot electron fraction permitted to identify the parametric instability stimulated Raman scattering as the main source of hot electron production [101].

Other methods of hot electron characterization which are based on Ly$_\beta$-satellite emission of highly charged ions are developed in [102].

12.5 Hollow Ions

12.5.1 General Considerations

Another type of radiation emission originating from autoionizing states is the so-called HI or hollow atom emission (sometimes called "hyper-satellites"). An HI is an ion, where one or more internal shells are entirely empty, whereas higher shells are filled with two or more electrons. Simple examples are the $K^0 L^X$ configurations for $X = 2$–8 (note that $N = 1$ corresponds to the H-like Ly$_\alpha$ transitions and $N = 2$ corresponds to the He-like $2l2l'$-satellites near Ly$_\alpha$). Other examples are the configurations $K^1 L^0 M^X$ for $X = 2$–18 (the L-shell does not contain any electrons, whereas the M-shell is filled with two or more electrons).

Traditionally, HI x-ray emission has been observed in ion beam surface interaction experiments [103,104] under low-density conditions: highly charged ions are extracted from a heavy particle accelerator at definite energies and are then brought into interaction with matter. However, recently also in a few visible and IR laser-produced plasma experiments, observation of HI emission has been claimed [53,105–107], as well as in high current dense Z-pinch plasmas [108]. Although HI emission has been identified on the basis of atomic structure calculations, the origin of their high intensities in dense laser-produced plasmas and Z-pinch plasmas are not yet well understood.

In [53,54], charge exchange between intermixing inhomogeneous dense plasmas has been proposed to explain the HI emission. However, up to now no *ab initio* atomic kinetic models were developed to provide proof of any reasonable production mechanisms except transient three-body recombination [26] and charge exchange [50]. We note that usual collisional-radiative simulations indicate that HI emission in dense laser-produced plasmas should be quite small, that is, below Bremsstrahlung. Current achievements in the understanding of HI emission in dense plasmas concerns essentially atomic structure calculations and corresponding identification of configurations. However, agreement with observed HI emission (intensities) has only been achieved by means of numerous free intensity parameters (including LTE assumptions [107]) to increase considerably (many orders of magnitude) the population of the HI configurations with respect to atomic ground and singly excited states. The origin of these large parameters itself remained not well justified.

It is therefore of fundamental interest to develop *ab initio* HI population kinetics to shed more light into this mystery. The first attempt in this direction has been made by the development of *ab initio* HI population kinetics which is driven by intense radiation fields [72–74]. It was shown that effective HI x-ray emission can be excited up to an observable level by the initiation of an effective photo-ionization chain reaction [73]:

$$K^2L^N + h\nu \rightarrow K^1L^N + e, \tag{12.125}$$

$$K^1L^N + h\nu \rightarrow K^0L^N + e. \tag{12.126}$$

It is important to note that the fully time-dependent simulations carried out with the MARIA code [29–31] do not contain any free parameters. The excitation mechanism is therefore well identified: chain reaction-induced photoionization of the K-shell by intense radiation fields according to Equations 12.125 and 12.126. It should be noted that the laser pulse duration is typically of the order of 10–100 fs, and therefore, the second photoionization (Equation 12.126) is effective before the configuration K^1L^N is destroyed by autoionization.

12.5.2 X-Ray Transitions Originating from K^0L^N Configurations

HI transitions originating from the configurations K^0L^N of highly charged ions, that is, $K^0L^N \rightarrow K^1L^{N-1} + h\nu_{\text{hollow}}$ are of particular interest for dense plasmas research:

1. These transitions connect the most strongly bound states in atomic/ionic systems. Strongly coupled plasma effects can therefore be studied as perturbations to these stable transitions [73].

2. The response time of HI emission is of the order of some 10 fs and represents therefore a fast x-ray emission switch at times when the plasma density is highest [72].

3. Intensity contributions from the low-density long-lasting recombining regime are negligible [72–74].

4. The absorbing ground states are the autoionizing states K^1L^N which are weakly populated even in dense plasmas. Radiation transport effects will therefore be small even in ICF plasmas with above solid density compression [73].

5. The HI x-ray transitions $K^0L^N \rightarrow K^1L^{N-1} + h\nu_{\text{hollow}}$ are well separated from other transitions, and the identification of double K-hole configurations is therefore less complicated (for $N = 2$–5, the transitions are essentially located between the H-like Ly_α-line and the He-like He_α-line).

Despite these outstanding properties for advanced diagnostics, HI emission is rather complex: the large number of levels and transitions does not really permit *ab initio* simulations with an LSJ-split level structure to achieve spectroscopic precision. When employing usual reduction methods, for example, the superconfiguration method [109] or a hydrogen-like approximation, the number of levels is reduced to a manageable number, however, the number of transitions is also strongly reduced. This reduction considerably modifies the total contour of the HI transitions (e.g., important for Stark broadening analysis; see below) due to an average of transitions and other atomic data (transition probabilities, autoionizing rates, line center positions, etc.). It is therefore very difficult to obtain a spectroscopic precision (high-resolution analysis of the spectral distribution) with the traditional superconfiguration method.

This reduction problem of the traditional superconfiguration method has recently been solved by the virtual contour shape kinetic theory (VCSKT) [31]: all transitions are maintained in an LSJ-split manner and all level populations are non-LTE due to a virtual population kinetics. Moreover, the VCSKT permits an analytical solution that speeds up the simulations by orders of magnitude without loss of spectroscopic precision. The interesting reader is referred to [31] for further details.

The interest in HI kinetics driven by intense radiation fields [73] originates from the fact that free electron x-ray lasers are currently under construction worldwide to provide extremely intense radiation fields. This allows us to provoke chain reactions of photo-ionizations according to Equations 12.125 and 12.126: XFEL in Germany [110], SPring8 in Japan [111], and LCLS in US [112]. Due to the outstanding radiative properties for dense plasma research outlined above, HI x-ray emission is of great interest for the free electron x-ray laser community [113] as well as for the dense plasma physics community [114]. Moreover, recently a new heating mechanism, namely "Auger electron heating" has been identified via high resolution spectroscopy of hollow ion states [115].

The great advantage of the HI emission induced by intense radiation fields is that a low-temperature, high-density plasma (strongly coupled plasma), which usually does not emit useful x-ray transitions, is transformed via photoionization chain reactions to an x-ray emitting strongly coupled plasma. The intensities of the transitions $K^0L^N \to K^1L^{N-1} + h\nu_{\text{hollow}}$ are characteristic to the populations K^2L^N and allow even detailed studies of the ionic charge stage distribution [72–74] in dependence of the temperature.

12.5.3 Stark Broadening Analysis of HI X-Ray Emission

Let us now proceed to possible density diagnostics. Figure 12.26 shows detailed Stark broadening calculations (carried out with the PPP code) for the hollow Mg ion x-ray transitions $K^0L^N \to K^1L^{N-1} + h\nu$, $kT_e = 100\,\text{eV}$ for $N = 1$–5. Line intensities within one configuration K^0L^N have been calculated assuming a statistical population for all LSJ-split levels in order not to mask the Stark broadening with population effects for different plasma densities. All HI electric dipole transitions and all energy levels have been included in the simulations (note that the minimum number of levels/transitions are 17/48 for the $N = 2$ configuration, 34/246 for the $N = 3$ configuration, 60/626 for

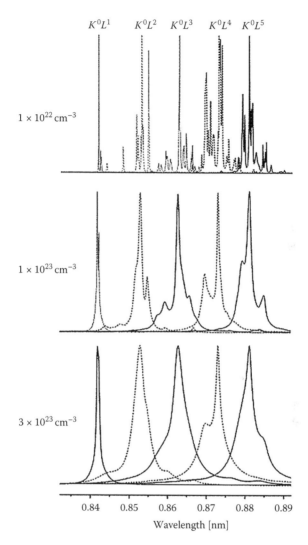

FIGURE 12.26

PPP Stark broadening simulations (normalized to peak) for x-ray transitions $K^0L^N \rightarrow K^1L^{N-1} + h\nu$ in hollow Mg ions, $kT_e = 100\,\text{eV}$.

the $N = 4$ configuration, and 65/827 for the $N = 5$ configuration, the number of Stark transitions is of the order of 10^6). Transitions from different charge states have been normalized to maximum peak intensity.

It can be seen that strong changes in the total contours emerge for near-solid-density plasmas. For densities less than $10^{22}\,\text{cm}^{-3}$, numerous single transitions are resolved (upper spectrum in Figure 12.26). This low-density simulation indicates that the broadening of the total contour is not only determined by

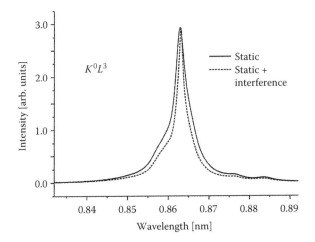

FIGURE 12.27
PPP Stark broadening simulations (normalized to peak) in intermediate coupling scheme showing the influence of interference effects for x-ray transitions $K^0L^3 \to K^1L^2 + h\nu$ in hollow Mg
ions. Interference effects lead to a considerable narrowing of the contour and also to a shift of the
contour maximum, $kT_e = 100\,\text{eV}$.

the Stark broadening of single transitions, but also by the oscillator strength
distribution over wavelengths in an even more important manner. Also in this
respect the VCSKT [31] provides the appropriate answer: all line transitions
are included in the simulations with their correct line center positions and
oscillator strengths distribution over wavelengths without any approximation (opposite to the traditional superconfiguration method where new artificial line center positions are calculated from certain averages of LSJ levels).

Let us finish the Stark broadening analysis of HI with a discussion of interference effects [2,4,5,66]. As the lower states of the HI configurations are
autoionizing states by itself (states K^1L^{N-1}), the number of lower levels is
also large and interference effects between upper and lower levels might be
important. Figure 12.27 compares Stark profile simulations for the HI x-ray
transitions $K^0L^3 \to K^1L^2 + h\nu$ with and without taking into account interference effects (intensities are normalized to peak). The simulations involve all
LSJ-split levels and all corresponding matrix elements in an intermediate
coupling scheme. It can clearly be seen that interference effects lead to a considerable "contour shape narrowing" as well as to a "peak center shift" of the
total contour.

12.6 Conclusion

The analysis of nonequilibrium phenomena in fusion and laser-produced
plasmas is a challenging field of research and corresponding high-resolution

x-ray spectroscopy provides a unique and independent characterization of the plasma. Dielectronic satellite and HI x-ray emission play a key role in the analysis of several important and complex phenomena such as, for example, suprathermal electrons, instabilities, particle transport, and strong coupling of particles. Fast time scales down to some 10 fs, an inherent spatial confinement of the emission to high-temperature and high-density areas, negligible overlap from low-density recombining regimes, and negligible opacity make dielectronic satellite and HI x-ray transitions an ideal plasma probe. Numerical simulations as well as numerous analytical approaches have been developed to characterize nonequilibrium phenomena. Particular emphasize has been paid to determine the electron temperature, electron density, hot electron fraction, ion acceleration, bulk electron temperature, inhomogeneities, particle and radiation transport. Numerous examples from fusion and laser-produced plasma experiments have been presented.

Acknowledgment

Fruitful and stimulating discussions with Prof. V. S. Lisitsa are greatly appreciated.

References

1. R. Loudon, *The Quantum Theory of Light* (2nd edition). Oxford Science Publications, Oxford, 1983.
2. H. R. Griem, *Plasma Spectroscopy*. New York: McGraw-Hill, 1964.
3. R. W. P. McWhirter, Spectral intensities. In R. H. Huddelstone and S. L. Leonard (Eds.), *Plasma Diagnostic Techniques*. New York: Academic Press, 1965.
4. H. R. Griem, *Spectral Line Broadening by Plasmas*. New York and London: Academic Press, 1974.
5. H. R. Griem, *Principles of Plasma Spectroscopy*. New York: Cambridge University Press, 1997.
6. I. I. Sobelman, *Theory of Atomic Spectra*. Oxford, UK: Alpha Science International Ltd, 2006.
7. D. Salzmann, *Atomic Physics in Hot Plasmas*. New York: Oxford University Press, 1988.
8. V. A. Boiko, A. V. Vinogradov, S. A. Pikuz, I. Yu. Skobelev, and A. Ya. Faenov, *J. Sov. Laser Res.* 6, 82, 1985.
9. C. de Michelis and M. Mattioli, *Nucl. Fusion* 21, 677, 1981.
10. V. S. Lisitsa, *Atoms in Plasmas*. New York: Springer, 1994.
11. L. A. Bureyeva and V. S. Lisitsa, A perturbed atom, *Astrophysics and Space Physics Reviews*, Moscow, 2000.
12. M. Alonso and E. J. Finn, *Physik III: Quantenphysik und Statistische Physik*. Amsterdam: Inter European Editions, 1974.

13. F. B. Rosmej, *J. Phys. B Lett.: At. Mol. Opt. Phys.* 28, L747, 1995.
14. A. Unsöld, *Physik der Sternatmosphären*. Berlin: Springer, 1955.
15. D. Mihalas, *Stellar Atmospheres* (2nd edition). San Francisco: W.H. Freeman, 1978.
16. D. Mihalas and B. Weibel-Mihalas, *Foundations of Radiation Hydrodynamics*. New York: Dover Pub. Inc., 1999.
17. F. E. Irons, *JQSRT* 22, 1, 1979.
18. F. E. Irons, *JQSRT* 22, 21, 1979.
19. F. E. Irons, *JQSRT* 22, 37, 1979.
20. F. E. Irons, *JQSRT* 24, 119, 1980.
21. G. Bekefi, *Radiation Processes in Plasmas*. New York: Wiley, 1966.
22. G. C. Pomraning, *The Equations of Radiation Hydrodynamics*. New York: Dover Pub. Inc., 1973.
23. J. Dawson and C. Oberman, *Phys. Fluids* 5, 517, 1962.
24. L. N. Tsintsadze, D. K. Callebaut, and N. L. Tsintsadze, *J. Plasma Phys.* 55, 407, 1996.
25. Yu. K. Zemtsov, A. Yu Sechin, and A. N. Starostin, *JETP* 83, 909, 1996.
26. F. B. Rosmej, A. Ya. Faenov, T. A. Pikuz, F. Flora, P. Di Lazzaro, S. Bollanti, N. Lizi et al., *JQSRT* 58, 859, 1997.
27. F. B. Rosmej, A. Schulz, K. N. Koshelev, and H.-J. Kunze, *JQSRT* 44, 559, 1990.
28. F. B. Rosmej and V. S. Lisitsa, *Plasma Physics Reports* 37, 521, 2011.
29. F. B. Rosmej, *J. Phys. B. Lett.: At. Mol. Opt. Phys.* 30, L819, 1997.
30. F. B. Rosmej, *Europhys. Lett.* 55, 472, 2001.
31. F. B. Rosmej, *Europhys. Lett.* 76, 1081, 2006.
32. F. B. Rosmej, V. S. Lisitsa, R. Schott, E. Dalimier, D. Riley, A. Delserieys, O. Renner et al., *Europhys. Lett.* 76, 815, 2006.
33. W. Engelhardt, *Course on Diagnostics for Fusion Reactor Conditions*, Vol. 1, Varenna, Italy, 1982, EUR 8351-IEN, p. 1.
34. D. Pasini, M. Mattioli, A. W. Edwards, R. Gianella, R. D. Gill, N. C. Hawkes, G. Magyar et al., *Nucl. Fusion* 30, 2049, 1990.
35. R. A. Hulse, *Nucl. Technol. Fusion* 3, 259, 1983.
36. F. B. Rosmej, V. S. Lisitsa, D. Reiter, M. Bitter, O. Herzog, G. Bertschinger, and H.-J. Kunze, *Plasma Phys. Control. Fusion* 41, 191, 1999.
37. F. B. Rosmej and V. S. Lisitsa, *Phys. Lett. A* 244, 401, 1998.
38. F. B. Rosmej, R. Stamm, and V. S. Lisitsa, *Europhys. Lett.* 73, 342, 2006.
39. V. A. Abramov, V. S. Lisitsa, and A. Yu. Pigarov, *JETP Lett.* 42, 356, 1985.
40. V. A. Shurygin, *Plasma Phys. Rep.* 30, 443, 2004.
41. F. B. Rosmej, R. Stamm, and V. S. Lisitsa, Convergent coupling of helium to the H/D background in magnetically confined plasmas, In *Symposium on the 18th International Conference on Spectral Line Shapes (ICSLS)*, Auburn, USA, AIP 874, 276, 2006.
42. P. Beiersdorfer, M. Bitter, M. Marion, and R. E. Olson, *Phys. Rev. A* 72, 032725, 2005.
43. P. Beiersdorfer, A. L. Osterheld, T. W. Phillips, M. Bitter, K. W. Hill, and S. von Goeler, *Phys. Rev. E* 52, 1980, 1995.
44. F. B. Rosmej, *Rapid Commun. Phys. Rev. E* 58, R32, 1998.
45. F. B. Rosmej, *J. Phys. B Lett.: At. Mol. Opt. Phys.* 33, L1, 2000.
46. A. Escarguel, F. B. Rosmej, C. Brault, Th. Pierre, R. Stamm, and K. Quotb, *Plasma Phys. Control. Fusion* 49, 85, 2006.

47. F. B. Rosmej, N. Ohno, S. Takamura, and S. Kajita, *Contrib. Plasma Phys.* 48, 243, 2008.
48. L. A. Bureyeva, V. S. Lisitsa, D. A. Petrov, D. A. Shuvaev, F. B. Rosmej, and R. Stamm, *Plasma Phys. Rep.* 29, 835, 2003.
49. J. Rosato, H. Capes, L. Godbert-Moret, M. Kubiti, Y. Marandet, F. B. Rosmej, and R. Stamm, *Contrib. Plasma Phys.* 48, 153, 2008.
50. F. B. Rosmej, H. R. Griem, R. C. Elton, V. L. Jacobs, J. A. Cobble, A. Ya. Faenov, T. A. Pikuz et al., *Phys. Rev. E* 66, 056402, 2002.
51. P. Monot, P. D'Oliveira, S. Hulin, A. Ya. Faenov, S. Dobosz, T. Auguste, T. A. Pikuz et al., *Phys. Plasmas* 8, 3766, 2001.
52. F. B. Rosmej, D. H. H. Hoffmann, W. Süß, M. Geißel, O. N. Rosmej, A. Ya. Faenov, T. A. Pikuz et al., *Nucl. Instrum. Methods A* 464, 257, 2001.
53. F. B. Rosmej, A. Ya. Faenov, T. A. Pikuz, A. I. Magunov, I. Yu. Skobelev, T. Auguste, P. D'Oliveira et al., *J. Phys. B. Lett.: At. Mol. Opt. Phys.* 32, L107, 1999.
54. N. E. Andreev, M. V. Chegotov, M. E. Veisman, T. Auguste, P. D'Olivera, S. Hulin, P. Monot et al., *JETP Lett.* 68, 592, 1998.
55. O. Renner, T. Missalla, P. Sondhauss, E. Krousky, E. Förster, C. Chenais-Popovics, and O. Rancu, *Rev. Sci. Instrum.* 68, 2393, 1997.
56. O. Renner, F. B. Rosmej, E. Krouský, P. Sondhaus, M. P. Kalashnikov, P. V. Nickles, I. Uschmann, and E. Förster, *JQSRT* 71, 623, 2001.
57. O. Renner, P. Adámek, P. Angelo, E. Dalimier, E. Förster, E. Krousky, F. B. Rosmej, and R. Schott, *JQSRT* 99, 523, 2006.
58. P. Adamek, O. Renner, L. Drska, F. B. Rosmej, and J.-F. Wyart, *Laser Particle Beams* 24, 511, 2006.
59. X. Li, Z. Xu, and F. B. Rosmej, *J. Phys. B: At. Mol. Opt. Phys.* 39, 3373, 2006.
60. I. Yu. Skobelev, A. Ya. Faenov, B. A. Bryunetkin, and V. M. Dyakin, *JETP* 81, 692, 1995.
61. A. Ya. Faenov, S. A. Pikuz, A. I. Erko, B. A. Bryunetkin, V. M. Dyakin, G. V. Ivanenkov, A. R. Mingaleev, T. A. Pikuz, V. M. Romanova, and T. A. Shelkovenko, *Phys. Scripta* 50, 333, 1994.
62. F. B. Rosmej, D. H. H. Hoffmann, W. Süß, M. Geißel, A. Ya. Faenov, and T. A. Pikuz, *Phys. Rev. A* 63, 032716, 2001.
63. F. B. Rosmej, D. H. H. Hoffmann, W. Süß, M. Geißel, P. Pirzadeh, M. Roth, W. Seelig et al., *JETP Lett.* 70, 270, 1999.
64. F. B. Rosmej, D. H. H. Hoffmann, W. Süß, A. E. Stepanov, Yu. A. Satov, Yu. B. Smakovskii, V. K. Roerich et al., *JETP* 94, 60, 2002.
65. F. B. Rosmej, O. Renner, E. Krouský, J. Wieser, M. Schollmeier, J. Krása, L. Láska, et al., *Laser Particle Beams* 20, 555, 2002.
66. I. I. Sobelman, V. A. Vainshtein, and E. A. Yukov, *Excitation of Atoms and Broadening of Spectral Lines* (2nd edition), Berlin: Springer, 1995.
67. J. Rosato, R. Stamm, M. Koubiti, Y. Marandet, L. Mouret, F. B. Rosmej, and H. Capes, *J. Nucl. Mater.* 363–365, 421, 2007.
68. J. Rosato, H. Capes, L. Godbert-Moret, M. Kubiti, Y. Marandet, F. B. Rosmej, and R. Stamm, *Contrib. Plasma Phys.* 48, 153, 2008.
69. J. Rosato, H. Capes, Y. Marandet, F. B. Rosmej, L. Godbert-Mouret, M. Koubiti, and R. Stamm, *Europhys. Lett.* 84, 43002, 2008.
70. B. Talin, A. Calisti, L. Godbert, R. Stamm, R. W. Lee, and L. Klein, *Phys. Rev. A* 51, 1918, 1995.

71. A. Calisti, S. Ferri, C. Mossé-Sabonnadière, and B. Talin, *Pim Pam Pum,* Report Université de Provence, Marseille, 2006.

72. F. B. Rosmej, R. W. Lee, and D. H. G. Schneider, *High Energy Density Phys.* 3, 218, 2007.

73. F. B. Rosmej and R. W. Lee, *Europhys. Lett.* 77, 24001, 2007.

74. F. B. Rosmej, R. W. Lee, D. H. G. Schneider, and H.-K. Chung, Some 10 fs x-ray emission switches driven by intense x-ray free-electron laser radiation, *SPIE - Laser-Generated and Other Laboratory X-ray and EUV Sources and Applications III (OP321),* San Diego, USA, Invited talk, 2007.

75. F. B. Rosmej, A. Ya. Faenov, D. H. H. Hoffmann, T. A. Pikuz, W. Süß, and M. Geißel, Autoionization spectral line shapes in dense plasmas, In *Proceedings of the XVth International Conference on Spectral Line Shapes,* Berlin, 2000, Germany, Invited talk, AIP 559, 19, 2002.

76. A. H. Gabriel, *Mon. Not. R. Astron. Soc.* 160, 99, 1972.

77. F. B. Rosmej and J. Abdallah Jr, *Phys. Lett. A* 245, 548, 1998.

78. F. B. Rosmej, A. Ya. Faenov, T. A. Pikuz, F. Flora, P. Di Lazzaro, S. Bollanti, N. Lizi et al., *J. Phys. B Lett.: At. Mol. Opt. Phys.* 31, L921, 1998.

79. V. P. Shevelko and L. A. Vainshtein, *Atomic Physics for Hot Plasmas.* Bristol: IOP Publishing, 1993.

80. L. A. Vainshtein and V. P. Shevelko, *Program ATOM.* Preprint 43. FIAN, Moscow: Lebedev Physical Institute, 1996,

81. O. Renner, E. Krouský, F. B. Rosmej, P. Sondhauss, M. P. Kalachnikov, P. V. Nickles, I. Uschmann, and E. Förster, *Appl. Phys. Lett.* 79, 177, 2001.

82. F. B. Rosmej, A. Ya. Faenov, T. A. Pikuz, I. Yu. Skobelev, A. E. Stepanov, A. N. Starostin, B. S. Rerich et al., *JETP Lett.* 65, 708, 1997.

83. F. B. Rosmej, U. N. Funk, M. Geißel, D. H. H. Hoffmann, A. Tauschwitz, A. Ya. Faenov, T. A. Pikuz et al., *JQSRT* 65, 477, 2000.

84. F. B. Rosmej, A. Calisti, R. Stamm, B. Talin, C. Mossé, S. Ferri, M. Geißel, D. H. H. Hoffmann, A. Ya. Faenov, and T. A. Pikuz, *JQSRT* 81, 395, 2003.

85. N. Yamamoto, T. Kato, and F. B. Rosmej, *JQSRT* 96, 343, 2005.

86. S. Kienle, F. B. Rosmej, and H. Schmidt, *J. Phys. B: At. Mol. Opt. Phys.* 28, 3675, 1995.

87. R. C. Elton, J. A. Cobble, H. R. Griem, D. S. Montgomery, R. C. Mancini, V. L. Jacobs, and E. Behar, *JQSRT* 65, 185, 2000.

88. V. A. Vinogradov, I. Yu. Skobelev, and E. A. Yukov, *Sov. Phys. JETP* 45, 925, 1977.

89. V. J. Jacobs and M. Blaha, *Phys. Rev. A* 21, 525, 1980.

90. F. B. Rosmej, *JQSRT* 51, 319, 1994.

91. K. N. Koshelev and N. Pereira, *JAP* 69, R21, 1991.

92. E. V. Aglitskii, P. S. Antsiferov, and A. M. Panin, *Sov. J. Plasma Phys.* 11, 159, 1985.

93. R. U. Datla and H. R. Griem, *Phys Fluids* 22, 1415, 1979.

94. F. B. Rosmej, *Nucl. Instrum. Res. Phys. B* 98, 33, 1995.

95. L. A. Woltz, V. L. Jacobs, C. F. Hooper, and R. C. Mancini, *Phys. Rev A* 44, 1281, 1991.

96. N. C. Woolsey, A. Asfaw, B. Hammel, C. Keane, C. A. Back, A. Calisti, C. Mossé, et al., *Phys. Rev. E* 53, 6396, 1996.

97. F. B. Rosmej, A. Calisti, B. Talin, R. Stamm, D. H. H. Hoffmann, W. Süß, M. Geißel, A. Ya. Faenov, and T. A. Pikuz, *JQSRT* 71, 639, 2001.

98. F. B. Rosmej, D. H. H. Hoffmann, M. Geißel, M. Roth, P. Pirzadeh, A. Ya. Faenov, T. A. Pikuz, I. Yu. Skobelev, and A. I. Magunov, *Phys. Rev. A* 63, 063409, 2001.

99. K. Bergmann, O. N. Rosmej, F. B. Rosmej, A. Engel, C. Gavrilescu, R. Lebert, and W. Neff, *Phys. Rev. E* 56, 5959, 1997.

100. J. A. Koch, M. H. Key, R. R. Freeman, S. P. Hatchett, R. W. Lee, D. Pennington, R. B. Stephens, and M. Tabak, *Phys. Rev. E* 65, 016410, 2001.

101. S. H. Glenzer, F. B. Rosmej, R. W. Lee, C. A. Back, K. G. Estabrook, B. J. MacCowan, T. D. Shepard, and R. E. Turner, *Phys. Rev. Lett.* 81, 365, 1998.

102. F. B. Rosmej, R. Schott, E. Galtier, P. Angelo, O. Renner, F. Y. Khattak, V. S. Lisitsa, and D. Riley, *High Energy Density Phys.* 5, 191, 2009.

103. I. A. Armour, B. C. Fawcett, J. D. Silver, and E. Träbert, *J. Phys. B: At. Mol. Opt. Phys.* 13, 2701, 1980.

104. J. P. Briand, L. de Billy, P. Charles, S. Essabaa, P. Briand, R. Geller, J. P. Desclaux, S. Bliman, and C. Ristori, *Phys. Rev. Lett.* 65, 159, 1990.

105. A. M. Urnov, et al., *JETP Lett.* 67, 489, 1998.

106. A. Ya. Faenov, A. I. Magunov, T. A. Pikuz, I. Yu. Skobelev, S. A. Pikuz, A. M. Urnov, J. Abdallah et al., *Phys. Scripta* T80, 536, 1999.

107. J. Abdahllah Jr, I. Yu. Skobelev, A. Ya. Faenov, A. I. Magunov, T. A. Pikuz, F. Flora, S. Bollanti et al., *Quant. Electron.* 30, 694, 2000.

108. F. B. Rosmej, D. H. H. Hoffmann, W. Süß, M. Geißel, O. N. Rosmej, A. Ya. Faenov, T. A. Pikuz et al., *Nucl. Instrum. Methods A* 464, 257, 2001.

109. A. Bar-Shalom, J. Oreg, W. H. Goldstein, D. Shvarts, and A. Zigler, *Phys. Rev. A* 40, 3183, 1989.

110. XFEL—European Free Electron Laser, Germany. http://xfel.desy.de/

111. SSRL—Stanford Synchrotron Radiation Laboratory, USA. http://www-ssrl.slac.stanford.edu/lcls/

112. Spring8 Synchrotron, Japan. http://www.spring8.or.jp/

113. F. B. Rosmej and R. W. Lee, Hollow ion emission. In *XFEL Technical Design Report*, Chapter 6, pp. 251-253, DESY, 2006, http://xfel.desy.de/tdr/tdr

114. F. B. Rosmej, P. Angelo, and Y. Aouad, Contour shape analysis of hollow ion X-ray emission. In *Proceedings of the 19th International Conference on Spectral Line Shapes*, Madrid, Espagne, Invited talk, AIP Conference Proceedings 1058, American Institute of Physics, New York, 2008, Spectral Line Shapes, Vol. 15, p. 349, 2008.

115. E. Galtier, F. B. Rosmej, D Riley, T. Dzelzainis, F. Y. Khattak, P. Heimann, R. W. Lee et al. *Physical Review Letters* 106, 164801, 2011.

13

Short-Wavelength Free Electron Lasers

John T. Costello and Michael Meyer

CONTENTS

13.1 Introduction

The invention of conventional lasers [1] with their unique property to deliver intense monochromatic and coherent photon beams has had a dramatic impact on many scientific and technological domains. Current research in physics, chemistry, and biology is based, to a large extent, on the application of laser systems with extremely high spectral resolution (operating in continuous wave mode) or with high pulse energies and/or ultrashort duration in pulsed mode. However, the wavelength regime of conventional lasers is limited to the infrared and visible spectral ranges since the reflectivity of mirrors (needed for the optical cavity) drops drastically at very short wavelengths. Hence the main photon sources for extreme-UV (EUV) and x-ray radiation are generally synchrotron radiation facilities based on large electron storage rings [2]. The latest (third) generation synchrotron sources are very powerful tools for studying processes induced by the interaction of short-wavelength light with strongly bound inner-shell electrons and for unraveling the complex structure of solids or biomolecules in diffraction experiments. However, synchrotron radiation is produced by the emission of electron bunches circulating at relativistic velocities in the storage rings, and therefore the pulse duration is generally limited to several picoseconds and the pulse energies reside typically in the nanoJoule regime. It is against this background that a substantial research effort to develop intense EUV photon sources with laser-like properties, opening up exciting new research fields, has been undertaken during the past two decades.

A number of different approaches to constructing sources of laser-like radiation in the EUV have been pursued. The one closest to the original concept of the laser relies on generating a population inversion between highly charged ions (HCIs) via electron collisional excitation in a laser-produced plasma [3]. The development of these sources is ongoing, and laboratory-scale systems can provide, for selected wavelengths, a partially coherent EUV beam of low repetition rate (typically 10 Hz) and picosecond pulse duration. Alternatively, short-wavelength radiation can be produced by focusing intense femtosecond laser pulses into a gas jet [4] or onto a solid surface [5], resulting in the emission of a comb of short-wavelength lines corresponding to high harmonics of the drive laser wavelength. These high harmonic sources are widely used nowadays as small-scale coherent laboratory sources and have had a very strong impact on coherent [6] and dynamical studies in atoms including the possibility of producing and probing with attosecond EUV pulses [7]. Most recently, the two aforementioned techniques have been combined to produce fully coherent EUV beams by injection seeding a collisionally excited plasma amplifier with wavelength-matched EUV harmonics [8,9].

The most recent approach, the single-pass free electron laser (FEL), is based on the technology of electron accelerators. Laser-like radiation is produced by the self-amplified spontaneous emission (SASE) process [10], which starts from shot noise in an electron beam accelerated to relativistic energies. The general feasibility of the FEL concept was demonstrated in 1971, when Madey [11] showed that coherent radiation could be generated and amplified by stimulated emission when relativistic electrons propagated colinearly with the radiation field in the periodic magnetic structure of an undulator. The SASE principle is illustrated in Figure 13.1 [12].

Electrons are deflected periodically as they propagate through the undulator, undergoing a slalom-type motion, and in doing so, they exchange energy with the radiation field. If there are N undulator pole pairs with a magnetic period of λ_u, then one can write that

$$N_u \lambda_u = v_e t,$$

where v_e is the electron velocity and t the undulator traversal time. In the laboratory frame, the observer see the emitted radiation Doppler (wavelength)

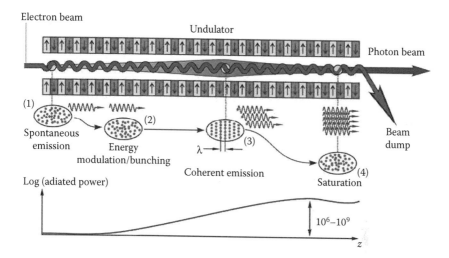

FIGURE 13.1
Schematic representation of the SASE process in an undulator. (From Berliner Elektronenspeicherring-Gesellschaft für Synchrotronstrahlung (BESSY). http://www.bessy.de/publicRelations/publications/files/sc.pdf. With permission.)

downshifted to λ_{FEL}, that is,

$$N_u \lambda_{FEL} = (c - v_e)t,$$

$$\therefore \lambda_{FEL} \approx \lambda_u \frac{c - v_e}{v_e} t \implies \lambda_{FEL} \approx \frac{\lambda_u}{2\gamma^2}.$$

The energy exchange is greatest along the undulator axis. A rigorous analysis shows that

$$\lambda_{FEL} = \frac{\lambda_u}{2n\gamma^2} \left(1 + \frac{K^2}{2}\right),$$

where $\gamma = E_e/m_0 c^2$ is the electron energy in units of the electron rest mass, n the harmonic number (the fundamental mode corresponds to $n = 1$), and K the undulator parameter. It can be computed from

$$K = \frac{eB_0 \lambda_u}{2\pi mc},$$

where B_0 is the peak magnetic field.

Referring to Figure 13.1, an electron beam of extremely high density passes through a long undulator. The primary spontaneous emission interacts with the electron bunch on its pathway through the undulator where it induces an energy modulation and concomitant spatial bunching of the electron beam so that the stimulated emissions from the microbunches add coherently. The process is regenerative and leads to an exponential growth in the energy of the radiation pulse in a single pass through the undulator. Saturation is reached when the microbunches are fully developed and no more electrons are available to increase the periodic density modulation. In order to obtain a sufficient gain within a short distance, the peak current has to be of the order of several kiloamperes and the beam diameter of the electron bunch has to be less than 100 μm over the undulator length. At the FEL facility in Hamburg (FLASH), the total length of the undulator is 27 m. For the upcoming x-ray FELs, undulators of more than 100 m are envisaged in order to achieve saturation at wavelengths below 1 nm.

13.2 FLASH: The Free-Electron LASer in Hamburg

13.2.1 The Development of Hamburg's Short-Wavelength FEL

FLASH was the first EUV FEL worldwide user facility to become operational. It has once been joined by the SCSS VUV FEL in Japan and the LCLS x-ray FEL in the United States. At the time of writing the FERMI EUV FEL in Trieste is generating light while construction of the Swiss FEL at the Paul Scherrer Institute is well advanced. Finally a test facility is operational at the MAXLAB in Sweden. Returning to the FEL at Hamburg, FLASH had its genesis as the Tesla Test Facility (TTF) at DESY in Hamburg where the centerpiece of the project was a vacuum-ultraviolet (VUV) FEL operating at the then record short wavelength of 98 nm, a milestone which was reported by DESY in 2000 [13]. It soon reached GW peak power levels [14] and was used to perform what was the first VUV multiphoton ionization (MPI) experiment on rare gas clusters [15]. In the succeeding phase of the project (TTF2), the design goal was a 1 GeV machine that could extend the shortest wavelength down to 6 nm. By 2005, the VUV FEL at Hamburg had become the EUV FEL at Hamburg and was rechristened "FLASH." It opened to users in autumn 2005 and since then it has hosted a wide variety of experiments in atomic and molecular physics, a number of which will be described in Section 13.2.5.

FLASH reported world record peak (up to 10 GW) and average (almost 100 mW) coherent power at a wavelength of 13.7 nm in 2007 [16]. The multinational collaboration, which included the authors of this chapter, also observed ~10 nJ pulses at the third (4.5 nm) and nJ level pulses at fifth (2.8 nm) harmonic wavelengths, the latter bringing FLASH into the biologically important "water window." Following a shut down for upgrade work in the summer of 2007, the FLASH team reported that they had reached the minimum design wavelength of 6 nm [17] in October 2007.

In Figure 13.2a, we show a bird's eye view of the FLASH experimental hall located at the end of the tunnel containing the linear accelerator (LINAC) and the undulator modules in which SASE occurs. The hall was part of EXPO 2000 and it housed a scientific exhibition during that event. Figure 13.2b shows an inside view of the experimental hall, whereas Figure

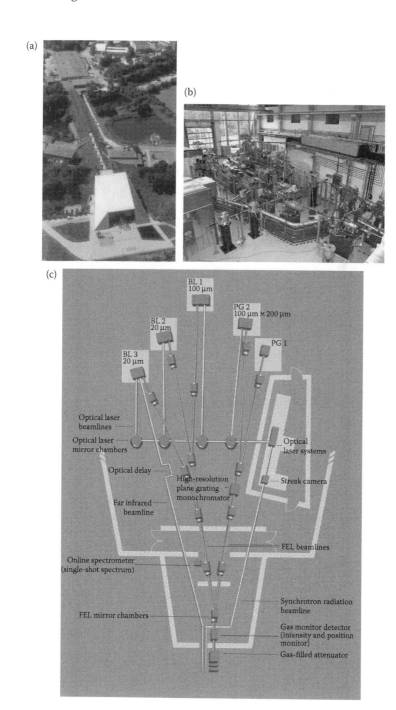

FIGURE 13.2
External view of the FLASH tunnel and experimental hall (a), a view from inside the experimental hall (b), and the beamline layout including the femtosecond laser hutch (c). (From Free Electron Laser in Hamburg, Germany. http://flash.desy.de. With permission.)

13.2c shows the beamline layout. There are five EUV beamlines available to users: BL1 (100 μm focus), BL2 (25 μm focus), BL3 (unfocussed), PG1 (a high-resolution monochromator), and PG2 (monochromator with 100 μm focus). Time-synchronized optical pulses from an ultrafast laser are delivered to each experiment via beamlines, which run alongside each EUV beamline for two-color experiments. The two-color facility is described in detail in Section 13.2.4.

13.2.2 Design and Operation

As indicated earlier, FLASH is based on a linear accelerator (LINAC) coupled to a series of long and precisely aligned undulators acting in concert as a quasi-single SASE-FEL device (Figure 13.3). Its design, operation, and performance have been described in a number of recent papers [16,18,19], while the FLASH website [20] contains a great deal of technical information, performance data, and advice for prospective users.

To ensure efficient SASE action and a concomitantly saturated output, as well as a coherent EUV beam, it is essential that the high-energy electron bunch train is of the highest possible quality; specifically it must comprise bunches of high peak current, ultrashort duration, and small momentum spread. The quality factor most frequently used to specify an electron beam is the emittance, which is effectively the beam size times the beam divergence (in units of mm mrad). Linear accelerators are known to produce beams, which match these essential criteria, and so FLASH is a LINAC-based system. Electron bunches of up to a few nanocoulomb are produced by a photoinjector system which comprises a Cs_2Te photocathode illuminated by 4 ps UV pulses (262 nm) that are obtained from an actively mode-locked laser driven off the FEL radio-frequency (RF) signal. The system is designed to deliver bunch trains (referred to as a macropulse) containing up to 800 bunches with a bunch-to-bunch separation of 1 μs at a macropulse repetition frequency of 10 Hz. Electron bunches are accelerated in a 1.5-cell, 1.3-GHz RF cavity powered by a 5 MW Klystron source which yields a peak accelerating field of 40 MV/m at a nominal 3 MW of power injected into the cavity.

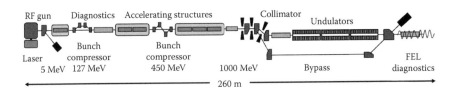

FIGURE 13.3
Schematic diagram of the EUV FEL FLASH at DESY, Hamburg, showing the accelerating structures and undulators. (From Free Electron Laser in Hamburg, Germany. http://flash.desy.de. With permission.)

The bunch train enters the first TESLA module which boosts the energy to 127 MeV before it encounters the first bunch compressor. As each electron bunch extends over a distance of a few millimeters, a considerable fraction of the RF wavelength of 23 cm, it experiences an accelerating gradient which impresses a position-dependent momentum spread, that is, each bunch becomes energy-chirped. To reduce the bunch width to the subpicosecond timescale, so that the peak current can exceed the 1 kA design threshold, it passes through a compressor comprising magnetic chicanes which slow the front edge (fast component) of the bunch while speeding up the bunch tail using RF phase control. Bunch lengths of <100 fs are possible by this action. The process is not completely unlike the chirped pulse amplification (CPA) technique of Strickland and Mourou [21] in which matched grating pairs are first used to introduce a differential path length between the red and blue components of an ultrashort optical pulse before amplifying it. The amplified pulse is recompressed by reversing the chromatically induced path length deviation in a second grating pair (the compressor). Similar effects can be observed in transmission optics where group velocity dispersion and self-phase modulation are used to affect CPA. Since mid-2007, the complete linear accelerator has comprised six 12-m long modules each housing eight 9-cell superconducting cavities which can accelerate bunch trains to 1 GeV and so FLASH can be tuned from 60 nm down to 6 nm. Modules 2 and 3 also induce a position-dependent momentum spread which is used to controllably compress the bunch yet again in a second compressor.

The core of the SASE-FEL, namely the undulator train, contains six undulators each 4.5 m long and separated by 0.6 m from each other. The FEL wavelength is tuned by changing the energy of the electron beam rather than the more traditional method of modulating the magnetic field via the undulator gap as happens in insertion devices at storage rings. This restriction arises because the undulator gap must be maintained in very precise alignment to ensure SASE, and regular changing of the gap is not conducive to this aim. In fact, the r.m.s. deviation of the electron beam from the ideal orbit plane must lie below 10 μm. Quadrupole magnetic focusing elements, located between each sub-undulator, are used to ensure the smooth transfer of the beam through the undulator chain.

The smooth operation and optimization of FLASH is completely dependent on innovative diagnostic systems that have been under continuous development since the initiation of the project in the late 1990s. As it is the key performance driver, diagnostic systems for the electron beam are distributed right along the FEL. The electron beam cross-section can be viewed directly by observing optical transition radiation produced as the beam traverses a series of thin Si wafers coated with aluminum with the induced radiation patterns being monitored by a distributed camera system [22]. Beam width measurements are complemented by data obtained from wire scanners passed through the beam at various locations along its length [23]. In this way, both beam diameter and divergence can be measured. The emittance at an operating wavelength of 32 nm is $2 \pm 0.4\pi$ mm mrad before the first compressor rising to $2.9 \pm 0.3\pi$ and $4.3 \pm 0.4\pi$ mm mrad for the horizontal and vertical emittances, respectively, of the spike produced after the compressor.

13.2.3 Radiation Properties and Key Performance Indicators

It is of course the radiation properties that are most of interest for atomic, molecular, and optical physics. As FLASH is basically a spontaneous radiation amplifier, no two output pulses have exactly the same characteristics. Hence, a huge effort has been made by the DESY team to develop a comprehensive and fully integrated photon diagnostics system which yields the energy and spectral distribution of each pulse. All diagnostic systems, including those brought to the end stations by users, are integrated into a single distributed data acquisition system called DOOCS (Distributed Object Oriented Control System). DOOCS provides a unique timestamp for each FLASH pulse which is added as a tag to all data files produced by the various DESY diagnostic systems. Users can develop their own codes to access the timestamps and so user-acquired experimental data can be normalized to each FEL pulse during analysis. Such comprehensive systems are absolutely crucial when using statistical light sources such as SASE FELs.

There has been a large number of studies published over the years coincident with the phased development of FLASH [24], most recently on its operation in the EUV [16,19]. The key diagnostics concern the measurement of the center wavelength and spectral distribution, achieved with the aid of a grazing incidence vacuum spectrometer located in the FEL tunnel [25] and also the pulse energy distribution obtained from either microchannel plates [26] or gas monitor detectors [27]. Given the critical importance of single-shot measurements in SASE FEL diagnostics, the readout system for the spectrometer comprises an EUV to visible light down converter imaged onto a gated CCD camera so that a single micropulse can be selected from each FLASH macropulse and its spectrum recorded. As FLASH produces trains of femtosecond pulses, each separated from the other by 1 μs, gated detectors are needed to extract data due to single pulses out of each train.

In Figure 13.4, we show a comparison of the peak brilliance of FLASH compared with third-generation synchrotron light sources and predicted values for two of the x-ray FELs currently under construction, namely European XFEL [28] and LCLS [29]. The solid circles indicate actual measured data and are in excellent agreement with the predictions of FEL models such as the FAST code [30]. Figure 13.4 also includes data not just for the fundamental wavelength (13.7 nm here) but also for the third and fifth harmonics. The harmonics are very useful as very-short-wavelength coherent radiation sources and in fact the fifth harmonic (2.7 nm) in Figure 13.4 lies in the "water window."

Harmonics have one other collateral benefit. When the electron beam traverses the undulator, it undergoes a slalom-type motion. Ideally this motion

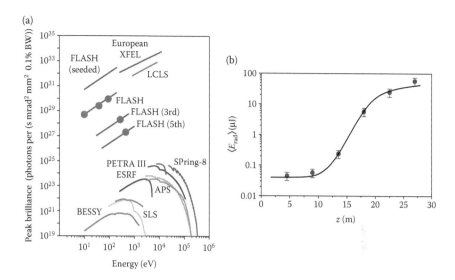

FIGURE 13.4

(a) Comparison of measured (solid circles) with predicted brilliant values (lines) for FLASH. Also shown are values for latest generation synchrotrons and next-generation x-ray FELs. (b) Growth in energy per pulse (μJ) as the number of undulators in the FEL chain is increased from one to the maximum number of six. The output clearly saturates as undulators 5 and 6 kick in. Measurements were made at 13.7 nm. (From W. Ackermann et al., *Nat. Photon. 1*, 336, 2007. With permission.)

should be confined to a single plane and so the SASE radiation should be close to 100% plane-polarized. In this case, only odd harmonics are observed with measurable intensity. Strong odd harmonic emission also signifies saturated output at the fundamental wavelength. On the other hand, deviation from in-plane motion results in elliptically polarized beams and the observation of increased intensity in even harmonics. Thus, harmonic analysis is a useful measure and diagnostic of FEL performance.

Fundamental and harmonic signals can be observed directly by optical spectroscopy or by photoelectron spectroscopy [31,32] where the kinetic energy of electrons ejected by the harmonic components of a beam results in well-defined photoelectron lines that are easily identified. In Figure 13.5, we show optical and photoelectron spectra [31] to illustrate the usefulness of both approaches in harmonic analysis.

The optical spectrum was obtained from a single micropulse from FLASH via a flat-field spectrometer equipped with a Harada grating [33] and a back-illuminated CCD camera. The photoelectron technique has the advantage that there are no order sorting issues (as with a grating spectrometer) and it can be used in a noninvasive way anywhere along the beamline, before or after an end station [32]. Apart from acting as performance markers of the FEL, it is important to have harmonic content analysis as these spectral components can give rise to signals which may be confused with two, three, or even higher

(a)

FLASH—Single Shot Spectrum May 2006.
Fundamental wavelength—25.5 nm.
EUV filter—1 μm polypropylene

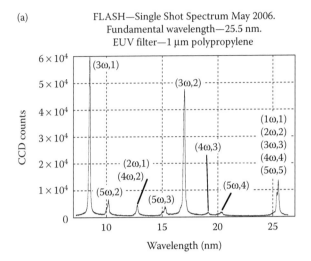

(b)

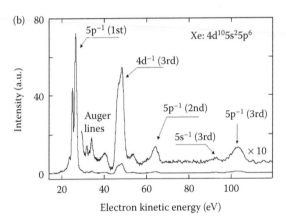

FIGURE 13.5
(a) Single-shot FLASH spectrum measured on a flat-field spectrometer with back-illuminated CCD readout (unpublished). Various harmonics appearing in different diffraction orders of the grating are labeled. The polypropylene filter removes almost all radiation at 25.5 nm and so the line at this wavelength position is mainly due to the third harmonic in third order. Measurements in 2006 showed that the even harmonics had all but disappeared and FLASH had reached optimal operation. (b) PES of Xe obtained at a FLASH wavelength of 32 nm (38.5 eV). Photolines due to photoionization by the second and third harmonics of FLASH are indicated. (From S. Duesterer et al., *Opt. Lett. 31*, 1750, 2006. With permission.)

photon processes. The key performance measures for FLASH are summarized in Table 13.1.

The average energy per pulse is 70 μJ (see also Figure 13.4b) but can fluctuate from 10 to >150 μJ between successive micropulses in the worst case. This is in fact a feature rather than a bug. By binning data according to FEL pulse energy, users can mine the data for quantities such as ion yield and plot them

TABLE 13.1

Key FLASH Parameters (Measured)

FLASH Parameter	Value
Wavelength range (fundamental)	6–60 nm (20–200 eV)
Spectral width (FWHM)	0.5–1%
Divergence	<0.1 mrad
Pulse energy	70 µJ (average) and 170 µJ (peak)
FEL harmonics (at 13.7 nm)	Third: 4.6 nm (250 nJ); fifth: 2.7 nm (10 nJ)
Pulse duration (FWHM)	10–50 fs
Peak power (fundamental)	1–10 GW
Average power (fundamental)	0.1 W (at 3000 pulses/s)
Photons per pulse	$\sim 10^{13}$

as a function of EUV intensity [34]. The pulse width cannot be measured directly as currently streak cameras are limited to a time resolution of 200 fs or so. Instead by comparing the statistical distribution of the radiation with that predicted by a SASE FEL model, the EUV pulse duration can be extracted. Most recently, we did this for FLASH operating at 13.7 nm in the well-known EUV lithography band [16]. In the exponential growth regime, the radiant output from a SASE-FEL model exhibits characteristics of a completely chaotic light source [35]. In this case, the probability distribution of the emitted pulse energies maps nicely onto a gamma distribution:

$$ p(E) = \frac{M^M}{\Gamma(M)} \left(\frac{E}{\langle E \rangle} \right)^{M-1} \frac{1}{\langle E \rangle} \left(-M \frac{E}{\langle E \rangle} \right), $$

where $\Gamma(M)$ is the gamma function and $M = 1/\sigma_E^2$ with $\sigma_E^2 = \langle (E - \langle E \rangle)^2 \rangle / \langle E \rangle^2$. The parameter M can be interpreted as the number of modes in the radiation pulse. Fitting this expression to the experimentally determined statistical distribution permits one to obtain a value for M and hence σ_E. The result is shown in Figure 13.6a for FLASH operating at 13.7 nm.

From the spectral distribution of the full FEL bandwidth, one can estimate the coherence time τ_c to be ~4 fs and so the pulse duration, given by $M\tau_c$ [35], becomes 8 fs (in the exponential gain regime). In saturated operation, the pulse energy distribution narrows and no longer approximates a gamma distribution. In this case, one has to rely on fitting the predictions of FAST FEL code [30] to experiment as shown in Figure 13.6b. Once fitted, one can use FAST to predict the expected shapes of FLASH pulses as shown in Figure 13.6c. It is clear that each pulse consists of one to three sub-10 fs spikes which is in close agreement with predictions based on FLASH as a statistical light source. We note here that proposals for attosecond pulse generation in EUV FELs have already been made [36] which will permit nonlinear processes to be driven at short wavelengths on characteristic atomic timescales.

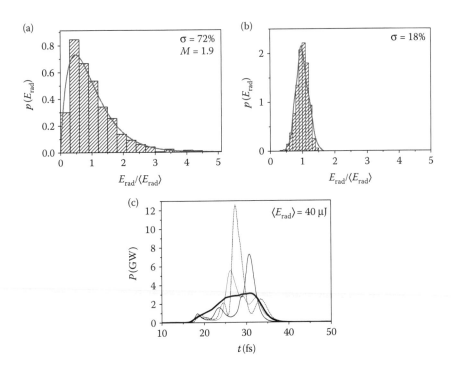

FIGURE 13.6
(a) Gamma distribution in the exponential gain region (solid line) fitted to the FLASH radiation pulse distribution which yields an M-value of ~ 2. (b) Radiation pulse distribution in the saturation regime compared to the prediction (solid line) of the FAST FEL code (Adapted from E. L. Saldin, E. A. Schneidmiller, and M. V. Yurkov, *Nucl. Instrum. Meth. A* 429, 233, 1999.) (c) Predicted temporal structure of the EUV pulse output from FLASH for an average energy per pulse of 40 µJ. Data obtained from FAST with measured radiation parameter values. The thick solid line represents the average pulse envelope, while the thin lines correspond to single shots. (Adapted from W. Ackermann et al., *Nat. Photon.* 1, 336, 2007.)

The transverse coherence has been modelled [37] and measured on a number of occasions using the traditional Young's double-slit experiment. The fringe pattern obtained at an operating wavelength of 100 nm [38] for a slit width of 100 µm and for a separation of 1 mm is shown in Figure 13.7a. The experiment was repeated for a number of targets of varying slit separation and the fringe visibility, defined as $(I_{Max} - I_{Min})/(I_{Max} + I_{Min})$, measured.

The results show that the SASE output is almost 100% coherent across the majority of its beam footprint. Since the SASE output beam builds up from shot noise in the FEL amplifier, it is not temporally coherent. However, recent experiments on laser plasma EUV amplifiers with high harmonics of an optical laser [8,9] have been extended to FEL amplifiers with some success [39]. Such systems hold out the promise of fully coherent EUV lasers (albeit at discrete wavelengths) which can bring the third space dimension into play so that applications requiring full spatial and phase coherence, even on relatively

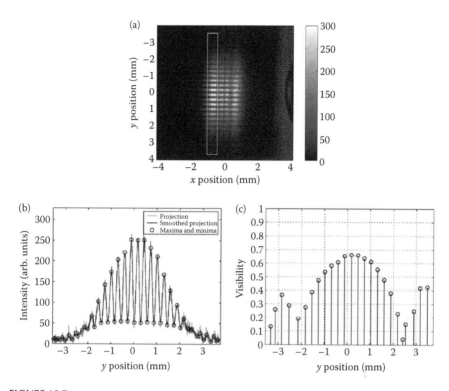

FIGURE 13.7
Recorded fringe pattern (a), corresponding lineout (b), and fringe visibility (c) for 100 μm slit width and separation of 1 mm. (Adapted from R. Ischebeck et al., *Nucl. Instrum. Meth. A 507*, 175, 2003.)

short timescales, such as multipulse holographic imaging and microscopy at the nanometer scale can at least be contemplated. At the time of writing, a project to seed FLASH with high harmonics of an optical laser has been sanctioned. Looking further into the future, FLASH is a milestone on the road to the European XFEL or x-ray FEL at Hamburg, a European project. The XFEL project status at the time of writing is summarized at the end of this chapter.

13.2.4 Two-Color Pump–Probe Facility

Combining short, intense optical pulses from an independent ultrashort pulse laser with comparable EUV pulses from FLASH is a particularly important task as it opens up a plethora of applications in pump and probe experiments and the study of coherent or dressed states of matter. For this reason, an additional pump–probe facility comprising two different laser systems has been installed in the experimental hall at FLASH. The optical laser can be used either to prepare the species before ionization by the EUV pulses or to characterize the final products of photoionization or photodissociation

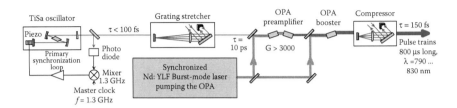

FIGURE 13.8
Layout of the optical laser system at FLASH.

processes. The simultaneous action of both pulses enables the investigation of two-photon (resonant) excitations and to study the dynamics of photoinduced processes under the influence of a strong electromagnetic field. In addition, the possibility to control the temporal delay between the two pulses opens access to time-resolved experiments.

At present, two optical laser systems are available for user operation at FLASH. The first one provides high-energy (up to 10 mJ) 800-nm pulses with a duration that can be varied from 120 fs to 4 ps at a repetition rate of up to 10 Hz, that is, one pulse per bunch train (macropulse) of the FEL. The second system (Figure 13.8) comprises three modules but can reproduce the overall bunch train structure of the FLASH, that is, up to 800 pulses per macropulse with a spacing of 1 μs between the individual pulses.

Referring to Figure 13.8, the output of a Nd:YLF-based burst mode laser, which is a modified copy of the photocathode laser, is frequency-doubled to deliver 523-nm pulses with energies of up to 0.4 mJ with a pulse duration of 12 ps. These pulses are used to amplify the ultrashort pulses of a Ti:Sapphire laser (50 fs, 3 nJ, 108 MHz) in an optical parametric amplifier. The final result is a bunch train of ultrashort (120 fs) near-infrared (800 nm) laser pulses of about 500 μJ pulse energy.

Both lasers (femtosecond oscillator and picosecond pump) are electronically synchronized to the FEL pulses, that is, to the RF source (1.3 GHz, 108 MHz, and 9 MHz) driving the electron accelerator. These reference frequencies are delivered to the laser hutch by a 300-m long temperature-stabilized cable. The repetition rate of the femtosecond oscillator is continuously compared to these frequencies, and possible deviations are directly corrected by changing the length of the cavity with a piezoelectric controlled back reflector. This procedure provides synchronization with respect to the RF of better than 150 fs r.m.s. for an interval of up to several minutes. The high-energy laser pulse train, at the macropulse repetition frequency of 10 Hz, is obtained with the use of a regenerative amplifier (Coherent Hidra) seeded by one micropulse picked from each bunch train.

For time-resolved pump–probe experiments, the determination of the jitter and drift of the optical pulses with respect to the FEL pulses is essential. For this reason, different approaches have been used to measure (on a shot-to-shot basis) the relative delay between the optical and the EUV pulses. Long-term drift is monitored by an optical streak camera located in the laser hutch [40]. To obtain a timing reference for the FEL pulse, the optical portion of a synchrotron radiation pulse, produced when the electron beam exiting the final undulator is deflected by a dipole magnet into a beam dump (Figure 13.3), is guided by a series of broadband mirrors to the entrance slit of a streak camera. A small part of the optical laser pulse is deflected onto the camera slit to register its arrival time. The relative jitter between both pulses is determined from the readout of the streak camera trace.

In Figure 13.9a, a typical image from the streak camera is displayed. The two main features correspond to the optical laser and synchronized dipole radiation pulses. The timing trace corresponds to a full vertical bin of the image data. The software determines the difference in arrival times from the sweep

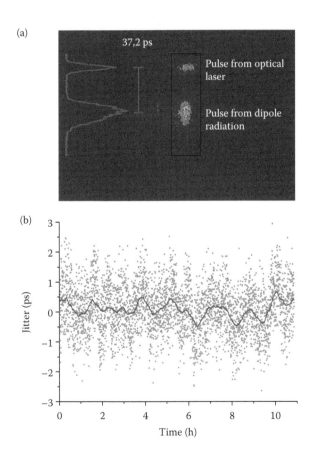

FIGURE 13.9
(a) Image of the laser pulse and the dipole radiation pulse as reference for the FEL arrival time on the streak camera. (b) The relative timing jitter between both pulses was determined by analyzing ca. 4000 images over 11 h for the graph. The resulting jitter was 800 fs r.m.s. (Adapted from P. Radcliffe et al., *Nucl. Instrum. Meth. Res. A 583*, 516, 2007.)

speed and peak positions (37.2 ps in this case). The procedure was repeated every 10 s for 11 h to obtain the trace in Figure 13.9b.

Even better temporal resolution on a shot-to-shot basis can be achieved with the aid of an electro-optical sampling system [41], which determines the jitter between the optical laser pulse and the FEL electron bunch directly. In this system, some of the optical laser pulses are sent into the accelerator tunnel by a glass fiber. Here, the laser pulses pass through an electro-optically active crystal located near to the electron bunch train (Figure 13.10).

When the electron bunch comes near to the crystal, its electric field induces a change in the birefringence inside the crystal, which alters the polarization state of the laser. Using a polarizer located behind the crystal, this phase effect is transformed into an intensity signal, which is monitored by a CCD camera. At a 45° angle of incidence, the optical laser pulses cross the crystal at different locations depending on their arrival times; hence a mapping of time to space is achieved. The current time resolution is better than 100 fs. The time resolution of pump-and-probe experiments is therefore no longer determined by the timing jitter window (500 fs) but rather by the accuracy with which the actual jitter can be measured for each individual FEL-optical pulse pair.

Other measurements for the arrival time of both the optical and the EUV pulse in the experimental chamber have been developed during the first years of operation at FLASH. These methods are based on obtaining a cross-correlation signal between both pulses. For experiments in the gaseous phase, measurements of two-color above threshold ionization signals have been shown to provide an accuracy of better than 50 fs for the relative delay by

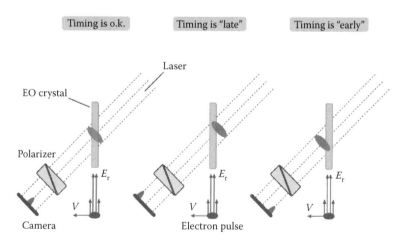

FIGURE 13.10
Schematic representation of the electro-optical sampling system which measures the temporal delay on a shot-to-shot basis between each optical laser pulse and the electron bunch giving rise to a synchronized EUV FEL pulses. (Adapted from Hamburger Synchrotronstrahlungslabor HASYLAB at Deutsches Elektronen-Synchrotron (DESY). http://hasylab.desy.de/facilities/flash/facility_information/laser_and_fel_timing)

comparing the experimental data to theoretical predictions. The measurements rely on the fact that the simultaneous interaction of both pulses leads to additional lines in the electron spectrum or the so-called sidebands. The intensity of these sidebands is very sensitive to the temporal overlap between both pulses, since it is determined directly by the optical field present at the atomic side when photoionization by the EUV photons takes place. A change in the relative position of the very short EUV pulse (10–20 fs) with respect to the much broader (>120 fs) optical pulse is converted into an intensity variation of the sidebands. In a same way, the optical reflectivity of a GaAs surface is strongly changed by the EUV pulse. This variation can also be monitored for each individual shot and provides a relative measure of the temporal delay between the optical and the EUV pulse. These processes are discussed in more detail in the following section.

13.2.5 Experiments on Atoms and Ions

Although FLASH has hosted a broad range of experiments including coherent imaging [42] and EUV laser–matter interactions [43], the focus of this section will be exclusively on atoms, ions, and molecules. For neutral and lowly ionized matter, the key effect of its interaction with EUV radiation is *photoionization* which has been a treasure trove (at times a Pandora's box) of intriguing and occasionally bemusing, but always beautiful, many-electron interactions and quantum interferences. Experiment and theory have worked side by side in a symbiotic drive to deliver complete experiments along with a deep understanding of the role of many-electron effects in photoionization [44]. In all of these studies, the EUV field has been weak and so each atom absorbs one, and only one, photon.

The competing field of ionization of atoms and molecules in intense infrared and visible laser fields has revealed many new dynamical processes such as multiphoton, above-threshold, and optical-field ionization [45] as well as short-wavelength coherent radiation via high harmonic generation (HHG) [46]. Of course, the interesting question arises, what happens if the driving field itself is of very high optical frequency—say it lies in the EUV?

In this regard, the opening up of phase 2 of the EUV Free Electron Laser at DESY-Hamburg, now known as "FLASH," heralds a new era in the study of intense EUV laser–matter interactions and indeed for the photoeffect itself. The EUV output of FLASH possesses unprecedented intensity and coherence, bringing nonlinear optics and spectroscopy into a regime where inner-shell electrons of neutral species and valence electrons of HCIs can mediate the fundamental photon–matter interaction. For the first time, atomistic systems, bathed in intense and coherent ultrahigh frequency electromagnetic fields, can be uniquely studied using photon, photoion, and photoelectron spectroscopy. Experiments on dilute samples and/or species with ultralow photoionization cross-sections, few photon processes in few electron systems, two-color (EUV+IR/EUV+EUV) pump–probe experiments on a femtosecond (even attosecond) time scale are all made possible with the advent of ultrafast EUV pulses and soon with x-ray pulses from the European XFEL [28] and LCLS [29] x-ray FELs.

A Web site on atomic, molecular and optical physics experiments at FLASH has been established [47], and a recent progress report has been published [48]. However, FLASH, as a high flux and intense EUV source, also offers the opportunity to perform laser spectroscopy on resonance transitions of HCIs. To date, one experiment has been performed and we begin this section with a focus on it.

13.2.5.1 EUV Laser-Induced Fluorescence of Li-Like Fe^{23+}

Subtle quantum electrodynamics (QED) effects have provided a rich and fertile ground for the most stringent tests of theory for many years, and laser spectroscopy has proven itself equal to the challenge of the required measurement accuracy. Continuous improvement in allied techniques have placed it at the heart of ultraprecise measurement, for example, the 1S to 2S transition in atomic hydrogen is known to an accuracy of better than 10^{-14} [49]. QED effects scale strongly with nuclear charge Z, for example, the Lamb shift scales as Z^4, and so one- or few-electron HCIs make ideal targets for QED studies as they lead to a significant enhancement of the salient effects, thereby reducing the need for a very narrow bandwidth from the laser system. However, it is clear from Figure 13.11a that resonance transitions for even Li-like ions move rapidly into the EUV with increasing charge state [50] and so FELs operating at fundamental photon energies in the 10–300 eV range are indeed timely. This trend is even more pronounced, of course, for He- and H-like HCIs where only x-ray lasers operating in the keV range will suffice.

Not only does moving further along an isoelectronic sequence provide a wavelength challenge that only EUV and x-ray lasers can meet, but the target ion flux available becomes much diminished in typical HCI sources such as EBITs [51]. Hence additionally, one requires the high flux that only a SASE FEL can provide. As a result, the future of QED research in ions is intimately connected to the development of EUV and x-ray FELs. Figure 13.11b shows the very first laser-induced fluorescence (LIF) measurement on the resonance $2S_{1/2} \rightarrow 2P_{1/2}$ transition of Fe^{23+}, the highest energy, direct laser-excited transition probed to date [50]. The center (nominal) transition energy was 48.613 eV. As the inherent bandwidth of the FEL is quite large (0.5% or 0.25 eV), in order to obtain high spectral resolution, the Heidelberg group used the PG2 monochromator beamline [52] operating with a photon bandwidth of ca. 0.025 eV. They compromised on available spectral resolution in order to obtain good flux of 3×10^{12} photons/s. The impressive measurement accuracy already achieved (22 ppm) equals current best conventional methods and will be significantly improved with the increased FLASH repetition rate in the near future. The relative accuracy already beats theoretical estimates for the dominant QED contributions and points to a bright future for fundamental QED tests on HCI systems at FLASH.

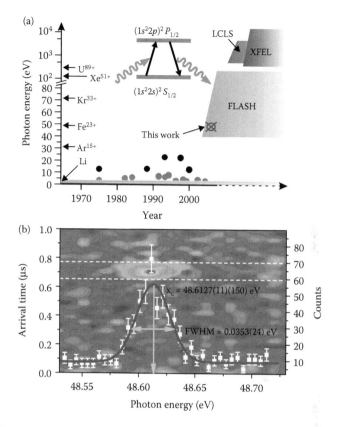

FIGURE 13.11

(a) $2S_{1/2} - 2P_{1/2}$ transition energy scaling along the Li-like isoelectronic sequence. (b) EUV LIF signal arising from direct excitation of the $2s-2p$ transition of Fe^{23+}. (Adapted from S. Epp et al., *Phys. Rev. Lett. 98*, 183001, 2007.)

13.2.5.2 Photoionization in Intense Laser Fields

Photoionization pathways for atoms and molecules in laser fields depend on a combination of ionization potential, laser photon energy, and laser intensity. Theoretical descriptions can be made within a perturbative framework or may require the full solution of the time-dependent Schrödinger equation depending on whether the particular experiment resides within the multiphoton (MPI), tunnel (TI) or the related over the barrier (OBI) ionization regimes [45]. A simple but instructive parameter (γ) that determines in which regime a particular experiment is likely to reside is due to Keldysh [53]:

$$\gamma = \sqrt{\frac{IP}{2U_p}},$$

where

$$U_P \text{ (eV)} = 9.3 \times 10^{-14} I(\text{W cm}^{-2})\lambda^2(\mu\text{m}).$$

Here IP is the ionization potential of the target atom or molecule and U_p the so-called ponderomotive potential (the cycle-averaged kinetic energy of a liberated electron in the laser field). As the laser wavelength moves into the EUV, the ponderomotive potential drops quadratically and so the Keldysh parameter increases linearly. Hence, EUV FELs will tend to favor multiphoton effects that are readily described by perturbation methods (high γ).

One way to look at this result is as follows. When the photon energy lies well above the ionization potential, the tunnel ionization rate will be less than the laser frequency and so the most favored route to ionization becomes multiple photon absorption. From another viewpoint, as the laser photon energy increases, fewer photons are needed to ionize the atoms and so the ionization probability becomes quite high at relatively low intensity (at least compared to visible/IR lasers). Taking the case of He, for a laser intensity of 10^{15} W/cm^2, $\gamma = 0.45$ for a 800-nm Ti–Sapphire laser (tunnel ionization regime), whereas for FLASH operating at 8 nm, $\gamma \sim 45$ (very clearly the MPI regime).

The result is hugely important since it means that EUV FELs bring high-intensity laser–matter interaction back into a space where few photon processes reign and accurate perturbative methods may be used to treat the problem. As one might expect, there has been a surge in theoretical activity since the late 1990s when firm design parameters for EUV/x-ray FELs started to appear. However, pioneering papers on intense VUV–atom interactions due to Lambropoulos and Zoller [54] appeared as early as 1981. The papers on double ionization of He in intense EUV fields by Nikolopoulos and Lambropoulos [55], Ivanov and Kheifets [56] and Feist et al. [57] illustrate nicely the state of play in theory and make a good starting point for a journey through the recent history of this archetypal problem.

13.2.5.3 *Few Photon Multiple Ionization (FPMI) of Rare Gas Atoms*

As expected, this particular domain has received much attention given the promise for exciting new results, which was brought into sharp relief by Phase 1 experiments on comparisons between free atoms and clusters [15]. Sorokin et al. [34] have performed systematic measurements on the photoion spectra of He and Ne under high irradiance, up to 10^{14} W cm^{-2} (Figure 13.12).

They chose two photon energies, namely 38.4 and 42.8 eV, which lie just below and just above the Ne$^+$ ionization threshold, respectively. They did so, as the key objective was to separate sequential ionization (SI) from the so-called direct or nonsequential ionization (NSI) pathways in Ne. As the experimental apparatus could be moved with high precision through the optimum FLASH beam focus and the pulse energies could be measured

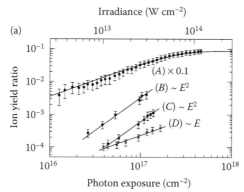

[a] Valid for a photon pulse duration of 25 fs.

FIGURE 13.12
(a) Ion yield ratios as a function of irradiance and photon density at 42.8 eV (A—Ne^{2+}/Ne^{+}), (C—Ne^{3+}/Ne^{2+}), (D—He^{2+}/He^{+}) and for 38.4 eV (B—Ne^{2+}/Ne^{+}). Ratio A implies sequential two-photon double ionization of Ne via Ne^{+} as the production mechanism, B is the result of one-photon ionization of Ne followed by two-photon single ionization of Ne^{+}, while C (Ne^{3+}) requires the additional step of direct two-photon single ionization of Ne^{2+} and D, two-photon double ionization of He. (b) Direct and sequential two-photon ionization cross-sections for He and Ne (and its ions). (Adapted from A. A. Sorokin et al., *Phys. Rev. A 75*, 051402(R), 2007.)

accurately using the gas monitor detector developed by this group [27], they were able to measure ionization yields as a function of EUV intensity (Figure 13.12). Using an analysis similar to that done earlier on N_2 ionization [58], Sorokin et al. were able to derive two-photon ionization cross-section for Ne, Ne^{+}, and Ne^{2+} (Table 1 of Ref. [34]). Their value of $1.6 \pm 0.6 \times 10^{-52}$ cm^4 s at 42.8 eV for two-photon double ionization of He is in good agreement with an earlier estimate of 1×10^{-52} cm^4 s using focused HHG radiation at 41.8 eV [59] and also with a recent theoretical value [60].

Moshammer et al. [61] brought the Heidelberg reaction microscope [62] to FLASH to obtain impressive data and insights into the dynamics of few photon multiple ionization of Ne and Ar. The microscope directs electrons and ions onto horizontally opposed position-sensitive detectors. From the position and time of flight (TOF) for each hit, the full momentum vectors of the photoionization products can be reconstructed. The FLASH photon energy

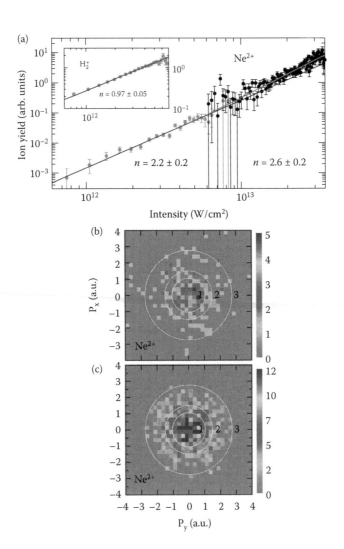

FIGURE 13.13
(a) Ne^{2+} ion yield as a function of intensity at 38 eV showing NSI to dominate at low intensity. Recoil ion momentum distributions (b) for Ne^{2+} at $I < 3 \times 10^{12}$ W cm^{-2} and (c) high intensity ($I > 2 \times 10^{13}$ W cm^{-2}). The z-axis corresponds to the laser propagation direction. The EUV field is polarized along the y-axis. The accumulation of events around the zero momentum point in (b) shows that the two electrons are emitted into opposite directions with roughly equal energies. The appearance of real events in zone 3 (c) indicates the addition of three-photon double ionization (3PDI) process at higher intensities of the FEL. Events are integrated along z-axis. (Adapted from R. Moshammer et al., *Phys. Rev. Lett. 98*, 203001, 2007.)

of 38 eV meant that sequential double ionization of Ne was forbidden. The ion yield scaling with focused intensity (Figure 13.13a) showed that nonsequential two-photon double ionization (NSTPDI), evidenced by the quadratic dependence of the Ne^{2+} yield on intensity, dominates for $I < 6 \times 10^{12}$ W cm^{-2}.

For higher intensities, a second process, three-photon double ionization (3PDI), is superimposed on NSTPDI. Here three photons are involved, one photon for the first step (Ne $\rightarrow$ Ne$^+$) followed by two photons for the second step (Ne$^+$ $\rightarrow$ Ne^{2+}). Recoil ion momentum distributions for Ne^{2+} are also shown in Figure 13.13 for low (b) and high (c) FLASH intensity. For low intensity, the concentration of events around the zero momentum origin leads to the conclusion that electrons are ejected into opposite directions (hemispheres) with roughly equal energies. This result stands in stark contrast to the one-photon double ionization case where such a dynamical decay pattern is suppressed for equal energy sharing.

A most intriguing and as yet incompletely understood result from FLASH has been published recently by Sorokin et al. [63]. Although, conventional wisdom would suggest few-photon few-electron ionization processes to be favored in the interaction of intense EUV radiation with atoms, these authors have observed Xe ions of charge up to 21 times ionized in a very simple experiment shown in Figure 13.14a.

The unfocussed beam from BL3 is directed onto a Mo–Si multilayer mirror operated at normal incidence with a reflectivity of 68% at 13.3 nm ($h\upsilon \sim 93$ eV). This configuration permitted the production of a small focal spot ($\sim$2.5 μm) and correspondingly an unprecedented irradiance of $\sim$8 $\times$ 10^{15} W/cm^2 to be achieved. This intensity is approximately an order of magnitude higher than possible to date with grazing incidence mirrors at BL2. One can see that the dominant charge state increases with EUV intensity rising from 2$^+$ at 2.5 $\times$ 10^{12} W/cm^2 to 11$^+$ at almost 8 $\times$ 10^{15} W/cm^2. The highest charge state observed was 21$^+$. This is to be compared with a maximum charge state of 8$^+$ in the case of 800 nm ($h\upsilon \sim 1.55$ eV) radiation from a Ti–Sapphire laser system operated at the same intensity [64]. These ion yields were well described by simple ADK tunnel ionization theory [65] with SI accounting for the ramp up to eight times ionized species. For the 93 eV case, the situation appears quite different. The ponderomotive potential amounts to much less than 1 eV and the Keldysh parameter lies in double figures, suggesting that the ionization process at FLASH is multiphoton in nature.

Sorokin et al. proposed a detailed accounting process for the observation of HCIs via multistep multiphoton ionization. However, the attainment of Xe^{21+} requires the absorption of some 5 keV of energy per atom or 57 FLASH photons within the 10–20 fs duration of each FLASH pulse. Certainly this result begs the question, what will happen at x-ray FELs where the photon energies will reach up to 10 keV and down to few femtosecond pulse durations?

13.2.5.4 Two-Color Photoionization Experiments on Atoms

As noted earlier, a femtosecond EUV/optical laser facility at DESY was built in the frame of a Framework 5, EU-Research and Technology Development grant by a collaboration involving Orsay (LIXAM), DCU (NCPST), Lund (LLC),

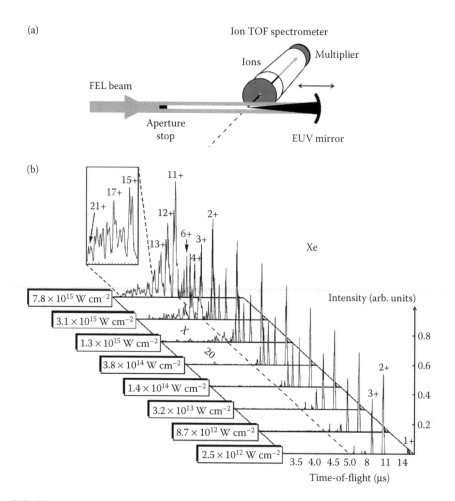

FIGURE 13.14

(a) Experimental setup for the study of the formation of HCIs in the interaction of intense EUV radiation with dilute targets. (b) Xe^{n+} ($1 \leq n \leq 21$) ion yield as a function of intensity at a photon energy of 93 eV. (Adapted from A. A. Sorokin et al., *Phys. Rev. Lett. 99*, 213002, 2007.)

MBI, BESSY, and DESY. The facility provides high-energy pulses of up to 10 mJ at the FEL macropulse repetition frequency of 5 or 10 Hz. It also provides 120 fs/20 μJ 800-nm pulses and 12 ps/400 μJ 523-nm pulses synchronized to the micro- and macropulse timing structure of FLASH. The facility can be used to perform both time-resolved pump–probe experiments and to study two-color coherent processes. We have built onto it a two-color photoionization experiment to determine FLASH harmonic content [32], study interference free sidebands [66], and determine timing jitter between EUV and FEL pulses on a femtosecond timescale (250 fs r.m.s.) [67]. The layout of the experiment [40] is illustrated in Figure 13.15.

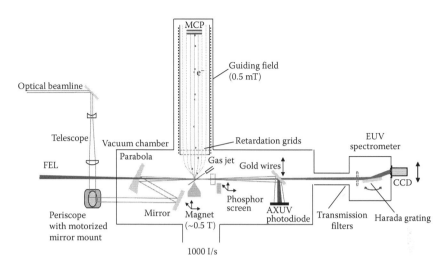

FIGURE 13.15
Detailed layout of the two-color photoelectron experiment at the EUV FEL FLASH. (Adapted from P. Radcliffe et al., *Nucl. Instrum. Meth. Res. A 583*, 516, 2007.)

The converging EUV beam is transported through a hole in the parabolic mirror used to focus the optical laser. The collinear EUV and 800 nm beams are brought to a common focus at the intersection point of a gas jet with the entrance aperture of a TOF electron spectrometer. The TOF is of the magnetic bottle type, which ensures 4π collection efficiency. The output from the TOF is fed to a LeCroy 8600A fast digital oscilloscope. A photoelectron spectrum (PES) trace is obtained and stored for each FLASH shot. The delay between these pulses is adjusted via an optical delay line (increment of 3.3 fs/micron). Obtaining spatial and temporal overlap of the beams is a nontrivial task as the experimental chamber is situated some 70 m from the exit port of the final undulator and 20 m from the optical laser. We use phosphor screens and fast (picosecond) photodiodes to obtain approximate spatial ($<10\,\mu m$) and temporal (20 ps) overlap. We have found the HS3 photodiode from International Radiation Detectors, Inc. (80 ps rise time) to offer the best compromise between responsivity at EUV/NIR wavelengths and time resolution while possessing very good radiation hardness. To obtain the absolute time delays, we must find the overlap point of the EUV and 800 nm pulses not in pico-time, but in femtotime—this point will become clear later when we come to illustrate some of the results obtained to date.

We focus in this section on a *coherent* process which can be driven by intense bichromatic laser fields, namely laser-assisted photoionization (LAP) [68,69]. In LAP experiments, when the EUV and optical laser fields overlap in space and time, a photoelectron is born into the optical dressing field with which it can exchange (i.e., gain or lose) one or more quanta of optical energy. The resultant PES displays sidebands adjacent to the main photoelectron line separated by the optical photon energy. Sideband intensity depends on the EUV photon energy and dressing field strength and so can be used to monitor the relative time delay between overlapping pulses [70].

Many such experiments have been performed using a single optical laser split into two parts. One part is focused into a rare gas jet to generate a comb of EUV high order (odd) harmonics (HOHs), (..., H_{2N-1}, H_{2N+1}, ...). The harmonics are then focused into a second gas jet where they produce a corresponding comb of photoelectron lines separated by twice the optical laser photon energy. The other part of the beam is overlapped in space/time with the EUV harmonics so that photoelectrons sense an intense dressing field with which they can exchange optical photons (hv). The sidebands so induced are disposed either side of the main photoelectron lines at $\pm(2N+1)hv$. In Figure 13.16, a typical LAP spectrum is shown where the vertical axis corresponds to time delay between EUV HOHs and optical pulses, while the horizontal axis corresponds to the photoelectron kinetic energy [6].

Referring to Figure 13.16, the intense vertical bars correspond to the main photoelectron lines produced under direct ionization by the HOH light source ..., H_{2N-1}, H_{2N+1}, ... The weaker interspersed lines are separated by the optical laser photon energy and appear only when the EUV and optical pulses are overlapped in a window of ± 30 fs determined by the optical pulse duration in this particular experiment. The sideband signal is proportional to the degree of overlap of the EUV and optical fields and is hence a measure of the cross-correlation between them. At very high intensities, the sidebands become subject to a strong ponderomotive shift and are even streaked [70]. However, the sidebands are also complicated by interference effects. Imagine that each photoelectron can absorb (or emit) just one optical photon. Then, each sideband results from the superposition of two sidebands—the upper sideband of one main line (say $H_{2N-1} + hv$) lies at the same photon energy as lower sideband of the next main photoline ($H_{2N+1} - hv$).

However, the measured signal is not simply the sum of the two sideband intensities. As the EUV harmonics are derived from a single driving laser pulse, they are phase coherent (in fact, if the phases can be locked with almost zero phase shift, the system approximates a mode locked laser and a train of attosecond pulses results). Hence the free electron waves or main photolines which the HOH produce are phase coherent, as are, in turn, the sidebands since they are formed by exchanging quanta with the dressing laser field. So for any single sideband, we cannot distinguish between the free electron waves which form it and in addition both these waves are phase coherent—in effect we have a classic two-slit experiment and we add the probability amplitudes rather than the free wave intensities. At sufficient dressing field intensities, a free electron can exchange n photons with the dressing field and so we have, in effect, an n-slit experiment.

Since FLASH could produce high flux at a single photon energy, it presented us with the first opportunity to study this effect in the interference free regime.

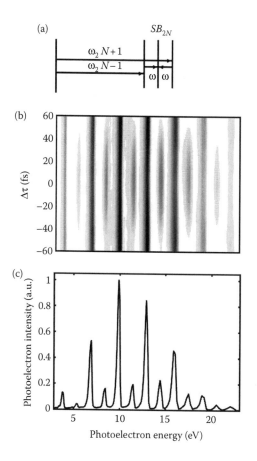

FIGURE 13.16

Coherent photoelectron sidebands formed by dressing photoelectrons generated by a comb of high harmonics with an intense near-infrared laser field. (a) Schematic representation of the formation of a single sideband, labeled SB_{2N}, via stimulated photon emission from a photoelectron formed by harmonic $N + 1$ or photon absorption by an electron created by harmonic $N - 1$, (b) variation of the photoelectron spectrum as a function of the time delay between the extreme ultraviolet harmonic comb and the NIR dressing field. The sidebands appear between the harmonic photolines for time delays close to zero. (c) A lineout of panel (b) at close to 0 fs showing the main photolines and smaller interleaved sidebands. (Adapted from P. O'Keeffe et al., *Phys. Rev. A 69*, 051401(R), 2004.)

In our first experiment [66], we operated with a long (12 ps) 523 nm pulse so that we did not have to worry about relative jitter between the EUV and optical laser pulses. In a more recent experiment [67], we have been able to carry out a comprehensive study of these sidebands with femtosecond pulses. Theoretical studies by Maquet and Taïeb [71] employ either the soft photon approximation (SFA) or the time-dependent Schrödinger equation. The so-called SFA requires the electron kinetic energy to greatly exceed the dressing field photon energy, as is the case in what follows where electrons liberated from rare gas atoms by EUV photons ($hv \sim$ 38–93 eV) are dressed by green ($hv \sim$ 2.4 eV) and near-infrared ($hv \sim$ 1.5 eV) or NIR laser fields.

In Figure 13.17, we show a comparison between theory and experiment for the case of He where the main 1 s^{-1} photoline was produced by FLASH operating at a wavelength of 25.5 nm (48.5 eV). The sidebands are observed to straddle this line at $\pm$1.55 eV (800 nm) energy separation as expected. From a fit of the theoretical trace to the experimental spectrum, we were able to derive EUV and optical laser intensities.

In Figure 13.18, we show a PES for the case of Xe irradiated by superposed 13.7 and 800 nm radiation fields. Sidebands of up to order 4 corresponding to a gain of up to four photons from the dressing laser field (800 nm) are observed. By analyzing Xe sideband numbers and intensities and comparing them with theory, we have been able to determine the relative delay between the FEL and optical laser.

The schematic in Figure 13.19 illustrates the basic principle. In this figure, we show how the number of sidebands generated in the case of Xe for 13.7 nm/800 nm irradiation depends on the relative delay between the two pulses. Hence simple sideband counting may be used to measure the relative delay between the pulses.

In Figure 13.20, we show the absolute jitter between the EUV and optical laser which is $\sim$500 fs (full-width at half-maximum (FWHM)). The trace is obtained by sweeping the optical time delay and measuring the average sideband intensity at each delay.

Cunovic et al. have developed a photoelectron imaging system to directly measure and display pulse and timing jitter using the LAP technique—the so-called time–space mapping [72]. The experimental scheme and cross-correlation trace are shown in Figure 13.21. Two other schemes utilizing space

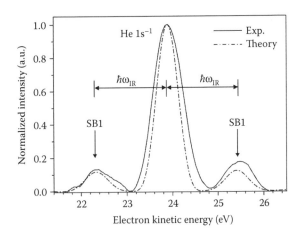

FIGURE 13.17
Comparison between measured and computed photoelectron sideband spectra for He. (Adapted from P. Radcliffe et al., *Appl. Phys. Lett.* 90, 131108, 2007.)

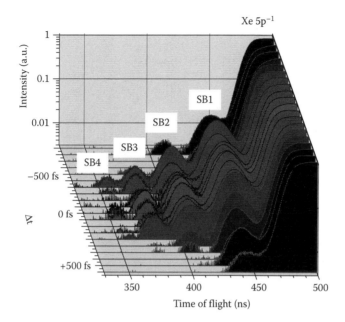

FIGURE 13.18
Multiple sidebands in Xe formed by 13.7 nm EUV and 800 nm laser fields. (Adapted from
P. Radcliffe et al., *Nucl. Instrum. Meth. Res. A 583*, 516, 2007.)

time mapping of FEL pulse modulated optical reflectivity have also been
demonstrated [73,74].

Coming back to our own work, it is instructive to look at the simple analytic
expression for sideband intensity derived by Maquet and Taïeb [71] in the
so-called SFA [75]:

$$B(\alpha_0 k_n) = \int_0^\pi \sin\theta \, d\theta \frac{d\sigma}{d\theta} J_n^2(\alpha_0 k_n \cos\theta),$$

where α_0 depends on the optical laser field strength and k_n is the momentum transferred to the nth band photoelectron. One can see that the total
sideband intensity depends on the differential cross-section, where θ is the
angle between the EUV and optical fields. Hence the sidebands should exhibit
dichroism, that is, sensitivity to the polarization direction of the optical field
relative to the EUV field. Figure 13.22a illustrates the process schematically,
while we show the results of He in Figure 13.22b [76].

The two-color process in Figure 13.22a we refer to as two-color above
threshold ionization of TC-ATI. At low optical fields, we see just one sideband disposed on either side of the diminished main photoline. For higher
optical fields, we see multiple sidebands. The orientation of the EUV and
optical field directions are shown schematically. Figure 13.22b shows a series
of photoelectron spectra of He taken at different EUV-optical field vector

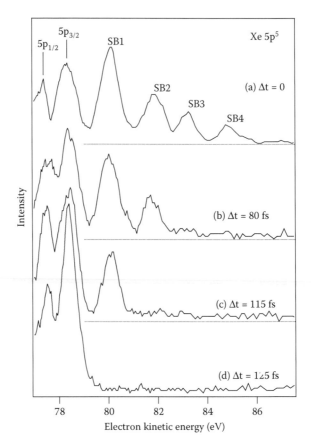

FIGURE 13.19
Time delay analysis derived from a study of Xe sidebands. (Adapted from P. Radcliffe et al., *Nucl. Instrum. Meth. Res. A 583,* 516, 2007.)

orientations where a wave structure, reminiscent of Malus' law, can be seen. This is clear evidence of atomic dichroism, which has already been seen, in two-color resonant excitation at synchrotrons. However, this is the first observation in the nonresonant case and was made possible only due to the high flux available from FLASH. The dichroic signal is illustrated more clearly in Figure 13.23, which shows line outs from two-color ATI photoelectron spectra of He compared with theory.

Full details of the analysis are available in Meyer et al. [76]. In summary, the comparisons show that in the low-field case, the ratio of s to d electron emission is ~3 with the SFA yielding a value of 3.2, while time-dependent second-order perturbation theory gives 2.93. The SFA is remarkably accurate at all optical field levels as can be seen from Figure 13.23 and indicates a single active electron approach is sufficient to reproduce TC-ATI spectra across a wide range of experimental conditions.

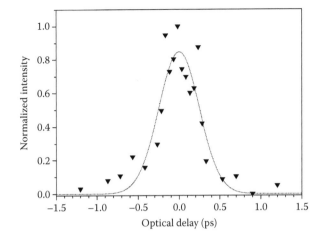

FIGURE 13.20
EUV-optical pulse jitter obtained by recording sideband at various nominal delays. (Adapted from P. Radcliffe et al., *Appl. Phys. Lett. 90*, 131108, 2007.)

There are many more exciting experiments which remain to be done such as EUV laser-driven autoionizing states [54], optical laser coupling of two (or more) autoionizing states [77], and indeed a two-color process where both EUV and optical fields are coupled together to drive a single autoionizing state into resonance. These experiments, possible only at intense EUV FELs

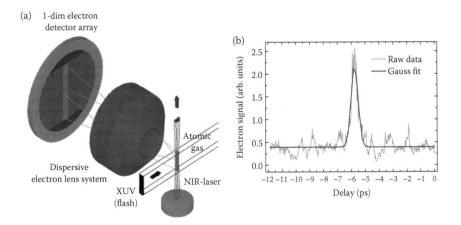

FIGURE 13.21
(a) Experimental setup for laser-assisted photoelectron imaging and EUV-optical pulse jitter measurement. Photoelectrons from a directed krypton gas jet produced at different points along the optical laser focal waist, corresponding to different delay times, are imaged to different points on an MCP. (b) Trace obtained showing the EUV-optical pulse delay distribution (temporal jitter). (Adapted from S. Cunovic et al., *Appl. Phys. Lett. 90*, 121112, 2007.)

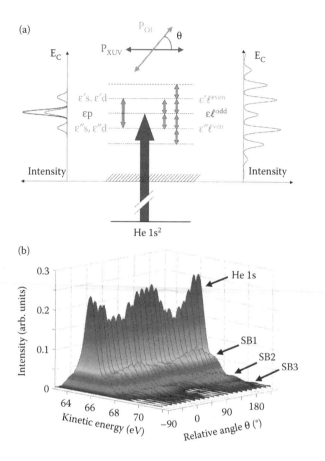

FIGURE 13.22

(a) Two-color above threshold ionization process for low (left) and high (right) optical laser intensities showing redistribution of electrons from the main line in the center to the sidebands. P_{XUV} refers to the polarization direction of the FEL beam while POL refers to the polarization direction of the optical laser. (b) Photoelectron spectra of He as a function of the angle between the EUV and optical laser field polarization vectors. He 1s refers to the main photoelectron line while SB1, SB2, and SB3 refer to the first, second, and third sidebands respectively. (Adapted from M. Meyer et al., *Phys. Rev. Lett.* 101, 193002, 2008.)

will open up photoionization to new parameter spaces and challenge the state-of-the-art in experiment in addition to testing the limits of current theory.

13.3 Future Prospects and XFEL

The success of the TTF project and, in particular, of the FLASH machine which is operating as planned have triggered also the extension of the project at

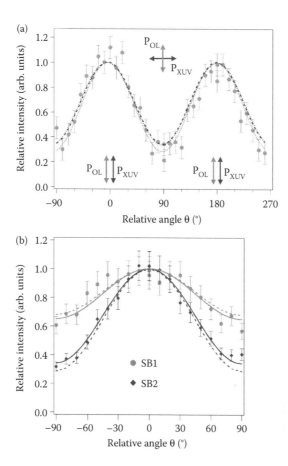

FIGURE 13.23

(a) Polarization dependence of the first sideband of He for low optical laser intensity (8×10^{10} W cm^{-2}). The solid curve is a fit to the experimental data. The theoretical curves (dashed and dotted lines) obtained from the SFA and time-dependent second-order perturbation theory are almost identical. (b) Polarization dependence of the first and second sidebands of He for high optical laser intensity (6×10^{11} W cm^{-2}). The solid curves are fits to the experimental data. The theoretical (dashed) curves are obtained from the SFA only. (Adapted from M. Meyer et al. *Phys. Rev. Lett.* 101, 193002, 2008.)

Hamburg to even shorter wavelengths. Worldwide there are presently three different projects scheduled for completion within the next 5 years, Spring-8 X-FEL in Japan [78], LCLS in the USA [29], and the European XFEL in Germany [28]. The XFEL in Hamburg is scheduled to be operational in 2015 and will deliver intense femtosecond-photon pulses in the wavelength region down to 0.1 nm (~13 keV). In order to achieve lasing at these short wavelengths, an installation of much larger dimension has to be constructed (Figure 13.24).

In fact, the overall length of the XFEL project is about 3.4 km. The purpose of the facility is to generate extremely intense (peak brilliance ~10^{33} photons/s/mm^2/mrad^2/0.1% BW), ultrashort (<100 fs) pulses of spatially

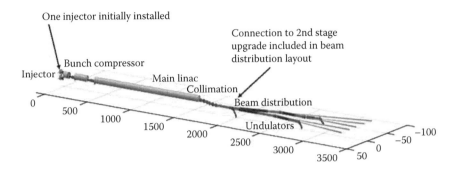

FIGURE 13.24
3D sketch of the accelerator complex for the XFEL in Hamburg. (Adapted from European Free Electron Laser, Hamburg, Germany. http://xfel. desy.de/tdr/tdr)

coherent x-rays, and to exploit them for revolutionary scientific experiments in a variety of disciplines spanning physics, chemistry, materials science, and biology. The basic process for generating the x-ray pulses remains SASE, whereby electron bunches are accelerated to high energy (up to 20 GeV) through a superconducting linear accelerator, and conveyed to long (up to ~200 m) undulators, where the coherent x-ray pulses are generated.

In the first phase, three photon beamlines behind SASE undulators are planned for the different applications which can be divided for the soft x-ray regime into investigations of nonlinear processes, high-field effects in small quantum systems, and coherent scattering experiments for structure determination of nanosystems and biomolecules. In the hard x-ray regime, priority will be given to femtosecond diffraction experiments, investigations of high-energy density matter, ultrafast coherent diffraction imaging of single

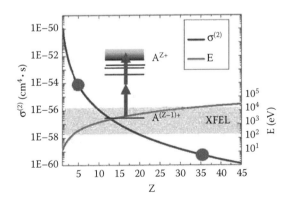

FIGURE 13.25
Cross-section and photon energies for the two-photon ionization process of H-like ions. (Adapted from European Free Electron Laser, Hamburg, Germany. http://xfel.desy.de/tdr/tdr)

particles, clusters, and biomolecules as well as structure determinations of nanodevices and dynamics at the nanoscale. The three SASE beamlines can be classified with respect to their characteristics. SASE 1 will deliver FEL radiation around 12 keV with high coherence spontaneous radiation, SASE 2 FEL radiation in the photon energy range between 3 and 12 keV and high time-resolution spontaneous radiation, and finally SASE 3 FEL radiation in the range 0.25 and 3 keV optimized for high flux and high-resolution applications.

From this new user facility, novel results of fundamental importance can be expected in materials physics, plasma physics, planetary science and astrophysics, chemistry, structural biology, and biochemistry, with significant impact on technologies such as nuclear fusion, catalysis, combustion (and their environmental aspects), as well as on biomedical and pharmaceutical technologies. As a final illustration of studies on highly charged samples, we mention the two-photon ionization of hydrogen-like ions (Figure 13.25) [79].

With increasing atomic number, the photon energies move rapidly into the keV range where only x-ray FELs can deliver the photon flux necessary to drive two-photon processes.

Acknowledgments

We sincerely thank many collaborators for their contributions to our two-color photoionization experiments at FLASH, especially Stefan Düsterer, Wen Bin Li, Paul Radcliffe, Harald Redlin, Mikhail Yurkov, Josef Feldhaus (all DESY), Ciaran Lewis (QUB), Denis Cubaynes (LIXAM), Eugene Kennedy (DCU), and the many postgraduate and postdoctoral members of the LIXAM and DCU teams who have carried out the experiments with us. Our work could not be possible without the support of the highly professional machine and photon diagnostics groups at FLASH. The DCU team is supported by Science Foundation Ireland (SFI Grant No. 07/IN.1/I1771), IRCSET, EO COST Action MP0601 and the HEA PRTLI IV Progammes of the Second National Development Programme (NDP2). The Orsay-DCU collaboration is supported by the Ulysses France-Irl. Exchange programme. The assistance of Paddy Hayden with the figures is greatly appreciated.

References

1. T. H. Maiman, *Nature* 187, 493, 1960.
2. H. Wiedemann, *Synchrotron Radiation*. Berlin and Heidelberg: Springer, 2002.
3. H. Daido, *Rep. Prog. Phys.* 65, 1513, 2002.

4. F. Krausz and T. Brabec, *Rev. Mod. Phys.* 72, 545, 2000.
5. B. Dromey et al., *Nat. Phys.* 2, 456, 2006.
6. P. O'Keeffe et al., *Phys. Rev. A* 69, 051401(R), 2004.
7. M. Hentschel et al., *Nature* 414, 509, 2001.
8. Y. Wang et al., *Nat. Photon.* 2, 94, 2008.
9. J. T. Costello, *Nat. Photon.* 2, 67, 2008.
10. S. V. Milton et al., *Science* 292, 2037, 2001.
11. J. Madey, *J. Appl. Phys.* 42, 1906, 1971.
12. Berliner Elektronenspeicherring-Gesellschaft für Synchrotronstrahlung (BESSY). http://www.bessy.de/publicRelations/publications/files/sc.pdf
13. J. Andruszkow et al., *Phys. Rev. Lett.* 85, 3825, 2000.
14. V. Ayvazyan et al., *Phys. Rev. Lett.* 88, 104802, 2002.
15. H. Wabnitz et al., *Nature* 420, 482, 2002.
16. W. Ackermann et al., *Nat. Photon.* 1, 336, 2007.
17. European Free Electron Laser, Hamburg, Germany. http://cerncourier.com/cws/article/cern/29710
18. J. Feldhaus et al., *J. Phys. B: At. Mol. Opt. Phys* 38, S799, 2005.
19. V. Ayvazyan et al., *Eur. Phys. J. D* 37, 297, 2006.
20. Free Electron Laser in Hamburg, Germany. http://flash.desy.de
21. D. Strickland and G. Mourou, *Opt. Commun.* 56, 219, 1985.
22. L. Catani, A. Cianchi, G. Di Pirro, and K. Honkavaara, *Rev. Sci. Instrum.* 76, 073303, 2005.
23. G. Schmidt, U. Hahn, M. Meschkat, and F. Ridoutt, *Nucl. Inst. Meth. A* 475, 545, 2001.
24. K. Tiedkte et al., *AIP Conference Proceedings*, 705, 588, 2004.
25. P. Nicolosi et al., *J. Electron Spectrosc. Relat. Phenom.* 144–147, 1055, 2005.
26. A. Bytchkov et al., *Nucl. Instrum. Meth. A* 528, 254, 2004.
27. M. Richter et al., *Appl. Phys. Lett.* 83, 2970, 2003.
28. European Free Electron Laser, Hamburg, Germany. http://www.xfel.eu
29. The Linear Coherent Light Source at SLAC, Stanford, USA. http://lcls.slac.stanford.edu
30. E. L. Saldin, E. A. Schneidmiller, and M. V. Yurkov, *Nucl. Instrum. Meth. A* 429, 233, 1999.
31. S. Duesterer et al., *Opt. Lett.* 31, 1750, 2006.
32. M. Wellhofer et al., *J. Inst.* 3, P02003, 2008.
33. N. Nakano, H. Kuroda, T. Kita, and T. Harada, *Appl. Opt.* 23, 2386, 1984.
34. A. A. Sorokin et al., *Phys. Rev. A* 75, 051402(R), 2007.
35. E. L. Saldin, E. A. Schneidmiller, and M. V. Yurkov, *Opt. Commun.* 148, 383, 1998.
36. E. L. Saldin, E. A. Schneidmiller, and M. V. Yurkov, *Opt. Commun.* 239, 161, 2004.
37. E. L. Saldin, E. A. Schneidmiller, and M. V. Yurkov, *Nucl. Instrum. Meth. A* 507, 106, 2003.
38. R. Ischebeck et al., *Nucl. Instrum. Meth. A* 507, 175, 2003.
39. G. Lambert et al., *Nat. Phys.* 4, 296, 2008.
40. P. Radcliffe et al., *Nucl. Instrum. Meth. Res. A* 583, 516, 2007.
41. Hamburger Synchrotronstrahlungslabor HASYLAB at Deutsches Elektronen-Synchrotron (DESY). http://hasylab.desy.de/facilities/flash/facility_ information/laser_and_fel_timing
42. H. N. Chapman et al., *Nat. Phys.* 2, 839, 2006.
43. J. Chalupsky et al., *Opt. Express* 15, 6036, 2007.

44. U. Becker and D. A. Shirley (Eds.), *VUV and Soft X-ray Photoionization*. New York: Plenum Press, 1996.
45. M. Protopapas et al., *Rep. Prog. Phys.* 60, 389, 1997.
46. P. Agostini and L. F. DiMauro, *Rep. Prog. Phys.* 67, 813, 2004.
47. Hamburger Synchrotronstrahlungslabor HASYLAB at Deutsches Elektronen-Synchrotron (DESY). http://hasylab.desy.de/science/user_collaborations/amopflash
48. J. T. Costello, *J. Phys. Conf. Ser.* 88, 012057, 2007.
49. M. Niering et al., *Phys. Rev. Lett.* 84, 5496, 2000.
50. S. Epp et al., *Phys. Rev. Lett.* 98, 183001, 2007.
51. W. Hu et al., *Can. J. Phys.* 86, 321, 2008.
52. M. Martins et al., *Rev. Sci. Instrum.* 77, 155108, 2006.
53. L. V. Keldysh, *Sov. Phys. JETP* 20, 1307, 1965.
54. P. Lambropoulos and P. Zoller, *Phys. Rev. A* 71, 379, 1981.
55. L. A. A. Nikolopoulos and P. Lambropoulos, *J. Phys. B: At. Mol. Opt. Phys* 40, 1347, 2007.
56. I. A. Ivanov and A. S. Kheifets, *Phys. Rev. A* 75, 033411, 2007.
57. J. Feist et al., *Phys. Rev. A* 77, 043420, 2008.
58. A. A. Sorokin et al., *J. Phys. B: At. Mol. Opt. Phys* 39, L299, 2006.
59. H. Hasegawa et al., *Phys. Rev. A* 71, 023407, 2005.
60. E. Foumouo et al., *Phys. Rev. A* 74, 063409, 2006.
61. R. Moshammer et al., *Phys. Rev. Lett.* 98, 203001, 2007.
62. J. Ullrich et al., *Rep. Prog. Phys.* 66, 1463, 2003.
63. A. A. Sorokin et al., *Phys. Rev. Lett.* 99, 213002, 2007.
64. K. Yamakawa et al., *Phys. Rev. Lett.* 92, 123001, 2004.
65. M. V. Ammosov, N. B. Delone, and V. P. Krainov, *Zh. Eksperiment. I Teoretich. Fiz.* 91, 2008, 1986.
66. M. Meyer et al., *Phys. Rev. A* 74, 011401(R), 2006.
67. P. Radcliffe et al., *Appl. Phys. Lett.* 90, 131108, 2007.
68. T. E. Glover et al., *Phys. Rev. Lett.* 76, 2468, 1996.
69. J. M. Schins et al., *Phys. Rev. A* 52, 1272, 1995.
70. E. S. Toma et al., *Phys. Phys. A* 62, 0618015(R), 2000.
71. A. Maquet and R. Taieb, *J. Mod. Opt.* 54, 1847, 2007.
72. S. Cunovic et al., *Appl. Phys. Lett.* 90, 121112, 2007.
73. T. Maltezopoulos et al., *New J. Phys* 10, 033026, 2008.
74. C. Gahl et al., *Nature Photonics* 2, 165, 2008.
75. N. M. Kroll and K. M. Watson, *Phys. Rev. A* 8, 804, 1973.
76. M. Meyer et al., *Phys. Rev. Lett.* 101, 193002, 2008.
77. S. I. Themelis, P. Lambropoulos, and M. Meyer, *J. Phys. B: At. Mol. Opt. Phys.* 37, 4281, 2004.
78. Spring-8 Compact SASE Source, Japan. http://www-xfel.spring8.or.jp
79. European Free Electron Laser, Hamburg, Germany. http://xfel.desy.de/tdr/tdr

14

QED Theory of Highly Charged Ions

O. Yu. Andreev and L. N. Labzowsky

CONTENTS

14.1 Introduction

During the past 30 years or so, the quantum electrodynamics (QED) theory of atoms and ions became a routine procedure for the description of the properties of electronic shells in atomic systems. This can be explained by the outstanding progress in the experimental investigations of two kinds of the "new" atomic systems: highly charged ions (HCIs) and superheavy atoms. While for the description of the superheavy atoms, that is, the systems containing more than 100 electrons, the sophisticated methods of the relativistic many-body theory (such as the superposition of configurations or the coupled clusters) are of the primary importance, the theory of ions with few electrons and the nuclear charge $Z \geq 50$ can be developed on the basis of QED perturbation theory. The formulation of the QED perturbation theory for the tightly bound electrons is not trivial, and unlike the QED theory of the loosely bound electrons in the light atoms cannot be essentially similar to the free electron QED. During the last decades, a number of general methods based on the S-matrix or the Green function were applied for the QED description of the tightly bound electrons in HCIs: adiabatic S-matrix approach [1], two-time Green's function method (TTGFM) [2], covariant evolution operator (CEO) approach [3], and the line profile approach (LPA) [4]. In this chapter, we will describe the recent status of the QED theory of HCIs. As a theoretical tool, we will use exclusively S-matrix methods [1,4], but the practical results obtained by all the methods will be discussed and compared.

14.2 S-Matrix Formulation of the Bound-State QED

Within the QED approach, we consider HCIs as a system of the tightly bound electrons, moving in the strong field of a nucleus (Furry picture of QED) and interacting with each other via the exchange of the photons. The Hamiltonian of this system in the second quantization representation looks like

$$\hat{H} = \hat{H}_0 + \hat{H}_{\text{int}}, \tag{14.1}$$

$$\hat{H}_0 = \hat{H}_0^e + \hat{H}_0^\gamma, \tag{14.2}$$

$$\hat{H}_0^e = \sum_{E_s > 0} \hat{n}_s^{(+)} E_s + \sum_{E_s < 0} \hat{n}_s^{(-)} E_s, \tag{14.3}$$

$$\hat{H}_0^\gamma = \sum_{k,\lambda} \omega \, \hat{n}_{k,\lambda}, \tag{14.4}$$

$$\hat{H}_{\text{int}} = \int d^3 r \, \hat{j}^\mu(x) \hat{A}_\mu(x), \tag{14.5}$$

$$\hat{j}^\mu(x) = e \, \hat{\bar{\psi}}(x) \gamma^\mu \, \hat{\psi}(x), \tag{14.6}$$

where $\gamma^\mu = (\beta, \beta\alpha)$ are the Dirac matrices. The charge of an electron is $e = -|e|$. We use the relativistic units ($\hbar = c = 1$, $\alpha = e^2/\hbar c$) and a pseudo-euclidean metric in 4-space with the sign convention for the metric tensor: $(g_{\mu\nu}) = \text{diag}(1, -1, -1, -1)$. Einstein sum convention is implied. The space and momentum 4-vectors are $x^\mu = (t, r)$ and $k^\mu = (\omega, k)$, respectively. In Equations 14.1 through 14.5, $\hat{H}_0$ is the Hamiltonian of the noninteracting fields and $\hat{H}_{\text{int}}$ the interaction Hamiltonian. The operators $\hat{n}_s^{(\pm)}$ are the operators of the number of the electrons in the states with the positive (negative) energies

$$\hat{n}_s^{(+)} = \hat{a}_s^{(+)} \hat{a}_s, \tag{14.7}$$

$$\hat{n}_s^{(-)} = \hat{b}_s^{(+)} \hat{b}_s, \tag{14.8}$$

where $\hat{a}_s^{(+)}$, $\hat{a}_s$ and $\hat{b}_s^{(+)}$, $\hat{b}_s$ are the creation and annihilation operators for the electrons in the positive-energy and negative-energy states, respectively.

These operators satisfy the anticommutator relations. We denote by E_s, the electron energy in the state s.

The field operators $\hat{\psi}(x)$ and $\hat{\psi}^+(x)$ for the electron–positron quantized field are

$$\hat{\psi}(x) = \sum_{E_s>0} \hat{a}_s \psi_s(x) + \sum_{E_s<0} \hat{b}_s^+ \psi_s(x), \tag{14.9}$$

$$\hat{\bar{\psi}}(x) = \sum_{E_s>0} \hat{a}_s^+ \bar{\psi}_s(x) + \sum_{E_s<0} \hat{b}_s \bar{\psi}_s(x), \tag{14.10}$$

where $\bar{\psi} = \psi^+ \beta$. The one-electron functions $\psi(x)$ in Equations 14.9 and 14.10 satisfy the Dirac equation for electron in the field of the nucleus:

$$\hat{h}^D(r)\psi_s(r) = E_s \psi_s(r), \tag{14.11}$$

$$\hat{h}^D(r) = \alpha \hat{p} + \beta m + eV_c(r), \tag{14.12}$$

where $\hat{p}$ is the momentum operator for the electron and $V_c(r)$ is the Coulomb field of the nucleus (point-like or extended). The employment of the functions $\psi_s(r)$, instead of the plain waves, describing the free electrons means the introduction of the Furry picture of QED. The current of the electromagnetic field is defined by Equation 14.6.

In Equation 14.6, $\omega = |k|$ means the photon frequency, k is the wave vector, $\lambda = (0,1,2,3)$ is the photon polarization, $\lambda = 0$ corresponds to the scalar (Coulomb) photons, and $\hat{n}_{k,\lambda}$ are the operators of the number of photons given by

$$\hat{n}_{k,\lambda} = \hat{C}_{k,\lambda}^+ \hat{C}_{k,\lambda}, \tag{14.13}$$

where $\hat{C}_{k,\lambda}^+$ and $\hat{C}_{k,\lambda}$ are the operators of the photon creation and annihilation, respectively. These operators satisfy the commutation relations.

The field operator for the electromagnetic field (the quantized electromagnetic potential) $\hat{A}_\mu(x)$ looks like

$$\hat{A}_\mu(x) = \sum_{k,\lambda} \left(\frac{2\pi}{V\omega}\right)^{1/2} e_\mu^{(\lambda)}(\hat{C}_{k,\lambda}\, e^{-ik^\nu x_\nu} + \hat{C}_{k,\lambda}^+\, e^{ik^\nu x_\nu}), \tag{14.14}$$

where $e_\mu^{(\lambda)}$ are the components of the polarization 4-vector and V is the normalization volume.

The operator $\hat{H}$ in the second quantization representation is acting on the state vector $\Phi(t)$. Then, defining the energy of the system of noninteracting electrons as

$$E^{(0)} = \langle \Phi(t)|\hat{H}_0|\Phi(t)\rangle, \tag{14.15}$$

we find for the N-electron atom

$$E^{(0)} = \sum_{s=1}^{N} E_s. \tag{14.16}$$

Unlike the quantum mechanics, the Schrödinger equation in QED can be written only in the second quantization representation, since this equation should also describe the processes of the creation and annihilation of the particles. The role of the wave function plays the state vector $\Phi(t)$ that describes how many particles are present in which states at the time moment t. It is convenient to use the interaction representation where both operators and wave functions are time-dependent. Then the Schrödinger equation looks like

$$i\frac{\partial}{\partial t}\Phi(t) = \hat{H}_{int}(t)\Phi(t), \tag{14.17}$$

where $\hat{H}_{int}$ is the Hamiltonian (Equation 14.5) in the interaction representation. Introducing the evolution operator $\hat{S}(t, t_0)$ by the relation

$$\Phi(t) = \hat{S}(t, t_0)\Phi(t_0), \tag{14.18}$$

one obtains from Equations 14.17 and 14.18, an equation

$$i\frac{\partial}{\partial t}\hat{S}(t, t_0) = \hat{H}_{int}(t)\hat{S}(t, t_0), \tag{14.19}$$

which has to be solved with the initial condition

$$\hat{S}(t_0, t_0) = 1. \tag{14.20}$$

Iterative solution of Equation 14.19 leads to the perturbation theory expansion

$$\hat{S}(t, t_0) = 1 + \sum_{l=1}^{\infty} \hat{S}^{(l)}(t, t_0), \tag{14.21}$$

$$\hat{S}^{(l)}(t, t_0) = (-i)^l \int_{t_0}^{t} dt_1 \int_{t_0}^{t_1} dt_2 \dots \int_{t_0}^{t_{l-1}} dt_l \, \hat{H}_{int}(t_1), \dots, \hat{H}_{int}(t_l). \tag{14.22}$$

Then Equation 14.22 can be presented in a more symmetric form with the use of the chronological product (T-product) of the operators

$$T(\hat{A}(t_1)\hat{B}(t_2)) = \begin{cases} \hat{A}(t_1)\hat{B}(t_2), & t_1 > t_2, \\ \pm\hat{B}(t_2)\hat{A}(t_1), & t_1 < t_2. \end{cases} \tag{14.23}$$

The positive sign in Equation 14.23 corresponds to the boson operators and the negative sign corresponds to the fermion operators. Then

$$\hat{S}^{(l)}(t, t_0) = \frac{(-i)^l}{l!} \int_{t_0}^{t} dt_1 \int_{t_0}^{t} dt_2 \dots \int_{t_0}^{t} dt_l \, T(\hat{H}_{\text{int}}(t_1), \dots, \hat{H}_{\text{int}}(t_l)). \quad (14.24)$$

The introduction of the T-product in Equation 14.23 is due to the fact that the operators $\hat{H}_{\text{int}}(t)$ corresponding to the different time moments do not commute, in principle.

The evolution operator on the time interval $t = \pm\infty$ is called S-matrix:

$$\Phi(\infty) = \hat{S}(\infty, -\infty)\Phi(-\infty). \quad (14.25)$$

In the free-electron QED, it is assumed that at the time moments $t = \pm\infty$, the particles (electrons and photons) do not interact with each other. Then, the initial and final state vectors correspond to the noninteracting particles. Let the initial state be $\Phi(-\infty) = \Phi_a^0$. Then, according to Equation 14.25:

$$\Phi(\infty) = \hat{S}\Phi(-\infty) = \sum_b \langle b|\hat{S}|a\rangle \Phi_b^0, \quad (14.26)$$

where the short-hand notation $\hat{S} \equiv \hat{S}(\infty, -\infty)$ and $\langle b|\hat{S}|a\rangle = \langle \Phi_b^0|\hat{S}|\Phi_a^0\rangle$ are used. The matrix elements (transition amplitudes) fully describe any collision (scattering) process in the free-electron QED. To some extent, this also concerns the processes with the bound electrons. However, the evaluation of such important atomic characteristics as the energy levels (more exactly, the shifts of the energy levels due to the interactions with photons or other electrons) and transition probabilities requires special attention (see below).

The evaluation of the S-matrix elements is greatly simplified with the use of the Feynman graph techniques which is based on the Wick theorem. Define the contraction $\overline{\hat{A}\hat{B}}$ of the two operators $\hat{A}$ and $\hat{B}$ as

$$\overline{\hat{A}\hat{B}} = T(\hat{A}\hat{B}) - N(\hat{A}\hat{B}), \quad (14.27)$$

where N denotes the normal product: all the creation operators in $\hat{A}$ and $\hat{B}$ should be placed to the left of all the annihilation operators. Then, the Wick theorem claims that the T-product in Equation 14.24 can be presented as the sum of the N products with all the possible contractions.

The contraction of the two-electron operators $\hat{\psi}(x_1)$ and $\hat{\psi}^+(x_1)$ is also called the electron propagator $S(x_1, x_2)$:

$$\overline{\hat{\psi}(x_1)\hat{\psi}(x_2)} \equiv S(x_1, x_2). \quad (14.28)$$

The photon propagator is defined in a similar way:

$$\overline{\hat{A}_\mu(x_1)\hat{A}_\nu(x_2)} \equiv D_{\mu\nu}(x_1, x_2). \tag{14.29}$$

The propagators do not contain the creation and annihilation operators and are functions of the space and time variables. The photon propagator in the bound-electron QED is the same as in the free-electron QED:

$$D_{\mu\nu}(x_1, x_2) = \frac{i}{2\pi} \frac{\delta_{\mu\nu}}{r_{12}} \int_{-\infty}^{\infty} d\omega \, e^{i|\omega|r_{12} - i\omega(t_1 - t_2)}, \tag{14.30}$$

where $r_{12} = |\boldsymbol{r}_1 - \boldsymbol{r}_2|$.

In Equation 14.30, the photon propagator is given in the Feynman (covariant) gage. In the theory of atoms and HCIs, the noncovariant Coulomb gage is also used often. Within this gage, the propagator for the Coulomb (scalar) photons is

$$D^c_{\mu\nu}(x_1, x_2) = \frac{i}{r_{12}} \delta(t_1 - t_2)\delta_{\mu,0}\delta_{\nu,0} \tag{14.31}$$

and the propagator for the transverse photons is

$$D^t_{\mu\nu}(x_1, x_2) = \frac{1}{2\pi i}\left[\frac{\delta_{\mu\nu}}{r_{12}}\int_{-\infty}^{\infty} d\omega \, e^{i|\omega|r_{12} - i\omega(t_1 - t_2)} - \boldsymbol{\nabla}_{1\mu}\boldsymbol{\nabla}_{2\nu}\frac{1}{r_{12}}\right.$$
$$\left.\times \int_{-\infty}^{\infty} d\omega \, e^{-i\omega(t_1 - t_2)}\frac{e^{i|\omega|r_{12}} - 1}{\omega^2}\right](1 - \delta_{\mu,0})(1 - \delta_{\nu,0}). \tag{14.32}$$

The electron propagator in the bound-state QED is different from the electron propagator in the free-electron QED:

$$S(x_1, x_2) = \frac{i}{2\pi}\int_{-\infty}^{\infty} d\omega \, e^{-i\omega(t_1 - t_2)} \sum_n \frac{\psi_n(\boldsymbol{r}_1)\bar{\psi}_n(\boldsymbol{r}_2)}{\omega - E_n(1 - i0)}. \tag{14.33}$$

In Equation 14.33 the summation runs over the entire spectrum of the Dirac equation 14.11, including both positive-energy and negative-energy states.

The employment of the Feynman graph techniques with the electron propagators (Equation) is the first step toward the construction of the bound-state QED. The numerous examples of the bound-electron Feynman graphs will be given below.

The next step is the choice of a method for the evaluation of the energy level shift, assuming that the zeroth-order atomic energy is given by the Equation 14.16. The methods for the evaluation of this shift are different within adiabatic S-matrix approach, TTGFM, CEO approaches, and the LPA. In the following few sections, we will describe the energy shift in the adiabatic S-matrix approach.

14.3 Adiabatic *S*-Matrix Theory

The formula for the evaluation of the energy shift on the basis of the adiabatic *S*-matrix approach was derived by Gell-Mann and Low [5]. The main feature of the Gell-Mann and Low formalism is the adiabatic switching of the interaction. This switching is achieved by changing the interaction Hamiltonian $\hat{H}_{int}(t)$ to the operator

$$\hat{H}_{int}(t, \gamma) = e^{-\gamma|t|}\hat{H}_{int}(t). \tag{14.34}$$

where $\gamma > 0$ is the adiabatic parameter. Then at the time moments $t = \pm$, the interaction is switched off and at $t = 0$ it is fully switched on. Using the interaction Hamiltonian (Equation 14.34), one can perform the calculations similarly to the free-electron QED and finally put $\gamma = 0$ that means the restoration of the interaction for the whole time interval. The formula for the shift of the atomic energy level *a*, given by Gell-Mann and Low, looks like

$$\Delta E_a = \lim_{\gamma \to 0} i\gamma e \frac{(\partial/\partial e)\langle \Phi_a^0|\hat{S}_\gamma(0, -\infty)|\Phi_a^0\rangle}{\langle \Phi_a^0|\hat{S}_\gamma(0, -\infty)|\Phi_a^0\rangle}, \tag{14.35}$$

where $\hat{S}_\gamma(t, t_0)$ is the adiabatic evolution operator. This operator can be obtained from the ordinary one by replacing $\hat{H}_{int}(t)$ by $\hat{H}_{int}(t, \gamma)$ in Equations 14.21 and 14.22. The differentiation with respect to the electron charge *e* is equivalent to the multiplication by the order of the perturbation theory *l*: the dependence on *e* enters the evolution operator via the expression for the electron current (Equation 14.6). Sucher [6] derived a symmetrized version of the energy shift formula

$$\Delta E_a = \lim_{\gamma \to 0} \frac{1}{2} i\gamma e \frac{(\partial/\partial e)\langle \Phi_a^0|\hat{S}_\gamma(\infty, -\infty)|\Phi_a^0\rangle}{\langle \Phi_a^0|\hat{S}_\gamma(\infty, -\infty)|\Phi_a^0\rangle}. \tag{14.36}$$

Practically, for the evaluation of the energy level shift, the formulas 14.35 and 14.36 are fully equivalent. However, there are general problems with the renormalization of the evolution operator for the finite time intervals; these problems arise also for the time interval $(0, \infty)$. Therefore, it is preferable to use the symmetrized formula, Equation 14.36.

The third step to the construction of the general QED theory for the tightly bound electrons consists of practical implementation of the Gell-Mann–Low–Sucher formula 14.36, combined with the Feynman graph techniques in the Furry picture to the evaluation of the different corrections to the energy levels. This step was made in [7], where the first general QED approach for the tightly bound electrons in atoms and HCIs was developed.

According to Labzowsky [7], to calculate the energy level shift with the use of the adiabatic formula 14.36, one has to evaluate first the derivative with respect to the electron charge. Expanding the numerator and denominator in Equation 14.36, in power series in $\hat{H}_{int}$ (this means, in power series in e), differentiating and restricting expansion to terms of order e^4 one obtains [7]

$$\Delta E_a = \lim_{\gamma \to 0} \frac{1}{2} i\gamma \left\{ \langle \Phi_a^0 | \hat{S}_\gamma^{(1)} | \Phi_a^0 \rangle + [2\langle \Phi_a^0 | \hat{S}_\gamma^{(2)} | \Phi_a^0 \rangle - \langle \Phi_a^0 | \hat{S}_\gamma^{(1)} | \Phi_a^0 \rangle^2] \right.$$

$$+ [3\langle \Phi_a^0 | \hat{S}_\gamma^{(3)} | \Phi_a^0 \rangle - 3\langle \Phi_a^0 | \hat{S}_\gamma^{(2)} | \Phi_a^0 \rangle \langle \Phi_a^0 | \hat{S}_\gamma^{(1)} | \Phi_a^0 \rangle + \langle \Phi_a^0 | \hat{S}_\gamma^{(1)} | \Phi_a^0 \rangle^3]$$

$$+ [4\langle \Phi_a^0 | \hat{S}_\gamma^{(4)} | \Phi_a^0 \rangle - 4\langle \Phi_a^0 | \hat{S}_\gamma^{(3)} | \Phi_a^0 \rangle \langle \Phi_a^0 | \hat{S}_\gamma^{(1)} | \Phi_a^0 \rangle$$

$$+ 4\langle \Phi_a^0 | \hat{S}_\gamma^{(2)} | \Phi_a^0 \rangle \langle \Phi_a^0 | \hat{S}_\gamma^{(1)} | \Phi_a^0 \rangle^2 - 2\langle \Phi_a^0 | \hat{S}_\gamma^{(2)} | \Phi_a^0 \rangle^2$$

$$\left. - \langle \Phi_a^0 | \hat{S}_\gamma^{(1)} | \Phi_a^0 \rangle^4] + \cdots \right\}. \tag{14.37}$$

For the free atom, in the absence of the external fields, the energy corrections contain only S-matrix elements of even order. The reason is that the operators $\hat{H}_{int}$, acting on the state vector Φ_a^0, should enter pairwise in the matrix element to give finally the same state Φ_a^0. Then the expression 14.37 looks simpler:

$$\Delta E_a = \lim_{\gamma \to 0} \frac{1}{2} i\gamma \left\{ 2\langle \Phi_a^0 | \hat{S}_\gamma^{(2)} | \Phi_a^0 \rangle + [4\langle \Phi_a^0 | \hat{S}_\gamma^{(4)} | \Phi_a^0 \rangle - 2\langle \Phi_a^0 | \hat{S}_\gamma^{(2)} | \Phi_a^0 \rangle^2] + \cdots \right\}. \tag{14.38}$$

Equations 14.35 through 14.38 are valid for the nondegenerate state of an atom. For extension to the degenerate state, see [1]. However, it is necessary to stress that these equations remain valid for the important case when the degenerate states differ by symmetry.

Next follows the procedure for the explicit evaluation of the limit $\gamma \to 0$ in Equation 14.38. This procedure was also described in [7]. For this purpose, one has to insert the adiabatic factor $e^{-\gamma|t|}$ in every Feynman graph vertex and perform the integration over all the time variables. It is important to distinguish between the "reducible" and "irreducible" Feynman graphs (or S-matrix elements). Reducible diagrams can be defined as those which can be divided into two parts by one single cut across the internal electron lines. The evaluation of the reducible S-matrix elements by means of Equation 14.36 gives rise to the singular terms proportional to $1/\gamma$, $1/\gamma^2$, and so on. These singularities are cancelled explicitly by the counterterms that arise from the denominator (i.e., from the terms with negative signs in Equations 14.37 and 14.38). For "irreducible" matrix element, the limit $\gamma \to 0$ can be evaluated in a very general way [8] (see also [1]). The contribution of the "irreducible" matrix elements always corresponds to the terms which originate from the numerator in Equation 14.36, that is, the terms with positive signs in Equation 14.37. Then in the nth order of perturbation theory,

$$\Delta E_a^{(n)} = \lim_{\gamma \to 0} \frac{1}{2} i n\gamma \langle \Phi_a^0 | \hat{S}_\gamma^{(n)} | \Phi_a^0 \rangle. \tag{14.39}$$

The result of the evaluation of the limit $\gamma \to 0$ can be presented in the form

$$\Delta E_{a,\text{irr}}^{(n)} = \langle \Phi_a^0 | U^{(n)} | \Phi_a^0 \rangle, \tag{14.40}$$

where the amplitude U is connected with the S-matrix by the relation

$$\langle \Phi_a^0 | \hat{S}_\gamma^{(n)} | \Phi_a^0 \rangle = -2\pi i\, \delta(E_a^{(0)} - E_b^{(0)}) \langle \Phi_b^0 | U^{(n)} | \Phi_a^0 \rangle. \tag{14.41}$$

The formulas 14.40 and 14.41 enable one to perform the calculations for the tightly bound electrons in the lowest orders of the perturbation theory, not employing the general formulas 14.36 through 14.38. The first evaluation of the second-order interelectron interaction correction which required the use of the adiabatic formalism was performed in [7].

14.4 First-Order Radiative Corrections to the Energy Levels in One-Electron HCIs

There are two lowest order radiative corrections to the energy levels of one-electron ions: electron self-energy (SE) correction and vacuum polarization (VP) correction. Corresponding Feynman graphs are depicted in Figure 14.1. The electron SE correction for the tightly bound electrons was first evaluated in [9,10] for the K-shell electrons in mercury atom. Since the Feynman graph (Figure 14.1a) is irreducible, the employment of the formulas 14.40 and 14.41 for the energy level shift was justified. The graph (Figure 14.1a) is ultraviolet-divergent. For its renormalization, the potential expansion method was first applied by Brown et al. [9,10]. This method became later an universal tool for the renormalization of the divergent radiative corrections for the tightly

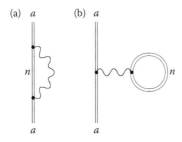

FIGURE 14.1
The first-order electron self-energy (a) and vacuum polarization (b) correction. The double solid line denotes the electron in the field of the nucleus (Furry picture), whereas the wavy line denotes the photon.

bound electrons. The method is based on the expansion of the bound-electron propagator (Equation 14.33) in powers of the nuclear potential:

$$\Delta E_a = \Delta E_a^a + \Delta E_a^b + \Delta E_a^c, \tag{14.42}$$

where $\Delta E_a^{a,b,c}$ correspond to the Feynman graphs (Figure 14.2a–c). For the tightly bound electrons, all the terms of the expansion (Equation 14.42) are, in principle, of the same order. The ultraviolet divergency is contained in the first two terms of the potential expansion $\Delta E_a^{a,b}$. The many-potential term ΔE_a^c is finite but the most difficult one to evaluate numerically. The renormalization of the divergent terms $\Delta E_a^{a,b}$ can be performed in the momentum space according to the standard procedures developed in free-electron QED. The renormalization algorithm requires the subtraction of two terms of the Taylor expansion of the free-electron self-energy operator $\Sigma(p)$ in the vicinity of the mass shell ($p^2 = m^2$):

$$\Sigma^{\text{ren}}(p) = \Sigma(p) - \left[\Sigma(p)\right]_{p^2=m^2} - (\not p - m)\left[\frac{\partial \Sigma(p)}{\partial p}\right]_{p^2=m^2}. \tag{14.43}$$

In Equation 14.43, $\Sigma^{\text{ren}}(p)$ denotes the renormalized expression for $\Sigma(p)$. Then,

$$\Delta E_a^a = \Delta E_a^{a,\text{ren}} + \Delta E_a^{a1} + \Delta E_a^{a2}, \tag{14.44}$$

where ΔE_a^{a1} and ΔE_a^{a1} correspond to the two subtracted terms in Equation 14.43. A regularized expression for the vertex (Figure 14.2b) is determined as

$$\Lambda_\mu^{\text{ren}}(p_1, p_2) = \Lambda_\mu(p_1, p_2) - \left[\Lambda_\mu(p_1, p_2)\right]_{p_1^2=p_2^2=m^2}. \tag{14.45}$$

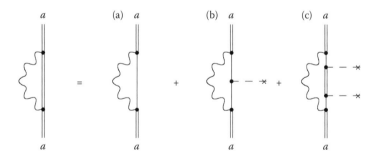

FIGURE 14.2
The potential expansion of the first-order electron self-energy correction. The double solid line denotes the electron in the field of the nucleus (Furry picture), whereas the wavy line denotes the photon, and the dashed line with the cross at the end corresponds to the Coulomb interaction with the nucleus. The external double lines correspond to the Dirac wave function for the electron in the atomic state a, the internal double lines correspond to the electron propagator. The ordinary internal solid line denotes the free electron propagator. The Feynman graphs (a)–(c) correspond to the "zero-potential," "one-potential," and "many-potential" contributions.

Hence,

$$\Delta E_a^b = \Delta E_a^{b,\text{ren}} + \Delta^{b1} E_a, \tag{14.46}$$

where ΔE_a^{b1} corresponds to the subtracted term in Equation 14.45.

Finally, the renormalized expression for the electron self-energy contribution results

$$\Delta E_a^{\text{ren}} = \Delta E_a^{a,\text{ren}} + \Delta E_a^{b,\text{ren}} + \Delta E_a^c. \tag{14.47}$$

To avoid the numerical evaluation of ΔE_a^c, Brown et al. [9,10] rearranged the terms in Equation 14.47, using the Ward identity [11]. From this identity, it follows that

$$\Delta E_a^{b1} = -\Delta E_a^{a2}. \tag{14.48}$$

The formula 14.47 takes the form

$$\Delta E_a^{\text{ren}} = (\Delta E_a - \Delta E_a^a) + \Delta E_a^{a2} + \Delta E_a^{a,\text{ren}}, \tag{14.49}$$

where the term ΔE_a^c is absent. Instead of this, one has to cancel the divergencies numerically in the expression $(\Delta E_a - \Delta E_a^a) + \Delta E_a^{a2}$. Later Mohr [12,13] was able to cancel the divergencies in Equation 14.49 analytically. The Mohr's method remains up to now one of the most accurate ones in the tightly bound electron QED theory (see the recent status of this method in [14]).

However, in the last decade, due to the development of the powerful numerical methods, including B-spline approach [15,16], the direct evaluation of Equation 14.47 became available. These numerical methods allow for the summations over the entire Dirac spectrum for arbitrary spherically symmetric potentials. This enables one to evaluate directly the many-potential term in the potential expansion. The direct potential expansion of the SE, analyzed also in [17], was later reconsidered in [18]. For the numerical calculations with the use of the direct potential expansion see, for example, [19,20].

Let us turn to the second lowest-order radiative correction described by the Feynman graph (Figure 14.1b). For the analysis, one can use again the potential expansion, also depicted in Figure 14.3. The first term of this expansion, corresponding to Figure 14.3a vanishes due to the Furry theorem [21]. The divergency is contained only in the one potential term (Figure 14.3b). After renormalization, this term reduces to the diagonal matrix element of the Uehling potential for the atomic electron [22]:

$$\phi_u(r) = \frac{2e^3 Z}{3\pi r} \int_1^\infty dx \, e^{-2mrx} \left(1 + \frac{1}{2x^2}\right) \frac{\sqrt{x^2 - 1}}{x^2}, \tag{14.50}$$

where Z is the nuclear charge and r the distance from the nucleus. Equation 14.50 represents a vacuum-polarization correction to the Coulomb interaction of the electron with the point-like nucleus. The Uehling term, being the

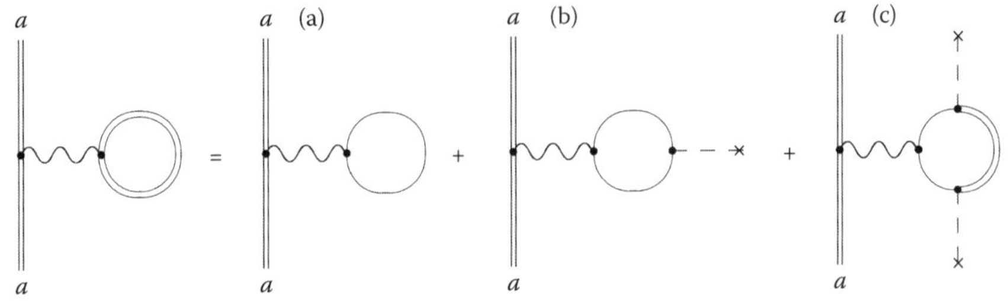

FIGURE 14.3

The potential expansion of the lowest order vacuum polarization correction. The notation are the same as in Figure 14.2. (a) Zero-potential, (b) one-potential, and (c) many-potential contributions.

fist term of potential expansion of the exact one-loop VP correction usually yields a fairly good approximation. Even for high Z values, the deviation from the full VP contribution does not exceed 10%. The first evaluation of the remainder contribution corresponding to the Feynman graph (Figure 14.3c) was performed by Wichmann and Kroll [23] for H-like ions with $70 \leq Z \leq 90$. Later this contribution (usually referred to as Wichmann–Kroll term) was calculated for the wide range of Z values, both for the point-like and extended nuclei [24–26].

14.5 Second-Order Radiative Corrections to the Energy Levels in One-Electron HCIs

The recent experimental accuracy achieved in the Lamb shift measurements in heavy H-like ions [27] requires to take into account in theoretical calculations the complete set of two-loop radiative corrections. The corresponding Feynman graphs are depicted in Figure 14.4. It is necessary to distinguish between the "irreducible" and "reducible" graphs, respectively. According to the definition given in Section 14.3, the graphs in Figure 14.4a, d, g, and h are reducible, whereas the other graphs are irreducible. A reducible graph can be divided again into two parts: irreducible part and reducible part. The irreducible part of a reducible diagram appears as the sum over intermediate electron states n in the eigenmode decomposition of the electron propagator (Equation 14.33), corresponding to the cutted electron line, however, restricted by the condition $n \neq a$. The remainder, which contains only the single "reference" state $n = a$, represents the reducible part of the reducible graph and is called the reference state correction (RSC).

The existence of the nonzero RSC is the only reason that does not allow to use Equations 14.40 and 14.41 for the evaluation of all the corrections in QED. In [7], it was proven explicitly that for the second-order Coulomb–Coulomb interelectron interaction, the RSC is exactly zero. The evaluation of the RSC within the adiabatic S-matrix approach appears to be rather cumbersome [1];

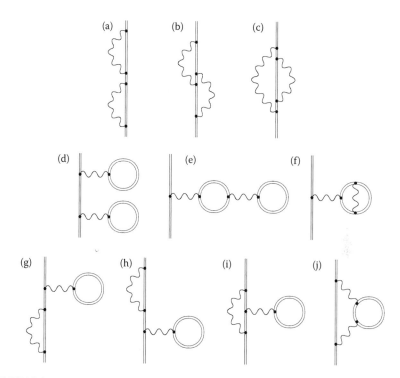

FIGURE 14.4
Two-loop radiative corrections for the H-like ions. The notation are the same as in Figures 14.2 and 14.3. The Feynman graphs (a)–(c) are called SESE graphs ("loop after loop," "cross loop," and "loop in loop," respectively), the graphs (d)–(f) are called VPVP graphs, and the graphs (g)–(j) are called SEVP graphs.

much simpler are the corresponding evaluations with the TTGFM [2] or the LPA [4].

The two-loop corrections presented in the first, second, and third rows in Figure 14.4 are often called SESE, VPVP, and SEVP corrections, respectively. These three groups of corrections represent three gauge-invariant subsets. Some of these corrections are individually gauge-invariant: for example, the irreducible contribution to the graph (Figure 14.4a), called "loop after loop" or SESE-lal-irr correction, turns out to be gauge-invariant in any covariant gauge. Its contribution to the Lamb shift for $1s_{1/2}$, $2s_{1/2}$, and $2p_{1/2}$ states of H-line HCIs was first calculated in [28] for high Z values. In [29], this calculation was extended to the entire range of nuclear charges $1 \leq Z \leq 92$.

The first complete evaluation of the SESE set of QED corrections was provided in [30]. The contribution VPVP (d) was evaluated in [31] for few heavy H-like ions and extended to larger number of Z in [32]. The corrections VPVP (e and f) were calculated in [33,34] in the Uehling approximation; later an exact value for the VPVP (e) contribution was obtained [35]. The SEVP (Figure 14.4g–i) contributions were evaluated in the Uehling approximation in [36] and in the exact form in [37]. The SEVP (Figure 14.4j) graph contribution has been calculated only within the Uehling approximation in [37,38].

14.6 Other One-Electron Corrections

To compare the one-electron corrections, cited above, with the experimental ones, it is necessary to take into account also the nuclear finite size, finite mass, and the nuclear polarization by atomic electrons.

The nuclear finite size correction within QED theory of atoms is usually included from the very beginning into the Dirac equation of the atomic electron via introducing the Coulomb interaction potential with the extended nucleus. As the most accurate fit for the nuclear charge density distribution, the Fermi formula is usually employed:

$$\rho(r) = \frac{N}{1 + \exp[(r - c)/a]}, \tag{14.51}$$

where the parameter c corresponds to the charge radius, the parameter a is a measure of the skin charge thickness, and N is the normalization constant. Then, according to the Poisson equation, the potential is

$$V(r) = \int \frac{\rho(r')}{|r - r'|} dr'. \tag{14.52}$$

The influence of the nuclear size corrections on the Dirac energy levels and on the QED corrections to the energy levels was studied in detail in [14,39].

The influence of the finite nuclear mass in the relativistic theory of the H-like atoms (the recoil correction to the energy levels) was obtained by Salpeter [40] utilizing the Bethe–Salpeter equation [41]. Actually Salpeter's result was semirelativistic since the expansion in terms of αZ (α is the fine structure constant) parameter was also employed. For the hydrogen atom, the use of the αZ expansion ($Z = 1$) is of course justified.

A generalization of the Bethe–Salpeter approach to all orders in αZ was presented in [42] (see also [1]). However, this work was restricted to the recoil correction, associated with the Coulomb interaction between the electron and the nucleus. In this approximation, the recoil correction to the level a of the H-line HCI looks like [1] (in atomic units)

$$\Delta E_a = \frac{1}{2M} \left(\hat{p}[P^{(+)} - P^{(-)}]\hat{p} \right)_{aa}, \tag{14.53}$$

where $\hat{p}$ is the momentum operator for the atomic electron, M is the mass of the nucleus, and $P^{(\pm)}$ are the projectors onto the positive (negative) spectrum of the Dirac equation for the atomic electron.

A closed expression for the relativistic recoil correction, including corrections due to the transverse photons which have the same order of magnitude

TABLE 14.1

Different Corrections to the Ground State of the H-like $^{238}U^{91+}$ Ion

Correction	Value	Reference
Nuclear size	198.81	
SE (Figure 14.1a)	355.05	[71]
VP (Figure 14.1b)	−88.60	[24]
SESE (Figure 14.4a^{red},b,c)	−0.90	[30]
SESE (Figure 14.4a^{irr})	−0.97	[28]
VPVP (Figure 14.4d)	−0.22	[32]
VPVP (Figure 14.4e,f)	−0.75	[32–34]
SEVP (Figure 14.4g,h,i)	1.14	[36]
SEVP (Figure 14.4j)	0.13	[37,38]
Recoil	0.46	[72]
Nuclear polarization	−0.20	[51]
Total binding energy shift	463.95	
Experiment	460.2 (4.6)	[73]

Note: All the values are given in eV

as the correction (Equation 14.53) in case of HCIs, were derived in [43] and in the complete form for the first time in [44,45]. In [43–45], a Green function approach was applied to the derivation of the energy shift. The first accurate calculations of the recoil correction for all Z values $1 \leq Z \leq 100$ were compiled in [46]. Recently, the recoil corrections in the H-like HCIs were reconsidered within the Bethe–Salpeter equation approach [47,48].

Nuclear polarization corrections in HCIs can be described correctly within a field-theoretical treatment based on the concept of effective photon propagators with nuclear polarization insertions [49–51].

The most recent results for the calculations of the different QED and non-QED corrections to the ground state of H-like uranium ions are given in Table 14.1 together with the latest experimental result.

14.7 Many-Electron Corrections

For the many-electron ions, apart from the one-electron corrections, the many-electron corrections represented by the many-electron Feynman graphs have to be considered.

In the first order of the QED perturbation theory, the many-electron corrections are represented by the interelectron interaction correction. The corresponding Feynman graph is depicted in Figure 14.5.

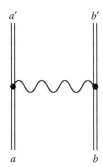

FIGURE 14.5
The lowest order correction for the interelectron interaction: one-photon exchange.

Many-electron corrections of the second order are given by the two-electron Feynman graphs (Figures 14.6 and 14.7) and by the three-electron Feynman graph in Figure 14.8. The graphs in Figures 14.6 and 14.8 represent the interelectron interaction corrections. The graphs in Figure 14.7 are called the screened self-energy and vacuum polarization corrections.

All the many-electron Feynman graphs of the first and second order have been rigorously calculated for the low-lying two- and three-electron configurations. The graphs in Figure 14.6 were first calculated for the ground state of the two-electron highly charged ions [52,53]. In the case of the three-electron configurations, the graphs in Figures 14.6 and 14.8 were first calculated for the $(1s)^2 2p_{1/2} - (1s)^2 2s$ splitting [54] (see also [55,56]). Calculation of the interelectron interaction corrections for the quasidegenerate energy levels of two-electron HCIs were first performed in [57,58]. The rigorous calculation of the self-energy and vacuum polarization screening correction (Figure 14.7) were first performed in [20,59] and in [60], respectively, for the low-lying three-electron configurations. The complete calculation of the contributions of the second-order Feynman graphs to the low-lying energy levels for two-electron HCIs is presented in [56].

At present, Feynman graphs of the third order of the perturbation theory are calculated in various approximations. Here we consider these graphs in

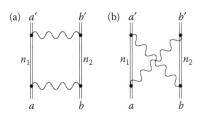

FIGURE 14.6
The second-order corrections for the interelectron interaction: two-photon exchange. Graphs (a) and (b) are called "box" and "cross" graphs, respectively.

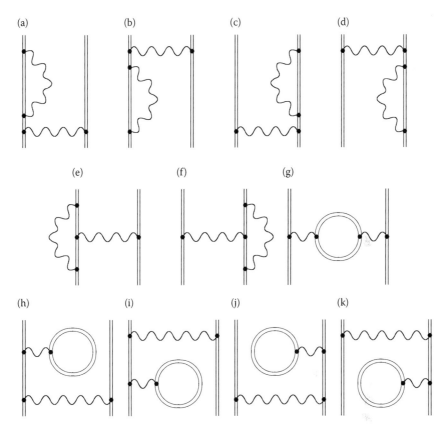

FIGURE 14.7
Feynman graphs representing the screened self-energy corrections (a–f) and the screened vacuum polarization corrections (g–k).

the frameworks of the relativistic many-body perturbation theory (RMBPT). It implies neglect of (1) retardation, (2) negative-energy part of the Dirac spectrum, and (3) "cross" graphs. One can show that item (3) follows from items (1) and (2). The third-order interelectron interaction corrections calculated in the frameworks of RMBPT can be considered as a dominant part of the corrections. The radiative corrections of the third order of the QED perturbation theory and the four-electron graphs are yet not considered. In this chapter, we consider the two- and three-electron graphs in the frameworks of RMBPT, that is, graphs in Figures 14.9 and 14.10. These corrections were first calculated in [55,61].

In Table 14.2, we collect different corrections to the splitting $2p_{1/2} - 2s_{1/2}$ in Li-like uranium. A part of radiative corrections of the second order for the $2p_{1/2} - 2s_{1/2}$ splitting is yet not calculated. This is the reducible part of the graph (Figure 14.4) and the corrections represented by the Feynman graphs (Figure 14.4b, c, and j).

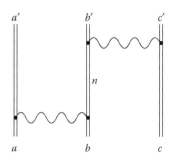

FIGURE 14.8
The three-electron Feynman graph representing second-order interelectron interaction correction.

Initially, the most popular method applied for QED calculations was the adiabatic S-matrix approach. This method allows one easily to derive formulas for numerical calculations of various corrections, however, only for relatively simple cases. The irreducible graphs (and irreducible parts of the reducible graphs) can be, in principle, evaluated by employing the adiabatic S-matrix approach considered in Section 14.3. The application of the adiabatic S-matrix approach for the reducible graphs becomes too complicated. This approach was also not applied for quasidegenerate levels. Accordingly, there were developed, more suitable QED approaches for description of HCIs. At present, the major part of the QED calculations is performed by employing the adiabatic S-matrix approach, TTGFM [2], CEO approach [3], and the LPA [4]. Calculations performed with the different QED approaches are equivalent.

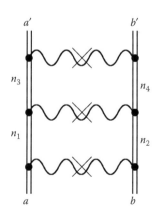

FIGURE 14.9
The three-photon exchange two-electron graph representing the third-order interelectron interaction correction. The cross at the wavy line (photon propagator) means the disregard of retardation.

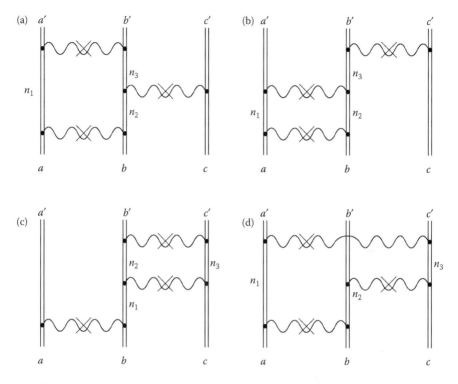

FIGURE 14.10
The three-photon exchange three-electron graphs representing the third-order interelectron inter-action correction (a through d). The cross at the wavy line (photon propagator) means the disregard of retardation.

14.8 Line Profile Approach

In this section, we would like to present the LPA. The basic ideas of the QED theory of the spectral line profile were first introduced in [62]. The application of the line profile theory to the evaluation of the energy level shifts was first discussed in [63] and later developed in [55]. The detailed description of the LPA is given in [4].

In the frameworks of the LPA, the energy levels of HCIs are associated with the resonances of a process of photon scattering on the ion. Accordingly, if we investigate a level A, we consider a resonant process

$$A_0 \to A \to A_0, \tag{14.54}$$

where an ion being in the ground-state A_0 absorbs a photon ω and goes to the excited-state A, and then decays back to the ground state.

TABLE 14.2

Different Corrections to the Splitting $2p_{1/2} - 2s_{1/2}$ in Li-like $^{238}U^{91+}$ Ion

	Correction	Value	Reference
Nuclear size		−33.35(6)	[74]
Interelectron interaction	1st order (Figure 14.5)	368.83	[55]
	2nd order (Figures 14.6 and 14.8)	−13.37	[55]
	3rd order (Figures 14.9 and 14.10)	0.17(2)	[55]
First-order radiative	SE (Figure 14.1a)	−55.87	[71]
corrections	VP (Figure 14.1b)	12.94	[37]
SE screening (Figure 14.7a–f)		1.52	[20]
VP screening (Figure 14.7g–k)		−0.36	[60]
Second-order radiative	SESE (Figure 14.4a^{irr})	0.10	[28]
corrections	VPVP (Figure 14.4d–f)	0.13	[32]
	SEVP (Figure 14.4g–i)	−0.21	[32]
Nuclear recoil		−0.07	[46]
Nuclear polarization		0.03	[51]
Total theory	QED	280.49(21)	
Experiment		280.59(9)	[75]

Note: All the values are given in eV

The energy level is characterized by two parameters: energy and width. They are assumed to be not dependent on the features of the experiment where they are measured. We note that the ion (a nucleus and a set of electrons) is not a conservative system because it interacts with the vacuum. Accordingly, the meaning of the energy should be explained.

The *S*-matrix of the scattering process, written as the series of the QED perturbation theory, contains terms singular at the position of the resonances. It is naturally regularized by the introduction of the level width. In order to make the energy of the level not depending on the features of the scattering process, we employ the resonance approximation, where we retain only the terms singular at the resonance. After the regularization, the absorption (or emission) probability at the position of the resonance ($\omega \approx E_A - E_{A_0}$), considered as a function of ω, can be well interpolated by the Lorentz contour. The Lorentz contour is described by the position of the resonance and the width. Accordingly, the energy of the level is defined by the position of the resonance. The width of the level is defined as the width of the corresponding Lorentz contour (A_0 is the ground state).

The LPA allows one to evaluate the absorption probability beyond the resonance approximation. The corresponding corrections to the energy are called the nonresonant corrections. These corrections are estimated for the resonant photon scattering processes [64,65]. They are found to be rather small; however, the nonresonant corrections determine the limits for the accuracy of the resonant frequency measurements [65]. For the two-photon $1s - 2s$ transition in hydrogen this limit is 10^{-5} Hz, and the accuracy of the corresponding experiment is 46 Hz [66]. For the one-photon $1s - 2p$ transition in hydrogen, the accuracy limit is 0.17 MHz, while the uncertainty of the measurement is about 6 MHz [67].

14.9 Evaluation of Energy

Let us consider a one-electron ion. In the lowest order of the QED perturbation theory, the S-matrix of the photon scattering process (Equation 14.54) is depicted in Figure 14.11 and reads

$$S_{a_0}^{(2)} = (-ie)^2 \int d^4 x_u \, d^4 x_d \, \bar{\psi}_{a_0}(x_u) \gamma^{\mu_u} S(x_u, x_d) \gamma^{\mu_d} \psi_{a_0}(x_d)$$

$$\times A_{\mu_u}^{*(k',\lambda')}(x_u) A_{\mu_d}^{(k,\lambda)}(x_d), \tag{14.55}$$

where $\psi_{a_0}(x) = \psi_{a_0}(r) \, e^{-i\varepsilon_{a_0} t}$ is a wave function of the electron in the ground state (the $1s$ electron from the Dirac spectrum), γ^μ is the Dirac matrix, $A_\mu^{(k,\lambda)}(x)$ is potential of the electromagnetic field (photon wave function), and $S(x, x')$ is the electron propagator defined by Equation 14.33.

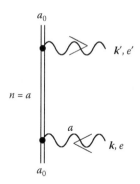

FIGURE 14.11
Feynman graph representing the photon scattering process on a one-electron ion. The wavy line with an arrow means the absorbed and emitted photons with momenta k, k' and polarizations e, e', respectively.

The S-matrix element after separating the time dependence in the integrand can be presented as

$$S_{a_0}^{(2)} = (-\mathrm{i})^2 \int \mathrm{d}t_u \, \mathrm{d}^3 r_u \, \mathrm{d}t_d \, \mathrm{d}^3 r_d \, \mathrm{d}\omega_n \, [\bar{\psi}_{a_0}(r_u) e\gamma^{\mu_u} A_{\mu_u}^{*(k',\lambda')}(r_u)]$$

$$\times \, e^{\mathrm{i}t_u(\varepsilon_{a_0}+\omega')} \, e^{-\mathrm{i}\omega_n(t_u - t_d)} \frac{\mathrm{i}}{2\pi} \sum_n \frac{\psi_n(r_u)\bar{\psi}_n(r_d)}{\omega_n - \varepsilon_n(1 - \mathrm{i}0)} \, e^{-\mathrm{i}t_d(\varepsilon_{a_0}+\omega)}$$

$$\times \, [e\gamma^{\mu_d} A_{\mu_d}^{(k,\lambda)}(r_d)\psi_{a_0}(r_d)], \tag{14.56}$$

where $\omega = |k|$ and $\omega' = |k'|$ are frequencies of the absorbed and emitted photons, respectively. The summation over n is extended over the entire Dirac spectrum for the bound electron, and ε_n are the Dirac electron energies. In the numerical calculations, the ion is inserted into a box of a finite radius; accordingly, all the Dirac spectrum becomes discrete. Moreover, the infinite Dirac spectrum is substituted by a finite one [15,16,68]. The size of the box and the number of eigenvectors in the approximate Dirac spectrum are chosen according to the necessary accuracy. We note that the index n at the frequency ω_n signifies that the frequency variable ω_n relates to the eigenmode decomposition (sum over n) of the (one-electron) Dirac propagator (see Equation 14.33), that is, ω_n is not a subject of the summation.

Integrating over time variables (t_u, t_d) and designating the expressions in the square brackets by

$$\Phi_{a_0}^{(0)}(r_d) = e\gamma^{\mu_d} A_{\mu_d}^{(k,\lambda)}(r_d)\psi_{a_0}(r_d), \tag{14.57}$$

$$\bar{\Phi}_{a_0}^{(0)}(r_u) = \bar{\psi}_{a_0}(r_u) e\gamma^{\mu_u} A_{\mu_u}^{*(k',\lambda')}(r_u), \tag{14.58}$$

we can write

$$S_{a_0}^{(2)} = (-\mathrm{i})^2 (2\pi)^2 \int \mathrm{d}^3 r_u \, \mathrm{d}^3 r_d \, \mathrm{d}\omega_n \, \bar{\Phi}_{a_0}^{(0)}(r_u)\delta(\omega_n - \omega' - \varepsilon_{a_0})$$

$$\times \frac{\mathrm{i}}{2\pi} \sum_n \frac{\psi_n(r_u)\bar{\psi}_n(r_d)}{\omega_n - \varepsilon_n(1 - \mathrm{i}0)} \delta(\omega + \varepsilon_{a_0} - \omega_n)\Phi_{a_0}^{(0)}(r_d). \tag{14.59}$$

The function $\Phi_{a_0}^{(0)}(r)$ can be considered as a "bare" vertex function, which describes a generic process of an absorption of a photon by an electron in the ground state. The function $\Phi_{a_0}^{(0)}(r)$ defines the features of the considered scattering process. In general, it is more complicated function than Equation 14.59. We employ the notation Φ_{a_0} for a generalized vertex function $\Phi_{a_0}^{(0)}$. The final expressions for the energy will not depend on the function $\Phi_{a_0}(r)$; accordingly, we will not specify its form. If it is necessary, the function $\Phi_{a_0}(r)$ can be evaluated by employing the perturbation theory. Also, the energy of the ground state may differ from the Dirac energy ε_{a_0}. We employ the notation

ϵ_{a_0} for the energy of the ground state a_0. Within the framework of the QED perturbation theory, we should consider $\epsilon_{a_0} = \varepsilon_{a_0} + O(\alpha)$.

For evaluation of the corrections to the wave function Φ_{a_0} and energy ϵ_{a_0} corresponding to the ground state, one has to consider graphs with multiple insertions, in particular, the insertions of the electron self-energy operator, into the outer electron lines of graph in Figure 14.11. The consideration of these insertions requires a regularization, because of the arising singularities. One way of regularization is introduction of an artificial state a_* "lower than ground" [4]. The partial width corresponding to the transition $a_0 \rightarrow a_*$ is considered as a regularization parameter which is set to zero at the end of the evaluation. Another way of regularization is application of the adiabatic approach (see Section 14.3). The arising singularities are combined into the additional factor $\exp[\Sigma_{a_0 a_0}/(i\gamma)]$ to the S-matrix element [69]. As the self-energy matrix element for the ground state $(\Sigma_{a_0 a_0})$ is real, this exponent does not contribute to the absolute value of the corresponding amplitude and, accordingly, to the line profile. The evaluation of the corrections to the wave function Φ_{a_0} and energy ϵ_{a_0} corresponding to the ground state is necessary for justification of the LPA backgrounds. Formally, the corrections to the ground state can be evaluated with the technique developed in the LPA for the excited states.

Considering the general case, we rewrite Equation 14.59 as

$$S_{a_0}^{(2)} = (-i)^2 (2\pi)^2 \int d^3 r_u \, d^3 r_d \, d\omega_n \, \bar{\Phi}_{a_0}(r_u)$$

$$\times \, \delta(\omega_n - \omega' - \epsilon_{a_0}) \frac{i}{2\pi} \sum_n \frac{\psi_n(r_u)\bar{\psi}_n(r_d)}{\omega_n - \varepsilon_n(1 - i0)} \delta(\omega + \epsilon_{a_0} - \omega_n)\Phi_{a_0}(r_d).$$

$$(14.60)$$

After performing spatial integration, we introduce the matrix

$$T_{n a_0} = - \int d^3 r \, \bar{\psi}_n(r)\Phi_{a_0}(r) \qquad (14.61)$$

and integrate over frequency ω_n in Equation 14.60. This yields the generic form

$$S_{a_0}^{(2)} = (-2\pi i)\delta(\omega - \omega') \sum_n \frac{T_{a_0 n}^* T_{n a_0}}{\omega + \epsilon_{a_0} - \varepsilon_n}. \qquad (14.62)$$

The amplitude of the scattering process is connected with the S-matrix element by the relation 14.41. We obtain an expression for the amplitude in terms of the matrices (Equation 14.61)

$$U_{a_0}^{(2)} = \sum_n \frac{T_{a_0 n}^* T_{n a_0}}{\omega + \epsilon_{a_0} - \varepsilon_n}. \qquad (14.63)$$

The index k at the amplitudes $U^{(k)}$ signifies the order of the perturbation theory (the power of $|e| = \sqrt{\alpha}$, the charge of the electron). It implies $U = \sum_{k=0}^{\infty} U^{(2k)}$ and $U^{(0)} = 0$.

Let us investigate the line profile as a function of ω at one of its maxima (the position of the resonance). We will be interested in the energy region close to the resonance value $\omega^{\text{res}} = -\epsilon_{a_0} + \epsilon_a + O(\alpha)$, where a one of the excited states of an ion and ϵ_a is the corresponding Dirac energy. In the resonance approximation, we retain only the term $n = a$ when summing over the Dirac spectrum Equation 14.63. This term is singular at $\omega = \omega^{\text{res}}$ and gives a dominant contribution. Then the resonant transition amplitude is given by

$$U_{a_0 a_0}^{(2)} = \frac{T_{a_0 a}^* T_{a a_0}}{\omega + \epsilon_{a_0} - \epsilon_a} = T^* D^{-1} T. \tag{14.64}$$

It is convenient to introduce the following generic notation:

$$T = T_{a a_0}, \tag{14.65}$$

$$D = \omega + \epsilon_{a_0} - V^{(0)}, \tag{14.66}$$

$$V^{(0)} = \epsilon_a. \tag{14.67}$$

The matrices T and T^* describe the process of absorption or emission, respectively. Accordingly, they define the process of scattering.

In order to obtain the Lorentz contour with a nonzero width we have to account for radiative corrections. For this purpose, we will consider graph (Figure 14.11), where the electron self-energy part is inserted into the internal electron line. For simplicity we neglect here the vacuum polarization. In the lowest order, this leads to the graph (Figure 14.12). The expression for the scattering amplitude in the resonance approximation takes the form

$$U_{a_0 a}^{(4)} = U_{a_0 a}^{(2)} \frac{V^{(2)}(\omega)}{\omega + \epsilon_{a_0} - \epsilon_a} = T^* D^{-1} \left[V^{(2)}(\omega) D^{-1} \right] T, \tag{14.68}$$

where

$$V^{(2)}(\omega) = e^2 \left(\widehat{\Sigma}^{\text{ren}}(\omega + \epsilon_{a_0}) \right)_{aa}. \tag{14.69}$$

The index $2k$ at $V^{(2k)}$ signifies the order of the perturbation theory (power of $\sqrt{\alpha}$) for the graphs which contribute to this function. It implies $V = \sum_{k=0}^{\infty} V^{(2k)}$. Here $\widehat{\Sigma}^{\text{ren}}(\omega)$ is the renormalized electron self-energy operator (see Section 14.4). Repeating these insertions up to higher orders (iterations), we can compose a geometric progression with the lth term

$$Q_l = U_{a_0 a}^{(2)} \left[\frac{V^{(2)}(\omega)}{\omega + \epsilon_{a_0} - \epsilon_a} \right]^l = T^* D^{-1} \left[V^{(2)}(\omega) D^{-1} \right]^l T. \tag{14.70}$$

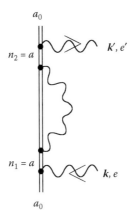

FIGURE 14.12
Feynman graph corresponding the electron self-energy insertion into the electron propagator in Figure 14.11.

This geometric progression is convergent for any ω except the interval near the position of the resonance $\omega \in [-\epsilon_{a_0} + \varepsilon_a - |V^{(2)}|, -\epsilon_{a_0} + \varepsilon_a + |V^{(2)}|]$. Applying the formula for convergent geometric progression, we arrive at

$$U_{a_0 a}(\omega) = \sum_{l=0}^{\infty} Q_l = \sum_{l=0}^{\infty} T^* D^{-1} \left[V^{(2)}(\omega) D^{-1} \right]^l T \tag{14.71}$$

$$= T^* \frac{1}{D - V^{(2)}(\omega)} T = \frac{T^* T}{\omega + \epsilon_{a_0} - V^{(0)} - V^{(2)}(\omega)}. \tag{14.72}$$

Formally, this corresponds to a partial resummation of an infinite number of diagrams of a particular class. The function $V^{(2)}$ defined by Equation 14.69 contains a nonzero imaginary part, and therefore, the resonance is shifted to the complex plane and Equation 14.72 is defined for all ω values. Accordingly, Equation 14.72 presents the analytic continuation of the expansion $\sum_{l=0}^{\infty} Q_l$ to the entire complex plane. This analytical continuation is considered as the regularization for the amplitude (Equation 14.71).

Taking the square modulus of the amplitude (Equation 14.71), integrating over the directions of the absorbed and emitted photons, and summing over the polarizations, we obtain the Lorentz profile for the absorption probability

$$dW(\omega) = \frac{1}{2\pi} \frac{\Gamma_{aa_0} \, d\omega}{(\omega + \epsilon_{a_0} - V^{(0)} - \text{Re}\{V^{(2)}(\omega)\})^2 + (\text{Im}\{V^{(2)}(\omega)\})^2}. \tag{14.73}$$

Here $dW(\omega)$ is the probability for absorption of a photon in the frequency interval $[\omega, \omega + d\omega]$ and Γ_{aa_0} is the partial width of the level a, connected with the transition $a_0 \rightarrow a$.

Having taken into account the graph (Figure 14.12), we improve the position of the resonance

$$\omega^{res} = -\epsilon_{a_0} + V^{(0)} + \text{Re}\{V^{(2)}(\epsilon_a - \epsilon_{a_0})\} + O(\alpha^2). \tag{14.74}$$

In the resonance approximation, the line profile (Equation 14.73) can be described by a Lorentz contour which is characterized by two parameters: the position of the resonance and the width. We define the energy shift of the excited level a via the shift of the resonance

$$\epsilon_a = \omega^{res} + \epsilon_{a_0} = V^{(0)} + \text{Re}\{V^{(2)}(\epsilon_a - \epsilon_{a_0})\} + O(\alpha^2) \tag{14.75}$$

and the width of the excited level as the width of the Lorentz contour at the point of the resonance

$$\Gamma_a = -2\,\text{Im}\{V(\omega^{res})\} = -2\,\text{Im}\{V^{(2)}(\epsilon_a - \epsilon_{a_0})\} + O(\alpha^2). \tag{14.76}$$

Here, Γ_a gives the total radiative (single-quantum) width of the level a.

We note that according to Equation 14.69,

$$V^{(2)}(\epsilon_a - \epsilon_{a_0}) = e^2 \left(\widehat{\Sigma}^{ren}(\epsilon_a)\right)_{aa}. \tag{14.77}$$

One can see that expressions for the energy and the width of the level a do not depend on the features of the scattering process.

For evaluation of the next order corrections to the energy, we have to consider the contribution of the Feynman graphs in Figure 14.4. For simplicity, we will consider only the graph Figure 14.4a. This graph is a reducible graph. In the frameworks of the LPA, we have to consider a graph (Figure 14.13). If $n_1 = n_3 = a$ and $n_2 \neq a$, the graph of Figure 14.13 can be viewed as a complicated insertion represented by Figure 14.4a in the graph (Figure 14.11) in the resonance approximation. We can write down the following expression for the scattering amplitude:

$$U^{(6)}_{a_0 a} = U^{(2)}_{a_0 a} \frac{V^{(4)}(\omega)}{\omega + \epsilon_{a_0} - \epsilon_a} = T^* D^{-1} \left[V^{(4)}(\omega)D^{-1}\right] T, \tag{14.78}$$

where

$$V^{(4)}(\omega) = e^4 \sum_{n \neq a} \frac{\widehat{\Sigma}^{ren}(\omega + \epsilon_{a_0})_{an}\,\widehat{\Sigma}^{ren}(\omega + \epsilon_{a_0})_{na}}{\omega + \epsilon_{a_0} - \epsilon_n}. \tag{14.79}$$

The singular term $n = a$ is not included here by definition. This term was taken into account in the geometric progression described above and represents

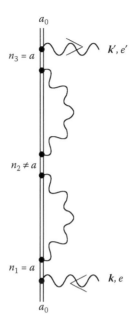

FIGURE 14.13
Feynman graph corresponding to the loop-after-loop self-energy insertion within the resonance approximation.

exactly the second term of this progression. Then repeating the evaluations leading to Equation 14.73 with

$$Q_l = U_{a_0a}^{(2)} \left(\frac{V^{(2)}(\omega) + V^{(4)}(\omega)}{\omega + \epsilon_{a_0} - \varepsilon_a} \right)^l \tag{14.80}$$

$$= T^* D^{-1} \left[(V^{(2)}(\omega) + V^{(4)}(\omega))D^{-1} \right]^l T, \tag{14.81}$$

we obtain the modified resonance condition

$$V^{(0)} + \text{Re}\{ V^{(2)}(\omega^{\text{res}}) + V^{(4)}(\omega^{\text{res}}) \} + O(\alpha^3) - \epsilon_{a_0} - \omega^{\text{res}} = 0. \tag{14.82}$$

Accordingly, we obtain the following expression for ω^{res}:

$$\omega^{\text{res}} = -\epsilon_{a_0} + V^{(0)} + \text{Re}\Big\{ V^{(2)}(\varepsilon_a - \epsilon_{a_0}) + V^{(4)}(\varepsilon_a - \epsilon_{a_0})$$

$$+ V^{(2)}(\varepsilon_a - \epsilon_{a_0}) \left[\frac{\partial}{\partial \omega} V^{(2)}(\omega) \right]_{\omega = \varepsilon_a - \epsilon_{a_0}} \Big\} + O(\alpha^3). \tag{14.83}$$

The term $V^{(4)}(\varepsilon_a - \epsilon_{a_0})$ is the contribution of the irreducible part of graph (Figure 14.4a). The term with the derivative defines the reducible part of the contribution of the graph (Figure 14.4a).

Up to now, the third-order corrections are considered only for the interelectron interaction. So, evaluating corrections of the third order we disregard the radiative corrections. Application of the LPA to the many-electron ions is performed similarly to the one-electron ion case. In this section, we will omit many derivations and present only the final results. For the details, we refer to a comprehensive review [4].

In the case of the many-electron ions, the matrix element $V^{(0)}$ is given by the sum of the Dirac energies of electrons defining the energy level under investigation. The matrix element $V^{(2)}(\omega)$ represents the one-photon exchange correction given by the graph in Figure 14.5. Important feature is that $V^{(2)}(\omega)$ in this case does not depend on ω. Accordingly, if we do not account for the radiative corrections, we can write $V^{(2)}(\omega) = V^{(2)}$. The matrix element $V^{(4)}(\omega)$ depends on ω (even if we disregard retardation). It represents the two-photon exchange corrections corresponding to the Feynman graphs in Figure 14.6. The next order corrections originate from the graphs in Figures 14.9 and 14.10 (we do not consider the four-electron graphs). We will designate the many-electron energies by a capital letter E.

Now the condition for ω^{res} looks like

$$V^{(0)} + \text{Re}\{V^{(2)}(\omega^{\text{res}}) + V^{(4)}(\omega^{\text{res}}) + V^{(6)}(\omega^{\text{res}})\} - E_{A_0} - \omega^{\text{res}} = 0. \qquad (14.84)$$

Resolving this equation, we obtain

$$\omega^{\text{res}} = -E_{A_0} + V^{(0)} + \text{Re}\left\{ V^{(2)}(E_{ab}^{(0)} - E_{A_0}) + V^{(4)}(E_{ab}^{(0)} - E_{A_0}) + V^{(6)}(E_{ab}^{(0)} - E_{A_0}) \right.$$

$$\left. + V^{(2)}(E_{ab}^{(0)} - E_{A_0}) \left[\frac{\partial V^{(4)}(\omega)}{\partial \omega} \right]_{\omega = E_{ab}^{(0)} - E_{A_0}} \right\} + O(\alpha^4), \qquad (14.85)$$

where $E_{ab}^{(0)} = \varepsilon_a + \varepsilon_b$, that is, sum of the Dirac energies. Here we utilized the fact that $\partial V^{(2)}(\omega)/\partial\omega = 0$ if we disregard radiative corrections. The term with derivative in equation together with $V^{(6)}$ represent the third-order (in α) corrections. We note that the term with derivative does not correspond to any Feynman graph and arises even in the case when the retardation is neglected.

When we derived formulas 14.74 and 14.85, we supposed that the employed perturbation theory series is well convergent. The interelectron interaction corrections are of order $1/Z$, where Z is the charge of the nucleus. Accordingly, the employed perturbation theory is plausible only for heavy ions. It is also necessary that the energy level under investigation is well isolated. However, for many-electron ions, the typical situation is the two (or more) closely lying energy levels. If the mixing of this levels is not prohibited by the symmetry, such levels are called quasidegenerate energy levels. The standard example is the two-electron configurations $(1s2p)^3P_1$ and $(1s2p)^1P_1$ which are strongly mixing in the region of middle and small Z [58].

We now turn to the application of the LPA to quasidegenerate levels. We consider levels that are not distinguishable by the symmetry and lying too close to each other for being treated separately by QED perturbation theory. Without loss of generality, we can restrict ourselves to two mixing configurations in two-electron ions.

Let us denote the considered two-electron wave functions (in the j–j coupling scheme) by $\Psi_1^{(0)}(r_1, r_2)$ and $\Psi_2^{(0)}(r_1, r_2)$. The energies corresponding to these wave functions are denoted by $E_1^{(0)}$ and $E_2^{(0)}$ and they are supposed to be close to the exact energies of the electron configurations under consideration. Employing the LPA, we will consider a scattering of a photon on a two-electron ion in its ground-state A_0. One may seek for the positions of resonances near the values $\omega_1^{\text{res}} = -E_{A_0} + E_1^{(0)} + O(\alpha)$ and $\omega_2^{\text{res}} = -E_{A_0} + E_2^{(0)} + O(\alpha)$. Here E_{A_0} is the energy of the ground state. Within the resonance approximation, we will have to retain two terms in the sum (Equation 14.63) corresponding to the basic functions $\Psi_1^{(0)}(r_1, r_2)$ and $\Psi_2^{(0)}(r_1, r_2)$. The scattering amplitude may be written in a matrix form as

$$U_{A_0} = T^+ D^{-1}[\Delta V D^{-1}]T, \tag{14.86}$$

where D is a 2×2 matrix, defined on the functions $\boldsymbol{\Psi}^{(0)} = (\Psi_1^{(0)}, \Psi_2^{(0)})$:

$$D = \omega + E_{A_0} - V^{(0)}, \tag{14.87}$$

$$V^{(0)} = \hat{h}^D(r_1) + \hat{h}^D(r_2). \tag{14.88}$$

Here $\hat{h}^D(r_1)$ and $\hat{h}^D(r_2)$ are the one-electron Dirac Hamiltonians (Equation 14.12) acting on the one-electron Dirac wave functions depending on r_1 or r_2, respectively. We also introduced a notation

$$\Delta V = V - V^{(0)} = V^{(1)} + V^{(2)} + V^{(3)} + \cdots . \tag{14.89}$$

Since the functions $\Psi_1^{(0)}$ and $\Psi_2^{(0)}$ are orthogonal, the matrix D is diagonal.

We compose a geometric matrix progression with the lth term

$$Q_l = T^+ D^{-1}[\Delta V D^{-1}]^l T \tag{14.90}$$

and perform the resummation in the usual manner. The expression for the amplitude reads

$$U_{A_0} = T^+[D - \Delta V]^{-1}T \equiv T^+ \frac{1}{D - \Delta V}T = T^+ \frac{1}{\omega + E_{A_0} - V}T. \tag{14.91}$$

Introducing the vector $\boldsymbol{\Psi} = (\Psi_1, \Psi_2)$ by means of the linear transformation $\boldsymbol{\Psi} = B\boldsymbol{\Psi}^{(0)}$, where the transformation matrix B is assumed to diagonalize the matrix $V = V^{(0)} + \Delta V$, that is, with $V^{\text{diag}} = B^+ V B$. The expression for the amplitude can now be written in the following form:

$$U_{A_0} = T^+_{A_0 \Psi_1} \frac{1}{\omega + E_{A_0} - [B^+ V B]_{\Psi_1 \Psi_1}} T_{\Psi_1 A_0}$$

$$+ T^+_{A_0 \Psi_2} \frac{1}{\omega + E_{A_0} - [B^+ V B]_{\Psi_2 \Psi_2}} T_{\Psi_2 A_0}$$

$$= T^+_{A_0 \Psi_1} \frac{1}{\omega + E_{A_0} - V^{\text{diag}}_{\Psi_1 \Psi_1}(\omega)} T_{\Psi_1 A_0}$$

$$+ T^+_{A_0 \Psi_2} \frac{1}{\omega + E_{A_0} - V^{\text{diag}}_{\Psi_2 \Psi_2}(\omega)} T_{\Psi_2 A_0}. \tag{14.92}$$

Accordingly, the positions of the resonances are determined by the equations

$$\omega^{\text{res}}_1 + E_{A_0} - \text{Re}\{V^{\text{diag}}_{\Psi_1 \Psi_1}(\omega^{\text{res}}_1)\} = 0, \tag{14.93}$$

$$\omega^{\text{res}}_2 + E_{A_0} - \text{Re}\{V^{\text{diag}}_{\Psi_2 \Psi_2}(\omega^{\text{res}}_2)\} = 0. \tag{14.94}$$

Hence, the energy of the configurations are

$$E_{\Psi_1} = \text{Re}\{V^{\text{diag}}_{\Psi_1 \Psi_1}(\omega^{\text{res}}_1)\}, \tag{14.95}$$

$$E_{\Psi_2} = \text{Re}\{V^{\text{diag}}_{\Psi_2 \Psi_2}(\omega^{\text{res}}_2)\}. \tag{14.96}$$

We can expand Equations 14.93 and 14.94 in a Taylor series around the values $\omega^{\text{res}}_1 = -E_{A_0} + E^{(0)}_1$ and $\omega^{\text{res}}_2 = -E_{A_0} + E^{(0)}_2$, respectively. As in the case of nondegenerate levels, this can be achieved up to any desired accuracy.

The LPA outlined above can be easily employed for an arbitrary number of degenerate levels. The generalization of the method to N-electron ions can also be achieved by the method similar to the one described at the end of the previous subsection.

14.10 Evaluation of Transition Probabilities

Let us describe the transition process, where an ion in the initial state (I) emits a photon (ω_0) and decays into the final state (F)

$$\text{I} \xrightarrow{\omega_0} \text{F}. \tag{14.97}$$

Within the framework of the LPA, each (excited) state of an ion is associated with the position of the resonance. Therefore, we will consider a more complicated process, which incorporates the process (Equation 14.97) as an intermediate one

$$A_0 \xrightarrow{\omega} I \xrightarrow{\omega_0} F \xrightarrow{\omega'} A_0, \tag{14.98}$$

that is, a transition from the state A_0 (we can assume it to be the ground state) to the state I with absorption of photon ω. Then, the state I decays to the state F with emission of photon ω_0 and, finally, the state F decays back to the state A_0 with emission of photon ω'. The initial state (I) is associated with the resonance near $\omega = -E_{A_0} + E_I^{(0)}$, where $E_I^{(0)}$ is the zeroth order energy of the state I (sum of the Dirac energies). The final state (F) is defined by the resonance near $\omega' = -E_{A_0} + E_F^{(0)}$.

In a way similar to the application of the LPA for the investigation of energy levels, the amplitude of scattering process (Equation 14.98) can be written as

$$U = T^+ \frac{1}{D(\omega') - \Delta V(\omega')} \Xi(\omega_0) \frac{1}{D(\omega) - \Delta V(\omega)} T, \tag{14.99}$$

where the matrix T describes the absorption of the photon ω by the ground state A_0. The diagonal matrix $D(\omega)$ and the matrix of the interaction operator $\Delta V(\omega)$ are defined by Equations 14.87 and 14.89, respectively.

The right denominator corresponds to the resonance associated with the state I and the left one refers to the resonance for the state F. Again the function $\Xi(\omega_0)$ stands for a complicated vertex insertion, which represents the emission of a photon ω_0, while the ion undergoes the transition (decay) from the state I to F. The transition matrix element involved in the vertex $\Xi(\omega_0)$ is calculated with the eigenvectors $\Psi_{I,F}$ corresponding to the states I and F; it is represented by the amplitude of the decay process (Equation 14.97)

$$U_{I \to F} = (\Xi(\omega_0))_{\Psi_F \Psi_I}. \tag{14.100}$$

The eigenvectors Ψ_I and Ψ_F and the vertex $\Xi(\omega_0)$ can be constructed by employing the perturbation theory order by order [4].

In the lowest order of the perturbation theory the vertex $\Xi(\omega_0)$ is given by $e\gamma^\mu A_\mu$, where e is the charge of the electron, γ^μ the Dirac matrix, and A_μ the photon wave function in the coordinate representation. For the second-order corrections to $\Xi(\omega_0)$, we refer to [4].

The transition probability for the decay I $\to$ F is given by the expression

$$w_{I \to F} = \int \frac{d^3 k_0}{(2\pi)^3} 2\pi |U_{I \to F}|^2 \delta(E_F + \omega_0 - E_I), \tag{14.101}$$

where $\omega_0 = |\mathbf{k}_0|$ is the frequency of the emitted photon and $\mathbf{k}_0$ the wave vector. The full transition probability is obtained after we have summed over the projections of the final state F and averaged over the projections of the initial state I:

$$W_{I \to F} = \frac{1}{2J_I + 1} \sum_{M_I M_F} w_{I \to F}. \tag{14.102}$$

In order to derive the amplitude defined in Equation 14.99, we have to construct the wave functions Φ_I and Φ_F. These functions are the eigenvectors of the matrix V, which was introduced at the end of Section 14.9. However, the diagonalization of the matrix V and evaluation of the corresponding eigenvectors is a serious task, because V is an infinite matrix. A possible solution of this problem is to replace it by a large but finite-dimensional matrix. Another option is the employment of perturbation theory. In this section, we will concentrate on the development of the proper perturbation theory.

The perturbation theory for the case of a nondegenerate level (as well as for the case of the fully degenerate levels) is well known [70]. Here we will introduce a perturbation theory for the case of quasidegenerate levels. Consider a set of N states with wave functions $\{\Psi\}$ defined, for example, in the j–j coupling scheme. Let us assume that these states are mixing with each other, that is, they may have the same symmetry and the corresponding energy levels are lying close to each other. Under such condition, the standard perturbation theory may not be applicable and has to be modified. Let $g = \{\Psi_{i_g}\}_{i_g=1}^{i_g=N}$ denote the set of these N states. We are interested in building an eigenvector corresponding to a state n_g ($\Psi_{n_g} \in g$); this eigenvector will be designated as Φ_{n_g}. We also suppose that all other states (not elements of the set g) are either not mixing with the state n_g or their energy levels are far enough from the level n_g; we suppose that the set g is large enough to embrace all the closely lying levels. Otherwise, we would have to enlarge the set g and apply again the perturbation theory developed below.

It is convenient to write down the infinite matrix V in a block form

$$V = \begin{bmatrix} V_{11} & V_{12} \\ V_{21} & V_{22} \end{bmatrix}, \tag{14.103}$$

where the subblock $V_{11} \equiv (V)_{i_g j_g}$ is constructed entirely on the states from the set g and the block V_{22} does not contain any of the states from the set g. From Section 14.1.8, we know that the matrix V can be written as a sum

$$V = V^{(0)} + \Delta V, \tag{14.104}$$

where the matrix $V^{(0)}$ is a diagonal one. The part ΔV contains the small parameter α and can be treated as a perturbation.

We can write the matrix V as

$$V = \begin{bmatrix} V_{11} & V_{12} \\ V_{21} & V_{22} \end{bmatrix} = \begin{bmatrix} V_{11}^{(0)} + \Delta V_{11} & \Delta V_{12} \\ \Delta V_{21} & V_{22}^{(0)} + \Delta V_{22} \end{bmatrix}. \tag{14.105}$$

The block V_{11} is a finite matrix and can be diagonalized numerically by means of the matrix B:

$$\tilde{V}_{11} = B^{\mathrm{T}} V_{11} B. \tag{14.106}$$

In general, the matrix V possesses complex elements and is symmetrical, that is, $V_{ij} = V_{ji}$. Accordingly, the matrix B is a complex orthogonal matrix, that is,

$$B^\mathsf{T} B = I = B^{-1} B, \tag{14.107}$$

where I is a unit matrix ($I_{ij} = \delta_{ij}$) of the proper dimension.

Compose a block diagonal matrix

$$A = \begin{bmatrix} B & 0 \\ 0 & I \end{bmatrix}, \tag{14.108}$$

which by definition is also an orthogonal matrix $A^\mathsf{T} = A^{-1}$.

Acting by the matrix A on the matrix V yields

$$\tilde{V} = A^\mathsf{T} V A = \begin{bmatrix} \tilde{V}_{11} & B^\mathsf{T} \Delta V_{12} \\ \Delta V_{21} B & V_{22} \end{bmatrix}. \tag{14.109}$$

Since by assumption the required state Ψ_{n_g} is not (or weakly) mixing with the states not included in the set g, the matrix $\tilde{V}$ can be diagonalized with the standard procedure [70]

$$\tilde{V}^{\mathrm{diag}} = \tilde{C}^\mathsf{T} \tilde{V} \tilde{C}, \tag{14.110}$$

where the matrix C can be constructed order by order. The zeroth and the first orders of the matrix C look like

$$\tilde{C}_{ij} = \tilde{C}_{ij}^{(0)} + \tilde{C}_{ij}^{(1)} = I_{ij} + \begin{bmatrix} 0 & \frac{(B^\mathsf{T} \Delta V_{12})_{ij}}{E_j - E_i} \\ \frac{(\Delta V_{21} B)_{ij}}{E_j - E_i} & \frac{(V_{22})_{ij}}{E_j - E_i} \end{bmatrix}. \tag{14.111}$$

The diagonalized matrices V and $\tilde{V}$ coincide, so we can write

$$V^{\mathrm{diag}} = \tilde{V}^{\mathrm{diag}} = (A\tilde{C})^\mathsf{T} V (A\tilde{C}). \tag{14.112}$$

Accordingly, an eigenvector Φ corresponding to a basic function Ψ can be defined as

$$\Phi = (A\tilde{C})\Psi. \tag{14.113}$$

Now we expand this expression for a state $\Phi_{n_g} \in g$ in the perturbation theory series

$$\Phi_{n_g} = A\tilde{C}\Psi_{n_g}^{(0)} = \sum_{k_g \in g} B_{k_g n_g} \Psi_{k_g}^{(0)} + \sum_{\substack{k \notin g \\ l_g \in g}} (\Delta V_{21})_{kl_g} \frac{B_{l_g n_g}}{E_{n_g}^{(0)} - E_k^{(0)}} \Psi_k^{(0)}. \tag{14.114}$$

Expression for ΔV_{21} is given by Equation 14.105. Summation over index k means the summation over all two-electron configurations (e.g., in j–j coupling scheme) including the negative part of the Dirac spectrum, which are not included in the set g. Note that the employment of the j–j coupling scheme is not obligatory.

In case, when the investigated state Ψ_{n_g} is a well-isolated and non-degenerated level, the set g consists only of this state $g = \{\Psi_{n_g}\}$. The matrix B reduces to a one-dimensional unit matrix.

The scheme, presented above, allows one for the evaluation of the transition probabilities between arbitrary quasidegenerate levels within rigorous QED approach.

References

1. L. Labzowsky, G. Klimchitskaya, and Yu. Dmitriev, *Relativistic Effects in the Spectra of Atomic Systems*. Bristol and Philadelphia: Institute of Physics Publishing, 1993.
2. V. M. Shabaev, Two-time Green's function method in quantum electrodynamics of high-z few-electron atoms. *Phys. Rep.* 356, 119, 2002.
3. I. Lindgren, B. Åsén, and S. Salomonson, The covariant-evolution-operator method in bound-state QED. *Phys. Rep.* 389, 161, 2004.
4. O. Yu. Andreev, L. N. Labzowsky, D. A. Solovyev, and G. Plunien, QED theory of the spectral line profile and its applications to atoms and ions. *Phys. Rep.* 455, 135, 2008.
5. M. Gell-Mann and F. Low, Bound states in quantum field theory. *Phys. Rev.* 84, 350, 1951.
6. J. Sucher, S-matrix formalism for level-shift calculations. *Phys. Rev.* 107, 1448, 1957.
7. L. N. Labzowsky, Electron correlation in the relativistic theory of atoms. *Zh. Eksp. Teor. Fiz.* 59, 167, 1970 [Sov. Phys. JETP 32, 94, 1970].
8. M. A. Brau, L. N. Labzowsky, In R. Damburg and O. Kukane (Eds.), Relativistic perturbation theory for atoms and ions. *Proceedings of the 6th Int. Conference on Atomic Physics*, New York: Plenum Press, p. 111, 1979.
9. G. E. Brown, J. S. Langer, and G. W. Schäfer, Lamb shift of a tightly bound electron. i. Method. *Proc. Roy. Soc. A* 251, 92, 1959.
10. G. E. Brown and D. F. Mayers, Lamb shift of a tightly bound electron. ii. Calculation for the k-electron in mercury. *Proc. Roy. Soc. A* 251, 105, 1959.
11. J. C. Ward, An identity in quantum electrodynamics. *Phys. Rev.* 78, 182, 1950.
12. P. Mohr, Self-energy radiative corrections in hydrogen-like systems. *Ann. Phys.* 88, 26, 1974.
13. P. Mohr, Numerical evaluation of the $1s_{1/2}$-state radiative level shift. *Ann. Phys.* 88, 52, 1974.
14. P. J. Mohr, G. Plunien, and G. Soff, QED corrections in heavy atoms. *Phys. Rep.* 293, 227, 1998.
15. W. R. Johnson, S. A. Blundell, and J. Sapirstein, Finite basis sets for the dirac equation constructed from b splines. *Phys. Rev. A* 37, 307, 1988.

16. V. M. Shabaev, I. I. Tupitsyn, V. A. Yerokhin, G. Plunien, and G. Soff, Dual kinetic balance approach to basis-set expansions for the dirac equation. *Phys. Rev. Lett.* 93, 130405, 2004.

17. L. N. Labzowsky, Lamb shift of the levels of the inner electrons in heavy atoms. *Zh. Eksp. Teor. Fiz.* 59, 2165, 1970 [Sov. Phys. JETP 32, 1171, 1971].

18. N. J. Snyderman, Electron radiative self-energy of highly stripped heavy atoms. *Ann. Phys.* 211, 43, 1991.

19. S. A. Blundell and N. J. Snyderman, Basis-set approach to calculating the radiative self-energy in highly ionized atoms. *Phys. Rev. A* 44, R1427, 1991.

20. V. A. Yerokhin, A. N. Artemyev, T. Beier, G. Plunien, V. M. Shabaev, and G. Soff, Two-electron self-energy corrections to the $2p_{1/2} - 2s$ transition energy in li-like ions. *Phys. Rev. A* 60, 3522, 1999.

21. W. H. Furry, A symmetry theorem in the positron theory. *Phys. Rev.* 51, 125, 1937.

22. E. A. Uehling, Polarization effects in the positron theory. *Phys. Rev.* 48, 55, 1935.

23. E. Wichmann and N. M. Kroll, Vacuum polarization in a strong coulomb field. *Phys. Rev.* 101, 843, 1956.

24. G. Soff and P. J. Mohr, Vacuum polarization in a strong external field. *Phys. Rev. A* 38, 5066, 1988.

25. G. Soff, Higher-order vacuum polarization charge densities for spherical symmetric external potentials. *Z. Phys. D* 11, 29, 1989.

26. N. L. Manakov, A. A. Nekipelov, and A. G. Fainstein, Polarization of the vacuum by strong Coulomb field and its contribution to the spectra of the multicharged ions. *Zh. Exsp. Teor. Fiz.* 95, 1167, 1989.

27. Th. Stöhlker, P. H. Mokler, F. Bosch, R. W. Dunford, F. Franzke, O. Klepper, C. Kozhuharov et al., 1s lamb shift in hydrogenlike uranium measured on cooled, decelerated ion beams. *Phys. Rev. Lett.* 85, 3109, 2000.

28. A. Mitrushenkov, L. Labzowsky, I. Lindgren, H. Persson, and S. Salomonson, Second order loop after loop self-energy correction for few-electron multicharged ions. *Phys. Lett. A* 200, 51, 1995.

29. S. Mallampalli and J. Sapirstein, Perturbed orbital contribution to the two-loop lamb shift in hydrogen. *Phys. Rev. Lett.* 80, 5297, 1998.

30. V. A. Yerokhin, A. N. Artemyev, V. M. Shabaev, M. M. Sysak, O. M. Zherebtsov, and G. Soff, Evaluation of the two-photon exchange graphs for the $2p_{1/2} - 2s$ transition in li-like ions. *Phys. Rev. A* 64, 032109, 2001.

31. H. Persson, I. Lindgren, S. Salomonson, and P. Sunnergren, Accurate vacuum-polarization calculations. *Phys. Rev. A* 48, 2772, 1993.

32. T. Beier, P. J. Mohr, H. Persson, G. Plunien, M. Greiner, and G. Soff, Current status of lamb shift predictions for heavy hydrogen-like ions. *Phys. Lett. A* 236, 329, 1997.

33. T. Beier and G. Soff, Källén-Sabry contribution to the lamb-shift in hydrogen-like atoms. *Z. Phys. D* 8, 129, 1988.

34. S. M. Schneider, W. Greiner, and G. Soff, The hyperfine structure of $^{209}_{83}\text{Bi}^{82+}$. *J. Phys. B* 26, L581, 1993.

35. G. Plunien, T. Beier, G. Soff, and H. Persson, Exact two-loop vacuum polarization correction to the lamb shift in hydrogenlike ions. *Eur. Phys. D* 1, 177, 1998.

36. I. Lindgren, H. Persson, S. Salomonson, V. Karasev, L. Labzowsky, A. Mitrushenkov, and M. Tokman, Second-order QED corrections for few-electron heavy ions: reducible breit-coulomb correction and mixed self-energy-vacuum polarization correction. *J. Phys. B* 26, L503, 1993.

37. H. Persson, I. Lindgren, L. N. Labzowsky, G. Plunien, T. Beier, and G. Soff, Second-order self-energy vacuum-polarization contributions to the lamb shift in highly charged few-electron ions. *Phys. Rev. A* 54, 2805, 1996.

38. S. Mallampalli and J. Sapirstein, Fourth-order vacuum-polarization contribution to the lamb shift. *Phys. Rev. A* 54, 2714, 1996.

39. T. Beier, P. J. Mohr, H. Persson, and G. Soff, Influence of nuclear size on QED corrections in hydrogenlike heavy ions. *Phys. Rev. A* 58, 954, 1998.

40. E. E. Salpeter, Mass corrections to the fine structure of hydrogen-like atoms. *Phys. Rev.* 87, 328, 1952.

41. E. E. Salpeter and H. A. Bethe, A relativistic equation for bound-state problems. *Phys. Rev.* 84, 1232, 1951.

42. L. N. Labzowsky, Nucleus recoil correction in the relativistic theory of the hydrogen atom. *Proceedings of the XVII All-Union Symposium on Spectroscopy, Part II*, Moscow, p. 89, 1972 (in Russian).

43. M. A. Braun, Recoil corrections in a strong nuclear field. *Zh. Eksp Teor. fiz.* 64, 413, 1973 [Engl. Transl.: Sov. Phys. JETP 67, 2039, 1973].

44. V. M. Shabaev, Mass corrections in a strong nuclear field. *Teor. Mat. Fiz.* 63, 394, 1985 [Engl. Transl.: Theor. Mat. Phys. 63, 588, 1985].

45. V. M. Shabaev, Nuclear recoil elect in relativistic theory of multiply charged ions. *Yad. Fiz.* 47, 107, 1988 [Engl. Transl.: Sov. J. Nucl. Phys. 47, 69, 1988].

46. A. N. Artemyev, V. M. Shabaev, and V. A. Yerokhin, Relativistic nuclear recoil corrections to the energy levels of hydrogenlike and high-z lithiumlike atoms in all orders in αZ. *Phys. Rev. A* 52, 1884, 1995.

47. G. S. Adkins and J. Saprstein, Recoil corrections in the hydrogen isoelectronic sequence. *Phys. Rev. A* 73, 032505, 2006.

48. G. S. Adkins and J. Saprstein, Recoil corrections in highly charged ions. *Phys. Rev. A* 76, 042508, 2007.

49. G. Plunien, G. Müller, W. Greiner, and G. Soff, Nuclear polarization contribution to the lamb shift in heavy atoms. *Phys. Rev. A* 39, 5428, 1989.

50. G. Plunien, G. Müller, W. Greiner, and G. Soff, Nuclear polarization in heavy atoms and superheavy quasiatoms. *Phys. Rev. A* 43, 5853, 1991.

51. A. V. Nefiodov, L. N. Labzowsky, G. Plunien, and G. Soff, Nuclear polarization effects in spectra of multicharged ions. *Phys. Lett. A* 222, 227, 1996.

52. S. Blundell, P. J. Mohr, W. R. Johnson, and J. Sapirstein, Evaluation of two-photon exchange graphs for highly charged heliumlike ions. *Phys. Rev. A* 48, 2615, 1993.

53. I. Lindgren, H. Persson, S. Salomonson, and L. Labzowsky, Full QED calculations of two-photon exchange for heliumlike systems: Analysis in the coulomb and feynman gauges. *Phys. Rev. A* 51, 1167, 1995.

54. V. A. Yerokhin, A. N. Artemyev, V. M. Shabaev, M. M. Sysak, O. M. Zherebtsov, and G. Soff, Two-photon exchange corrections to the $2p_{1/2} - 2s$ transition energy in li-like high- z ions. *Phys. Rev. Lett.* 85, 4699, 2000.

55. O. Yu. Andreev, L. N. Labzowsky, G. Plunien, and G. Soff, QED calculation of the interelectron interaction in two- and three-electron ions. *Phys. Rev. A* 64, 042513, 2001.

56. A. N. Artemyev, V. M. Shabaev, V. A. Yerokhin, G. Plunien, and G. Soff, QED calculation of the $n = 1$ and $n = 2$ energy levels in he-like ions. *Phys. Rev. A* 71, 062104, 2005.

57. I. Lindgren, B. Åsén, S. Salomonson, and A.-M. Mårtensson-Pendrill, QED procedure applied to the quasidegenerate fine-structure levels of he-like ions. *Phys. Rev. A* 64, 062505, 2001.

58. O. Yu. Andreev, L. N. Labzowsky, G. Plunien, and G. Soff, Calculation of quasidegenerate energy levels of two-electron ions. *Phys. Rev. A* 69, 062505, 2004.

59. V. A. Yerokhin, A. N. Artemyev, T. Beier, V. M. Shabaev, and G. Soff, Direct evaluation of the two-electron self-energy corrections to the ground state energy of lithium-like ions. *J. Phys B* 31, L691, 1998.

60. A. N. Artemyev, T. Beier, G. Plunien, V. M. Shabaev, G. Soff, and V. A. Yerokhin, Vacuum-polarization screening corrections to the energy levels of lithiumlike ions. *Phys. Rev. A* 60, 45, 1999.

61. O. M. Zherebtsov, V. M. Shabaev, and V. A. Yerokhin, Third-order interelectronic-interaction correction to the $2p_{1/2}$ – $2s$ transition energy in lithiumlike ions. *Phys. Lett. A* 277, 227, 2000.

62. F. Low, Natural line shape. *Phys. Rev.* 88, 53, 1952.

63. L. Labzowsky, V. Karasiev, I. Lindgren, H. Persson, and S. Salomonson, Higher-order QED corrections for multi-charged ions. *Phys. Scr. T* 46, 150, 1993.

64. L. N. Labzowsky, D. A. Solovyev, G. Plunien, and G. Soff, Asymmetry of the natural line profile for the hydrogen atom. *Phys. Rev. Lett.* 87, 143003, 2001.

65. L. Labzowsky, G. Schedrin, D. Solovyev, and G. Plunien, Theoretical study of the accuracy limits of optical resonance frequency measurements. *Phys. Rev. Lett.* 98, 203003, 2007.

66. M. Niering, R. Holzwarth, J. Reichert, P. Pokasov, Th. Udem, M. Weitz, T. W. Hänsch et al., Measurement of the hydrogen 1s- 2s transition frequency by phase coherent comparison with a microwave cesium fountain clock. *Phys. Rev. Lett.* 84, 5496, 2000.

67. K. S. E. Eikema, J. Waltz, and T. W. Hänsch, Continuous coherent lyman-α excitation of atomic hydrogen. *Phys. Rev. Lett.* 86, 5679, 2001.

68. C. F. Fischer and F. A. Parpia, Accurate spline solutions of the radial dirac equation. *Phys. Lett. A* 179, 198, 1993.

69. O. Yu. Andreev, L. N. Labzowsky, and G. Plunien, QED calculation of transition probabilities in two-electron ions. *Phys. Rev. A* 79, 032515, 2009.

70. L. D. Landau and E. M. Lifshitz, *Quantum Mechanics: Non-Relativistic Theory*. Vol. 3 (3rd ed.). Oxford: Pergamon Press, 1977.

71. P. J. Mohr and G. Soff, Nuclear size correction to the electron self-energy. *Phys. Rev. Lett.* 70, 158, 1993.

72. V. M. Shabaev, A. N. Artemyev, T. Beier, G. Plunien, V. A. Yerokhin, and G. Soff, Recoil correction to the ground-state energy of hydrogenlike atoms. *Phys. Rev. A* 57, 4235, 1998.

73. Th. Stöhlker, A. Gumberidze, D. Banas, H. F. Beyer, F. Bosch, S. Chafferjee, C. Kozhuharov et al., Recent developments for the investigation of ground-state transitions in heavy one-electron ions. *J. Phys. Conf. Ser.* 72, 012008, 2007.

74. T. Franosch and G. Soff, The influence of the nuclear shape and of the muonic vacuum polarization on strongly bound electrons. *Z. Phys. D* 18, 219, 1991.

75. J. Schweppe, A. Belkacem, L. Blumenfeld, N. Claytor, B. Feinberg, H. Gould, V. E. Kostroun et al., Measurement of the lamb shift in lithiumlike uranium (U^{89+}). *Phys. Rev. Lett.* 66, 1434, 1991.

15

Parity Nonconservation Effects in the Highly Charged Ions

Anastasiya Bondarevskaya and L. N. Labzowsky

CONTENTS

15.1 Introduction

The studies of the parity nonconservation (PNC) effects in atoms began in the 1970s as soon as the standard model (SM) of the electroweak interactions was introduced [1,2]. The first proposals for the atomic PNC experiments were provided in [3] (optical dichroism, Cs atom) and in [4] (optical rotation, Tl, Bi, and Pb atoms). The first observation of the PNC effect in atoms (optical rotation, Bi) was reported in Novosibirsk [5], and the most accurate recent experiment belongs to the Boulder group [6] (Cs atom, optical dichroism). The accuracy of the latter experiment reaches 0.5% that allows also for the observation of the small nuclear-spin-dependent PNC contribution caused by the anapole moment of the nucleus [7,8]. The atomic PNC experiments are indirect, that is, they require adequate theoretical description to extract the SM constants from the experimental data. To describe the PNC effect in the heavy many-electron atom with the required accuracy (better than 0.5%) appeared to be very hard problem. Due to the very short radius of the weak interactions, the PNC effects are proportional to the density of the valence electrons at the nucleus surface. Then the electron correlation between all 55 electrons in the Cs atom is involved. Moreover, the relativistic Breit interaction also appeared to be significant as well as quantum electrodynamical (QED) effects: vacuum polarization and electron self-energy corrections. The modern status of the problem is presented in detail in [9] (see also the most accurate latest calculation of QED corrections to the PNC effects in Cs in [10]). Still not all the electroweak radiative corrections to the PNC effects in heavy atoms are included, and the agreement of the atomic PNC experiments with the SM based on the high-energy determination of the free parameters requires further approval. The use of heavy atoms is necessitated by the Z^3 enhancement of the PNC effects with the growth in the nuclear charge Z.

It is, therefore, greatly attractive to use much simpler systems such as the highly charged ions (HCIs) with high Z values for the search of PNC effects in atomic physics. A number of proposals [11–18] were made during the last decades for the search of PNC effects in HCIs. In this chapter, we will discuss the recent status of the problem and the most prominent ways for its solution.

15.2 Relativistic Effective PNC Hamiltonian in Atoms

For the description of the PNC effects in heavy atoms and HCIs, it is necessary to use the fully relativistic theory for the electrons. However, the nucleons within the nuclei can be considered as nonrelativistic. This follows from the order of magnitude of the energies involved: these energies can be as large as the rest energy of the electrons $m_e c^2$, where m_e is the electron mass and c the speed of the light, but essentially lower than that of the nucleons. Within this approximation, the effective relativistic Hamiltonian of the interaction between the atomic electron and the nucleus is [9]

$$\hat{H}_W = -\frac{G_F}{2\sqrt{2}} Q_W \rho(r) \gamma_5. \tag{15.1}$$

Here $G_F = -1027 \times 10^{-5} m_p^{-2}$ is the Fermi constant, m_p the proton mass, Q_W the so-called weak charge of the nucleus given as

$$Q_W = -N + Z(1 - 4\sin^2\theta_W). \tag{15.2}$$

Here N and Z are the numbers of neutrons and protons in the nucleus and θ_W is the Weinberg angle (a free parameter of the SM). From the high-energy experiments it follows that $1 - 4\sin^2\theta_W \approx 0.08$. In Equation 15.1, enters also the nucleon density $\rho(r)$ within the nucleus; it is assumed that these densities are the same for protons and neutrons. Finally, γ_5 in Equation 15.1 is the Dirac pseudoscalar matrix. The value of Q_W given by Equation 15.2 is modified by the radiative corrections.

The Hamiltonian Equation 15.1 represents the nuclear-spin-independent part of the effective PNC interaction between the atomic electron and the nucleus. The dominant nuclear-spin-dependent part of this interaction, caused by the anapole moment of the nucleus, is [9]

$$\hat{H}_a = \frac{G_F}{\sqrt{2}} \kappa_a F_I(I)(\vec{\alpha}\vec{I})\rho(r),$$ (15.3)

where κ_a is the anapole moment of the nucleus, $\vec{\alpha}$ are the Dirac matrices, $\vec{I}$ is the nuclear spin, and $F_I(I)$ is given by

$$F_I(I) = (-1)^{I+1/2-l} \frac{I+1/2}{I(I+1)}.$$ (15.4)

Equation 15.3 and 15.4 are valid for the nuclei within one valence nucleon within a shell model for the nucleus and l is the orbital angular momentum of this nucleon. The anapole moment κ_a increases with atomic number and for the heavy nuclei is approximately equal to $\kappa_a \sim \alpha A^{2/3}$, where α is the fine structure constant and A the atomic number.

15.3 Mixing of Atomic States by the PNC Hamiltonian

The standard situation in atoms, favorable for the search of the PNC effects, is the admixture of the PNC E1 amplitude to the basic $M1$ amplitude. The relative weakness of the basic transition leads to the enhancement of the PNC effect. In one-electron ions, one can consider $ns \rightarrow 1s + \gamma(M1)$ transition as the basic one. In general, the amplitude corresponding to the $M1$ transition is

$$A^{(M)}_{njlm,n'j'l'm',JM} = \langle njlm|\vec{\alpha}\vec{A}^{(M)}_{JM}|n'j'l'm'\rangle,$$ (15.5)

where $njlm$ and $n'j'l'm'$ are the quantum numbers of the initial and final one-electron states, respectively. This set of quantum numbers includes the principal quantum number n, the total electron angular momentum j, and its projection m. The quantum number l in the relativistic theory does not relate to the orbital angular momentum which is not conserved, but fixes the parity of the state. In Equation 15.1, $\vec{\alpha}$ are the Dirac matrices and $\vec{A}^{(M)}_{JM}$ is the photon vector function for the magnetic type of photons with the angular photon momentum J and its projection M. One can find the explicit expression for the functions $\vec{A}^{(M)}_{JM}$ in many books on QED theory, for example, in [19,20].

The PNC interaction in atoms admixes to the $\vec{A}_{JM}^{(M)}$ amplitude the $\vec{A}_{JM}^{(E)}$ amplitude via the mixing of atomic states with opposite parity:

$$A_{njl,n'j'l',JM}^{(E)} = \sum_{n''j''l''m''} \frac{\langle njlm|\hat{H}_W|n''j''l''m''\rangle}{E_{n''j''l''} - E_{n'j'l'}}$$

$$\times \langle n''j''l''m''|\vec{\alpha}\vec{A}_{JM}^{(E)}|n'j'l'm'\rangle, \tag{15.6}$$

where $E_{n''j''l''} - E_{n'j'l'}$ is the energy interval between the mixed levels. In many cases (see below the situation in the two-electron ions), there is one level of opposite parity most close to $n'j'l'$. Then we can retain only one term in the sum in Equation 15.6.

The evaluation of the matrix element of the PNC interaction Equation 15.1 can be performed easily in a very general case. The wave function Ψ_{njlm} of the electron in an arbitrary central field can be expressed as

$$\Psi_{njlm} = \begin{pmatrix} \varphi_{njlm} \\ \chi_{njlm} \end{pmatrix}, \tag{15.7}$$

$$\varphi_{njlm} = g_{njl}(r)\Omega_{jlm}(\vec{n}), \tag{15.8}$$

$$\chi_{njlm} = if_{njl}(r)\Omega_{\tilde{j}lm}(\vec{n}), \tag{15.9}$$

where $\vec{n} \equiv \vec{r}/r$ and $\tilde{l} = 2j - l$. In Equations 15.6 through 15.9, the spherical spinors $\Omega_{jlm}(\vec{n})$ are introduced to obtain

$$\Omega_{jlm}(\vec{n}) = \sum_{m_l m_s} C_{jm}^{l1/2}(m_l m_s) Y_{lm_l}(\vec{v})\eta_{m_s}, \tag{15.10}$$

where η is the two-component spinor function.

Insertion of the wave functions Equation 15.6 into the matrix element of the operator $\hat{H}_W$ in Equation 15.6 after integration over the electron variables yields

$$\langle njlm|\hat{H}_W|n''j''l''m''\rangle \equiv iG_{njl;n''j''l''}\delta_{\tilde{l}l''}\delta_{jj''}\delta mm''$$

$$= -\frac{G_F}{2\sqrt{2}}Qwi \int [g_{njl}(r)f_{n''j''l''}(r) + f_{njl}g_{n''j''l''}(r)]$$

$$\times \rho(r)r^2 \, dr \, \delta_{\tilde{l}l''}\delta_{jj''}\delta mm''. \tag{15.11}$$

In the standard situation with mixing $ns_{1/2}$ and $n''p_{1/2}$ states

$$G_{n1/20;n''1/21} = -\frac{G_F}{2\sqrt{2}}Qwi \int [g_{n1/20}(r)f_{n''1/20}(r) + f_{n1/20}g_{n''1/20}(r)]\rho(r)r^2 \, dr. \tag{15.12}$$

In the nonrelativistic limit, only the components $g_{n1/20}(r)$ and $f_{n1/20}(r)$ are nonzero in the limit $r \to 0$, so that the mixing of $ns_{1/2}$ and $np_{1/2}$ is most effective for the enhancement of the PNC effect. It is also important that the matrix element Equation 15.11 appears to be purely imaginary.

With the PNC amplitude Equation 15.6 taken into account, the probability of the transition $ns \to n's + \gamma$ is

$$dW^J_{ns \to n's} = \pi \sum_{mm'} (|A^{(M)}_{njlm,n'j'l'm',JM}|^2 + 2\mathrm{Re}A^{(M)*}_{njlm,n'j'l'm',JM} A^{(E)}_{njlm,n'j'l'm',JM}). \qquad (15.13)$$

In Equation 15.13, we neglect the term quadratic in the PNC Hamiltonian. We are interested in the probability dW, not integrated over the directions of the photon emission and not averaged over the photon polarizations. Therefore, we change the photon quantum numbers from *JM* (total angular momentum and its projection) to $\vec{k}$ and $\vec{e}$ (photon momentum and polarization, respectively). We perform further evaluations in the dipole approximation for the mixing of *M1* and *E1* amplitudes. Then, neglecting retardation, we have

$$A^{(M)*}_{njlm,n'j'l'm',JM} = \langle njlm|\vec{\alpha}\vec{A}^{(E)}_{1M}|n'j'l'm'\rangle^* \Rightarrow \langle njm|\vec{\mu}(\vec{e} \times \vec{v})|n'j'm'\rangle^*, \quad (15.14)$$

$$\langle n\bar{j}lm|\vec{\alpha}\vec{A}^{(E)}_{1M}|n'j'l'm'\rangle \Rightarrow \langle n''jm|\vec{d}\,\vec{e}|n'j'm'\rangle, \qquad (15.15)$$

$$A^{(E)}_{njlm,n'j'l'm',JM} \Rightarrow \sum_{n''} \frac{iG_{njl,n''\bar{j}l}}{E_{n''\bar{j}l} - E_{njl}} \langle n''jm|\vec{d}\,\vec{e}|n'j'm'\rangle. \qquad (15.16)$$

Here μ and *d* are the magnetic and electric dipole moment operators of the atomic electron, respectively, $\vec{e}$ the polarization vector for the photon, and $\vec{v} \equiv \vec{k}/|\vec{k}|$.

For performing the angular integrations in Equations 15.14 through 15.16, we present the scalar products in spherical components

$$\vec{\mu}(\vec{e} \times \vec{v}) = \sum_q (-1)^q \mu^1_q (\vec{e} \times \vec{v})_{-q}, \qquad (15.17)$$

$$\vec{d}\vec{e} = \sum_{q'} (-1)^{q'} d^1_{q'} e_{-q'} \qquad (15.18)$$

and employ the Wigner–Eckart theorem:

$$\langle njm|\mu^1_q|n'j'm'\rangle^* = (-1)^{j'-m'} \begin{pmatrix} j' & 1 & j \\ \bar{m}' & q & m \end{pmatrix} \langle nj\|\mu^1\|n'j'\rangle, \qquad (15.19)$$

$$\langle n''jm|d^1_q|n'j'm'\rangle = (-1)^{j-m} \begin{pmatrix} j & 1 & j' \\ \bar{m} & q' & m' \end{pmatrix} \langle n''j\|d^1\|n'j'\rangle, \qquad (15.20)$$

where $\langle nj\| \cdots \|n'j'\rangle$ are reduced matrix elements, and $\bar{m} = -m$.

Performing now the summation over mm' in the interference term in Equation 15.13, we obtain

$$\sum_{mm'} \begin{pmatrix} j' & 1 & j \\ \bar{m}' & q & m \end{pmatrix} \begin{pmatrix} j & 1 & j' \\ \bar{m} & q' & m' \end{pmatrix} = \frac{1}{3}\delta_{qq'}, \tag{15.21}$$

and we find after summation over q,

$$\sum_{mm'} 2\,\mathrm{Re}\, A^{(M)*}_{njlm,n'j'l'm'} A^{(E)}_{njlm,n'j'l'm'}$$

$$= \frac{2}{3}\,\mathrm{Re}\, i((\vec{e}^* \times \vec{v})\vec{e}) \sum_{n''} \frac{G_{njl,n''jl}}{E_{n''jl} - E_{njl}} \langle nj\|\mu^1\|n'j'\rangle\langle n''j\|d^1\|n'j'\rangle. \tag{15.22}$$

Rearranging the vector–scalar product $(\vec{e}^* \times \vec{v})\vec{e} = (\vec{e} \times \vec{e}^*)\vec{v}$ and introducing the photon spin by the definition $\vec{s}_{ph} = i(\vec{e} \times \vec{e}^*)$, which is defined only for the circularly polarized photon, we arrive at the final expression for the transition probability

$$dW_{ns \to n's} = dW^{M1}_{ns \to n's}[1 + (\vec{s}_{ph}\vec{v})P_1], \tag{15.23}$$

where

$$P_1 = 2\sum_{n''} \frac{G_{ns,n''p}}{E_{n''p} - E_{ns}} \frac{\langle n''p\|d^1\|n's\rangle}{\langle ns\|\mu^1\|n's\rangle}. \tag{15.24}$$

The coefficient P is usually called "the degree of the parity violation." In case of the neutral Cs atom, for the valence electron $n = 7$, $n' = 6$, and $P \simeq 10^{-4}$.

When the levels $n''p$ and ns become too close, the width $\Gamma = \Gamma_{n''p} + \Gamma_{ns}$ should be introduced and Equation 15.24 should be replaced by

$$P_1 = 2\,\mathrm{Re}\sum_{n''} \frac{G_{ns,n''p}}{E_{n''p} - E_{ns} - \frac{i}{2}\Gamma} \frac{\langle n''p\|d^1\|n's\rangle}{\langle ns\|\mu^1\|n's\rangle}. \tag{15.25}$$

15.4 PNC Experiments with HCIs

PNC effects in atoms and HCIs can be observed as different types of asymmetries in atomic transitions. The most general expression for the one-photon transition probability of the type Equation 15.23 including different types of the PNC effects reads

$$dW_{ns-n's} = dW^{M1}_{ns-n's}[1 + (\vec{s}_{ph}\vec{v})P_1 + (\vec{\xi}\vec{v})P_2$$
$$+ (\vec{n}\vec{v})P_3 + (\vec{\xi}\vec{h})Q_1 + (\vec{e} \times \vec{v})\vec{\xi}Q_2], \tag{15.26}$$

where $\vec{\xi}$ denotes the polarization vector in case of the polarized ions, $\vec{h} = \vec{H}/|\vec{H}|$, $\vec{H}$ is the external magnetic field, P_i ($i = 1, 2, 3$) and Q_i ($i = 1, 2$) are the coefficients, and P_1 is given by Equation 15.24. In Equation 15.26, the terms with P coefficients correspond to the different PNC effects and the terms with Q coefficients are parity-conserving but can be used for the measurement of the ion beam polarization. Below we will consider the possibility of each type of experiments and discuss their advantages and disadvantages.

15.4.1 PNC Effects with Circularly Polarized Photons

The experiments with the constant P_1, that is, the observation of the circular dichroism in the transition $ns \rightarrow n's + \gamma$ is the standard way for the search of the PNC effects in neutral atoms. In HCIs the direct use of this approach is prevented by the lack of detectors for circularly polarized photons with energy of about 100 keV. However, in [15] it was proposed to use the relativistic Doppler effect for the excitation of $1s - 2s$ transitions in H-like ions via head-on collision with the laser photons. According to the estimates in [15] with the visible and near-UV lasers in existing storage rings, it is possible to excite $1s - 2s$ transitions in H-like HCIs up to $Z = 48$. For $Z = 10, P_1 \approx 4 \times 10^{-6}$ [15]. It is important that the circularly polarized laser is available in the desired frequency range.

15.4.2 Energy Level Crossings in Two-Electron Ions

In the neutral atoms, the essential enhancement of the coefficient P_1 occurs mainly due to the Z^3 dependence of the factor $G_{ns,n''p}$ in Equation 15.24 and due to the ratio of $E1/M1$ amplitudes. The energy denominator in Equation 15.23 does not play a serious role. This denominator becomes important in case of hydrogen atom: when we set $n'' = n$, the energy difference $E_{np} - E_{ns}$ represents the Lamb shift and becomes small: $E_{np} - E_{ns} \sim n^{-3} \times m\alpha(\alpha Z)^4 ru$. The PNC experiments with the hydrogen atom were discussed theoretically in [21]. In the HCIs, the Lamb shift grows rapidly and the magnitude of the P_1 coefficient drops down.

Another situation occurs in the He-like HCI where the levels with opposite parity may become very close to each other. This can be clearly seen when we consider the Z-dependence of the levels of the first excited configuration ($n = 1$ and $n' = 2$, where n and n' are the principal quantum numbers for the two electrons) of He-like HCIs. The picture of this dependence is given in Figure 15.1, where the results of the most accurate modern calculations [23] are used. The crossing (or near-crossing) of the levels with opposite parity occur at $Z = 32$ (Ge), $Z = 64$ (Gd) and $Z = 90$ (Th). Apart from this, the levels of the opposite parity 2^3P_0, 2^3P_1, and 2^1S_0 are very close to each other at $Z = 6$ (C). Though Z values are integers, the accuracy of these crossing can be very high. For example, the splitting $E(2^3P_0) - E(2^1S_0) = 0.004 \pm 0.74$ eV

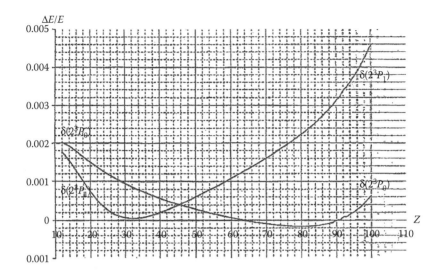

FIGURE 15.1
The ratios $\delta(2^3P_0) = (E(2^1S_0) - E(2^3P_0))/E(2^1S_0)$ and $(\delta)2^3P_1 = (E(2^1S_0) - E(2^3P_1))/E(2^1S_0)$ show the relative closeness for energy levels with opposite parity for He-like ions with different Z values [22]. The notation from the LS-coupling limit are used. The level crossings occur at $Z = 32$ (Ge), $Z = 64$ (Gd), and $Z = 92$ (U).

for $Z = 64$ [22], where the inaccuracy arises due to the higher-order inter-electronic interaction correlations, not included in the evaluation in [22]. This value should be compared to the energy interval $E(2^1S_0) - E(1^1S_0) = 55866.01$ eV.

For the first time, the proposal to use the crossings of the levels 2^1S_0 and 2^3P_1, and 2^1P_1 for $Z = 6$ and close to $Z = 30$ for the PNC experiments was made in [11]. The degree of the PNC effects was estimated as $P \approx 5 \times 10^{-4}f$, where f stands for the dimensionless constant, replacing the contribution of the anapole moment of the nucleus. Since the total angular momenta of the admixed levels are different, only the anapole moment interaction contributes to the level mixing. In 1974, when the paper [11] was published, it was not yet known that the anapole moment contribution dominates in the PNC nuclear-spin-dependent interaction between the atomic electron and the nucleus. The dominance of the anapole moment contribution was first established in 1980 [8]. The near-degeneracy of the levels 2^3P_0 and 2^1S_0 at $Z = 92$ (U) was first discussed in connection with PNC effects in [12], and then in [26]. In [12], the basic transition $2^3P_0 \rightarrow 1^1S_0$ was considered, which occurs as a two-photon $E1M1$ transition; in [13], the one-photon hyperfine-quenched $2^1S_0 \rightarrow 1^1S_0$ transition was chosen as a basic one. The levels 2^1S_0 and 2^3P_0 are mixed by the interaction 15.1. The degree of the PNC was calculated as $P_1 \approx 10^{-4}$ [13]. The same level mixing 2^3P_0 and 2^1S_0 with the transition $2^1S_0 \rightarrow 2^3S_1$ for $Z = 6$ was discussed in [23]. An original idea of how to avoid detection of the circular polarization of x-ray photons in PNC experiments with HCI was outlined in [14]. Here it was proposed to observe the two-photon emission from the 2^3P_0 level in He-like uranium, stimulated by the circularly polarized optical laser. The degree of the PNC effect was again $P = 3 \times 10^{-4}$.

15.4.3 PNC Effects with Polarized HCIs

PNC effects with polarized ion beams were considered in [16,17]. In [16], an experiment was proposed for the determination of Q_W in He-like Eu and in [17] a similar experiment was discussed where the anapole moment of the nucleus using He-like Ge ions could be measured. The coefficient P_2 in Equations 15.26 looks like

$$P_2 = \frac{3\lambda}{I+1}\varrho_1,$$ (15.27)

where λ is the degree of the ion beam polarization and I the nuclear spin. The measurement of the PNC effect in this case consists of registration the asymmetry of the photon emission with respect to the direction of the ion beam polarization. The idea of the experiment is based on the near-crossing of the 2^1S_0 and 2^3P_0 levels. The level scheme for the excited configurations in $^{151}_{63}\text{Eu}^{61+}$ ions ($Z = 63, I = 5/2$, abundance of the isotope 47,8) is shown in Figure 15.2. The basic transition is the hyperfine-quenched (HFQ) one-photon transition $2^1S_0 \rightarrow 1^1S_0 + \gamma(M1)$. The PNC-admixed transition is $1s2s^1S_0 + (\text{PNC mixing})1s2p^3P_0 + (\text{HF mixing})1s2p^3P_1 \rightarrow (1s^2)^1S_0 + \gamma(E1)$. The calculated asymmetry in the photon emission with respect to the ion beam polarization is [16]

$$P_2 \approx \lambda \times 10^{-4}$$ (15.28)

The level crossing actually occurs for Gd ($Z = 64$) where the spacing between the 2^1S_0 and 2^3P_0 levels is very small [22]: $\Delta E = (0.004 \pm 0.078)$ eV. In this case, according to Equation 15.25

$$\text{Re}\frac{1}{\Delta E - \frac{i}{2}\Gamma} = \frac{\Delta E}{(\Delta E)^2 + \frac{1}{4}\Gamma^2},$$ (15.29)

where Γ is defined by the width of the 2^3P_0 level: $\Gamma = 0.0016$ eV. The minimum value of the expression 15.29 corresponds to the spacing $\Delta E = 0.0078$ eV which gives $P_2^{\text{min}} \simeq \lambda \times 10^{-3}$, that is, 10 times larger than the value Equation 15.28 for Eu. The maximum value, corresponding to $\Delta E = \Gamma = 0.0016$ eV is $P_2^{\text{max}} = 0.052$.

The latter result is unprecedented for PNC effects in atoms and ions, though the big discrepancy between P_2^{min} and P_2^{max} does not allow us to draw definite conclusions for the weak interaction constants. However, the experimental situation in Gd^{62+} is less favorable than in Eu^{61+} where the level spacing is

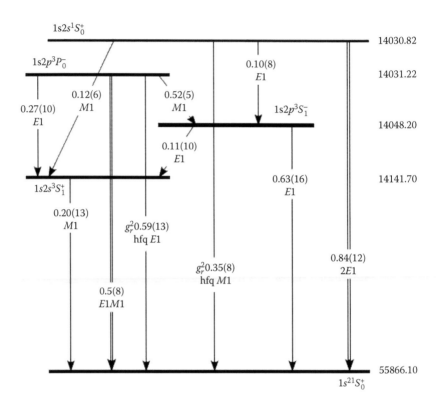

FIGURE 15.2
Energy level scheme of the first excites states of He-like europium. Numbers on the right-hand side indicate the ionization energies in eV. The partial probabilities of radiative transitions are given in s⁻¹. Numbers in parentheses indicate powers of 10. The double lines denote two-photon transitions.

much larger: $\Delta E = -0.224(69)$ [23]. The reason is that in Gd^{62+}, the lifetime of the 2^1S_0 level defined by the 2E1 two-photon transition to the ground state is about one order of magnitude smaller than the lifetime of the 2^3P_0 level, which is determined by the HFQ E1 transition to the ground state. This supplies a strong background from $2^3S_0 \to 1^1S_0 + \gamma$ transitions in experiments with Gd^{62+}: the HFQ $2^3P_0 \to 1^1S_0 + \gamma(E1)$ transition rate is five orders of magnitude larger than the basic HFQ $2^1S_0 \to 1^1S_0 + \gamma(M1)$ transition rate, and both transitions are not distinguishable in the x-ray spectra due to their close frequencies.

In Eu^{61+}, the situation is different. The weak asymmetry effect is smaller, but the 2^1S_0 level lives significantly longer than the 2^3P_0 level. The 2^1S_0 lifetime equals about 1.19 ps and corresponds to a typical decay length of about 0.1 mm in the laboratory. This enables one to "burn out" the 2^3P_1 level and to get rid of the parasitic $2^3P_0 \to 1^1S_0 + \gamma(E1)$ transition.

In case of mixing of the 2^1S_0 and 2^3P_0 states in two-electron HCIs, the formula 15.25 should be replaced by

$$P_1 = 2\,\mathrm{Re}\,\frac{G_{2^1S_0,2^3P_0}}{E(2^3P_0) - E(2^1S_0) - \frac{i}{2}\Gamma}\,\frac{\langle 2^3P_0\|A^{E1}_{HFQ}\|1^1S_0\rangle}{\langle 2^1S_0\|A^{M1}_{HFQ}\|1^1S_0\rangle}. \qquad (15.30)$$

Unlike Equation 15.25, we retained only one term 2^3P_0 in the sum over intermediate states: $\Gamma = \Gamma(2^3P_0) + \Gamma(2^1S_0)$. The reduced matrix elements of the HFQ transitions $\langle 2^1S_0\|A^{M1}_{HFQ}\|1^1S_0\rangle$ and $\langle 2^3P_0\|A^{E1}_{HFQ}\|1^1S_0\rangle$ are

$$\langle 2^1S_0\|A^{M1}_{HFQ}\|1^1S_0\rangle = \frac{\langle 2^1S_0\|\hat{H}_{HFS}\|2^3S_1\rangle}{E(2^1S_0) - E(2^3S_1)}\,\langle 2^3S_1\|A^{M1}\|1^1S_0\rangle, \qquad (15.31)$$

$$\langle 2^3P_0\|A^{E1}_{HFQ}\|1^1S_0\rangle = \frac{\langle 2^3P_0\|\hat{H}_{HFS}\|2^3P_1\rangle}{E(2^3P_0) - E(2^3P_1)}\,\langle 2^3P_1\|A^{E1}\|1^1S_0\rangle, \qquad (15.32)$$

where $\hat{H}_{HFS}$ is the operator of the hyperfine interaction and $\langle 2^3S_1\|A^{M1}\|1^1S_0\rangle$ and $\langle 2^3P_1\|A^{E1}\|1^1S_0\rangle$ are the ordinary reduced transition matrix elements in the frames of the relativistic theory [20].

15.4.4 Parity Violation Effects with HCIs in an External Magnetic Field

Now we turn to the next term in the right-hand side of Equation 15.26, which describes the asymmetry of the photon emission in the external magnetic field. This effect was considered theoretically in [24] for the hydrogen atom, but never before for HCIs. Following [24], we can write down the coefficient P_3 in Equation 15.26 for the case of the 2^1S_0 and 2^3P_0 level crossing in He-like HCIs like

$$P_3 = 2\,\mathrm{Re}\,\frac{[\Delta E_Z(H) - \Delta E_Z(-H)]G_{2^1S_0,2^3P_0}}{[E(2^3P_0) - E(2^1S_0) - \frac{i}{2}\Gamma]^2}$$

$$\times\,\frac{\langle 2^3P_0\|A^{E1}_{HFQ}\|1^1S_0\rangle}{\langle 2^1S_0\|A^{M1}_{HFQ}\|1^1S_0\rangle}, \qquad (15.33)$$

where $\Delta E_Z(H)$ is the distance between Zeeman sublevels with the same total angular momentum projection, belonging to the two levels mixed by the effective PNC Hamiltonian Equation 15.1. Unlike the situation considered in [24], we assume that $\Delta E_Z(H) \ll E(2^3P_0) - E(2^1S_0)$. We do not suppose that the Zeeman structure can be resolved in the experiment. Then the value of $\Delta E_Z(H)$ in Equation 15.33 is the weighted sum of the contributions of all Zeeman sublevels. The Zeeman splitting of the 2^1S_0 level occurs via the electron magnetic moment with the coefficient

$$\xi_S^2 = \left|\frac{\langle 2^1S_0|\hat{H}_{HFS}|2^3S_1\rangle}{E(2^1S_0) - E(2^3S_1)}\right|^2 \qquad (15.34)$$

and the Zeeman splitting of the 2^3P_0 level occurs via the electron magnetic moment with the coefficient

$$\xi_P^2 = \left| \frac{\langle 2^3P_0 | \hat{H}_{\text{HFS}} | 2^3P_1 \rangle}{E(2^3P_0) - E(2^3P_1)} \right|^2. \tag{15.35}$$

There is also direct Zeeman splitting of 2^1S_0 and 2^3P_0 levels via the nuclear magnetic moment, however this contribution cancels in $\Delta E_Z(H)$.

The Zeeman sublevels in case of zero electronic total angular momentum (i.e., for the states 2^1S_0 and 2^3P_0) can be labeled by the values of the nuclear angular momentum projection M_I. For different M_I values, the equality holds

$$\Delta E_Z(H, -M_I) = \Delta E_Z(-H, M_I). \tag{15.36}$$

Then the weighted sum mentioned above looks like

$$\Delta E_Z(H) - \Delta E_Z(-H) = \frac{1}{2I+1} \sum_{M_I} [\Delta E_Z(H, M_I) - \Delta E_Z(H, -M_I)]$$

$$= \frac{4\mu}{2I+1} H[\xi_{sS}^2 g(2^3S_1) - \xi_P^2 g(2^3P_1)], \tag{15.37}$$

where $g(2^3S_1)$ and $g(2^3P_1)$ are the bound electron g-factors for the He-like HCIs.

We define the g-factor for the two-electron ion in a homogeneous magnetic field $\vec{B}$, oriented along z-axis like

$$g(2^{2S+1}L_J) = \frac{2}{B_Z M} \langle 2^{2S+1}L_J | \vec{\alpha}_1 \vec{A}_1 + \vec{\alpha}_2 \vec{A}_2 | 2^{2S+1}L_J \rangle, \tag{15.38}$$

where we use the standard LS-coupling scheme notation for defining the different component of the multiplets in HCIs: L, S, and J are the values of the two-electrons spin, orbital momentum, and the total momentum, respectively, and M is the projection of the total momentum. In principle, in the HCI, the LS-coupling scheme breaks down but some of the LS states, including 2^3S_1 and 2^3P_1, which are of interest here, do not mix up with the other LS states, and we can use the corresponding notation.

Insertion of the vector potential $\vec{A}$ in Equation 15.38 in the form

$$\vec{A}_i = \frac{1}{2}(\vec{B} \times \vec{r}_i) \tag{15.39}$$

and integration over angles yields

$$g(2^3S_1) = -\frac{4}{3}(G_{1s1/2} + G_{2s1/2}), \tag{15.40}$$

$$g(2^3P_1) = -\frac{4}{3}(G_{1s1/2} + G_{2p1/2}), \tag{15.41}$$

where

$$G_{nlj} = \int_0^\infty r^3 g_{njl}(r) f_{njl}(r)\, \mathrm{d}r. \tag{15.42}$$

Substituting in Equation 15.42, the explicit expressions for the radial Dirac function for the point-like nucleus results:

$$G_{1s1/2} = -\frac{1}{4}(1 + 2\sqrt{1 - (\alpha Z)^2}), \tag{15.43}$$

$$G_{2s1/2} = -\frac{1}{4}\left(1 + 2\sqrt{\frac{1 + \sqrt{1 - (\alpha Z)^2}}{2}}\right), \tag{15.44}$$

$$G_{2p1/2} = \frac{1}{4}\left(-1 + 2\sqrt{\frac{1 + \sqrt{1 - (\alpha Z)^2}}{2}}\right). \tag{15.45}$$

The expressions 15.43 and 15.44 coincide with the ones given in [25].

For the most favorable case in Gd^{62+}, when $P_2^{max} = 0.052$ and for the magnetic field 5 T, we receive $\Delta E_Z(H) - \Delta E_Z(-H) = 0.14 \times 10^{-8}$ eV and $P_3^{max} = 0.14 \times 10^{-6}$. Thus, the PNC effect with the external magnetic field is relatively small even in the case of the full degeneracy of the 2^1S_0 and 2^3P_0 levels in the two electron HCIs.

15.5 Conclusions

There are few proposals [14,15] of how to measure the optical dichroism (coefficient P_1) avoiding the direct measurement in the x-ray region. A proposal to measure the asymmetry of the photon emission with respect to the direction of the ion beam polarization [17] (i.e., coefficient P_2) requires the production of the polarized beams. Schemes for the beam polarization production, preservation, and controlling were recently discussed in [26,27]. The observation of the asymmetry of the photon emission in an external magnetic field avoids many difficulties with the x-ray circular polarization measurement or with the necessity to produce the polarized ion beams. However, the magnitude of the PNC effect in the fields available at the moment is too small. The coefficients Q_1 and Q_2 in Equation 15.26 do not correspond to any PNC effects but, in principle, can be employed for the controlling of the ion beam polarization. The term with the coefficient Q_1 describes the dependence of the photon emission transition rate on the mutual orientation of the ion beam polarization and an external magnetic field. The term with the coefficient Q_2 describes the dependence of the photon emission on the mutual orientation of the ion beam polarization and the plane orthogonal to the photon polarization and the direction of the photon emission.

Acknowledgments

The authors are grateful to Guenter Plunien, Andrei Nefiodov, Anton Pro-zorov, Dieter Liesen, Fritz Bosch, Roger Hutton, and Yaming Zou for many helpful discussions. This work was supported by RFBR grant №080200026.

References

1. S. Weinberg, A model of leptons, *Phys. Rev. Lett.* 19, 1264, 1967.
2. A. Salam, Elementary particle theory, *Proceedings of the 8th Nobel Symposium*, New York: Wiley, 1968.
3. M. A. Bouchiat and C. Bouchiat, Parity violation induced by weak neutral currents in atomic physics, *J. Phys (Paris)* 35, 899, 1974.
4. I. B. Khriplovich, Feasibility of observing parity nonconservation in atomic transitions, *Pisma Zh. Eksp. Teor. Fiz.* 20, 686, 1974 [Engl. Transl. *Sov. Phys. JETP Lett.* 20, 315, 1974].
5. L. M. Barkov and M. S. Zolotorev, Observation of parity nonconservation in atomic transitions, *Pisma Zh. Eksp. Teor. Fiz.* 27, 379, 1978 [Engl. Transl. *Sov. Phys. JETP Lett.* 27, 357, 1978].
6. S. C. Bennett and C. E. Wieman, Measurement of the 6S → 7S transition polarizability in atomic cesium and an improved test of the standard model, *Phys. Rev. Lett.* 82, 2484, 1999.
7. Ya. B. Zeldovich, Electromagnetic interaction by parity violation, *Zh. Eksp. Teor. Fiz.* 33, 1531, 1957 [Engl. Transl. *Sov. Phys. JETP* 6, 1184, 1958].
8. V. V. Flambaum and I. B. Khriplovich, P-odd nuclear forces—a source of parity violation in atoms, *Zh. Eksp. Teor. Fiz.* 79, 1656, 1980 [Engl. Transl. *Sov. Phys. JETP* 52, 835, 1980].
9. J. S. M. Ginges and V. V. Flambaum, Violations of fundamental symmetries in atoms and tests of unification theories of elementary particles, *Phys. Rep.* 397, 63, 2004.
10. V. M. Shabaev, K. Pachucki, I. I. Tupitsyn, and V. A. Yerokhin, QED corrections to the parity-nonconserving 6s–7s amplitude in ^{133}Cs, *Phys. Rev. Lett.* 94, 213002, 2005.
11. V. G. Gorshkov and L. N. Labzowsky, Effects of parity nonconservation in heavy ions, *Zh. Eksp. Teor. Fiz. Pisma* 19, 768, 1974 [Engl. Transl. *Sov. Phys. JETP Lett.* 19, 394, 1974].
12. A. Schäfer, G. Soff, P. Indelicato, B. Müller, and W. Greiner, Prospects for an atomic parity-violation experiment in U^{90+}, *Phys. Rev. A* 40, 7362, 1989.
13. V. V. Karasiev, L. N. Labzowsky, and A. V. Nefiodov, Parity violation in heliumlike uranium, *Phys. Lett. A* 172, 62, 1992.
14. R. W. Dunford, Parity nonconservation in high-Z heliumlike ions, *Phys. Rev. A* 54, 3820, 1996.
15. M. Zolotorev and D. Budker, Parity nonconservation in relativistic hydrogenic ions, *Phys. Rev. Lett.* 78, 4717, 1997.

16. L. N. Labzowsky, A. V. Nefiodov, G. Plunien, G. Soff, R. Marrus, and D. Liesen, Parity-violation effect in heliumlike gadolinium and europium, *Phys. Rev. A* 63, 054105, 2001.
17. A. V. Nefiodov, L. N. Labzowsky, D. Liesen, G. Plunien, and G. Soff, Nuclear anapole moments from beams of highly charged ions, *Phys. Lett. B* 534, 52, 2002.
18. G. F. Gribakin, F. J. Currell, M. G. Kozlov, and A. I. Mikhailov, Parity nonconservation in electron recombination of multiply charged ions, *Phys. Rev. A* 72, 032109, 2005.
19. A. I. Akhiezer and V. B. Berestetskii, *Quantum Electrodynamics*, New York: Wiley, 1956.
20. L. Labzowsky, G. Klimchitskaya, and Yu. Dmitriev, *Relativistic Effects in the Spectra of Atomic Systems*. Bristol and Philadelphia: IOP Publishing, 1993.
21. A. N. Moskalev, Parity nonconservation effects due to neutral weak currents in mesic atoms, *Pisma Zh. Eksp. Teor. Fiz* [Engl. Transl. *Sov. Phys. JETP Lett.* 19, 216, 1974].
22. A. N. Artemyev, V. M. Shabaev, V. A. Yerokhin, G. Plunien, and G. Soff, QED calculation of the $n = 1$ and $n = 2$ energy levels in He-like ions, *Phys. Rev. A* 71, 062104, 2005.
23. G. von Oppen, Parity violation in two-electron systems, *Z. Phys. D* 21, 181, 1991.
24. Ya. I. Azimov, A. A. Anselm, A. N. Moskalev, and R. M. Ryndin, Some parity-nonconservation effects in emission by hydrogenlike atoms, *Zh. Eksp. Teor. Fiz.* 67, 17, 1974.
25. T. Beier, The g_j factor of a bound electron and the hyperfine structure splitting in hydrogenlike ions, *Phys. Rep.* 339, 79, 2000.
26. A. Prozorov, L. Labzowsky, D. Liesen, and F. Bosch, Schemes for radiative polarization of ion beams in storage rings, *Phys. Lett. B* 574, 180, 2003.
27. A. Prozorov, L. Labzowsky, G. Plunien, D. Liesen, F. Bosch, S. Fritzsche, and A. Surzhykov, Ion beam polarization in storage rings: production, controlling and preservation, *GSI Report* 2008-03.

Index

Printed and bound by CPI Group (UK) Ltd, Croydon, CR0 4YY

21/10/2024

01777042-0017